PROTEOMIC APPLICATIONS IN BIOLOGY

Edited by **Joshua L. Heazlewood**
and **Christopher J. Petzold**

Proteomic Applications in Biology
Edited by Joshua L. Heazlewood and Christopher J. Petzold
Subject Editors: Tsz-Kwong Man and Ricardo J. Flores

Published by InTech

Janeza Trdine 9, 51000 Rijeka, Croatia

Publishing Process Manager Martina Durovic

Technical Editor Teodora Smiljanic

Cover Designer InTech Design Team

Additional hard copies can be obtained from orders@intechopen.com

Proteomic Applications in Biology, Edited by Joshua L. Heazlewood and Christopher J. Petzold; Subject Editors: Tsz-Kwong Man and Ricardo J. Flores
p. cm.
ISBN 978-953-307-613-3

For EU product safety concerns:
IN TECH d.o.o., Prolaz Marije Krucifikse Kozulić 3, 51000 Rijeka, Croatia,
info@intechopen.com or visit our website at intechopen.com.

Meet the editors

Dr. Heazlewood received his doctorate in the area of plant molecular biology from La Trobe University, Australia. He undertook his post-doctoral research at The University of Western Australia where he used proteomic techniques to characterize plant mitochondria. He was awarded an Australian Research Council Postdoctoral Fellowship to continue this work and to expand investigations into post translation modifications using mass spectrometry. He received an Alexander von Humboldt Research Fellowship to work at the Max Planck Institute of Molecular Plant Physiology in Germany where he developed and employed phosphoproteomic techniques. Over the past few years Dr Heazlewood developed a series of plant proteomic resources that have been instrumental in advancing basic knowledge about protein function in plants. Dr Heazlewood was recently appointed Director of Systems Biology at the Joint BioEnergy Institute at the Lawrence Berkeley National Laboratory to investigate the biosynthesis and structure of plant cell walls employing proteomic and systems approaches.

Dr Petzold earned his doctorate in Analytical Chemistry from Purdue University with emphasis on gas-phase ion chemistry and mass spectrometric methods development. After graduating from Purdue, he performed his post-doctoral work at the University of California, Berkeley focusing on methods development and application of mass spectrometry to glycomics and lipidomics research applied to tuberculosis research. Dr Petzold joined Prof. Jay Keasling's research group in 2005 as part of the team tasked to study the impact of metabolic engineering on the host microbe via proteomics. He is currently a research scientist with Lawrence Berkeley National Laboratory in the Physical Biosciences Division and Deputy Director of Proteomics for the Joint BioEnergy Institute (JBEI). His research efforts at JBEI focus on applying quantitative proteomics research to metabolic engineering efforts and developing mass spectrometric solutions to problems facing biofuel production.

Contents

Preface

In the past decade the field of proteomics has expanded from a highly technical endeavor requiring dedicated practitioners to being widely utilized by researchers of diverse backgrounds for solving complex problems in biology. The approach is used broadly to complement basic molecular and biochemical research and encompasses basic protein identification, functional characterization, comparative approaches and large-scale analyses. During this period, technical improvements (*e.g.*, increased sensitivity, speed, mass accuracy, dynamic range) in mass spectrometric instrumentation provided a solid foundation for proteomic applications outlined in this collection.

This book is divided into four sections and is an interesting overview of the diversity and breadth of proteomic applications in biology. This is highlighted by the wide range of plant, animal and microbial species that are presented throughout the contents of this volume. Especially interesting are the variety of proteomic methods and analytical strategies employed by biological researchers. The technical side of proteomics has witnessed enormous changes in sample analysis strategies with many advanced users adopting 'shotgun' approaches for protein identification and quantification. In contrast, many of the studies outlined in this book employ established, but still powerful, two-dimensional gel electrophoresis strategies to array samples prior to protein identification and/or quantification. The advantages of such a visual media for assessment prior to identification are that they are useful and productive when access to high-end mass spectrometers is limited.

The first section of this book outlines the utilization of proteomics to address issues in agriculture. The increased pressure on food production in the coming decades due to expanding populations and the effects of climate change will require major re-assessments of how food is produced. It is extremely likely that proteomic approaches in combination with other analysis techniques will play an important role in the adaptation of this industry through agricultural research. The chapters outlined in this section cover the utilization and application of proteomics to assess the effects of unintended post-translational modifications on food-derived proteins, an overview of approaches and techniques to examine proteins from fruit, and the use of proteomic techniques to understand viral pathogenesis in cattle.

The second section focuses on the emerging role of metaproteomics to characterize and profile microbes and microbial communities. The majority of the Earth's biomass is comprised of microorganisms, many of which play essential roles in recycling elements such as carbon and nitrogen. While our knowledge of these complex communities is still very limited, recent advances in sequencing technologies, and now proteomics, enable detailed examination of these interesting populations. This section provides an overview of current proteomic approaches and methods being used to study microbial communities, demonstrating the utilization of these approaches to characterize a cold tolerant bacterial strain from the Antarctic.

The third section addresses the utilization of proteomics to study fundamental processes in plants. The development of novel approaches and processes using proteomics can be a direct driver for applications in other areas of biology and applied research. This section highlights the specific issues associated with undertaking proteomic analyses in plant biology, demonstrates and assesses its use in understanding the complex process of plant cell wall biosynthesis through a targeted characterization of xylan biosynthesis using multiple approaches by mass spectrometry and finally provides an overview of the subcellular isolation and proteomic characterization of the Golgi apparatus from plants and other species.

The final section highlights the use of comparative proteomics in biology to understand complex systems. One of the most significant contributions to the field of proteomics has been the development of reliable and reproducible quantitation techniques enabling comparative profiling of samples to be undertaken. This section examines the use of proteomics to identify protein markers to differentiate commercial tuna species, to identify proteins involved in pH adaptation in yeast strains, the characterization and identification of proteins in snake venom and lastly, the identification and characterization of a protein from onion involved in floral bolting.

Collectively these chapters outline studies and overviews that employ a wide variety of proteomic applications in biology. These approaches display extensive technical diversity and contrast the use of proteomics from the community level to the individual protein and together highlight the varied ways in which proteomics is being used to expand our understanding of biological systems.

Dr. Joshua L. Heazlewood
and **Dr. Christopher J. Petzold**
Joint BioEnergy Institute and Physical Biosciences Division
Lawrence Berkeley National Laboratory
Berkeley California,
USA

Part 1

Addressing Issues in Agriculture

1

Food Proteomics: Mapping Modifications

Stefan Clerens[1], Jeffrey E. Plowman[1] and Jolon M. Dyer[1,2,3]
[1]*Food & Bio-Based Products, AgResearch Lincoln Research Centre*
[2]*Biomolecular Interaction Centre, University of Canterbury*
[3]*Riddet Institute at Massey University, Palmerston North*
New Zealand

1. Introduction

Proteins are an essential element in the human diet. As food ingredients, they are primarily sourced from plants and animals; important sources include cereals, meat, poultry, fish and dairy. Proteomics offers a powerful new way to characterise the protein component of foods. Proteomics not only reveals which proteins are expressed in each tissue type, it also allows the investigation of differences in the protein composition of different tissues. In addition it has the power to track the proteome of tissues before and after harvest/slaughter, and to evaluate the effect of downstream treatments such as cooking or curing.

The proteomic evaluation of food proteins presents a unique set of challenges and opportunities. Muscle, milk and cereal proteomes are dominated by very abundant proteins, creating a dynamic range problem. In addition to post-translational modifications produced *in vivo,* food proteins are subjected to a wide range of post-harvest/post-slaughter environmental and processing insults prior to consumption. These modifications include side-chain oxidation, cross-link formation and backbone cleavage, and critically influence key food properties such as shelf-life, nutritional value, digestibility and health effects.

A profound understanding of proteomics, protein modifications and redox chemistry has allowed us to pioneer the application of redox proteomics to foods. This has led to the development of a unique proteomics damage scoring system, allowing a direct link between molecular-level understanding to intervention/mitigation at the processing level (Dyer et al., 2010). We anticipate that this ability will be pivotal in the development of next-generation food products.

This chapter outlines current achievements in the field of food proteomics. It deals with the full spectrum of protein-containing foods, including dairy, meat, seafood and cereal proteins. We focus on *ex vivo* protein modifications and their effects on foods. We devote attention to redox proteomics approaches applied to food, and pay special attention to the recent development of advanced redox proteomic-based approaches to evaluate and track food protein modifications. These approaches are illustrated in a case study that compares the protein damage level in a number of commercially available dairy products.

2. Proteins and nutrition

Proteins are key functional and structural components of all living cells and are an essential element in the human diet. The human body is capable of synthesising most of the amino acids from other precursors, but is unable to produce the nine essential amino acids (His, Ile, Leu, Lys, Met, Phe, Thr, Trp, Val), which must be supplied from the diet. Another six amino acids (Arg, Cys, Gln, Gly, Pro, Tyr) can be produced by the body, but may need a dietary source when endogenous production cannot meet metabolic requirements. Amino acids also act as precursors for many coenzymes, hormones, nucleic acids and other molecules, and can also be used as a metabolic fuel.

The total amount of protein in the body is fairly static, but individual proteins are constantly being degraded and re-synthesised. The rate of this turnover is affected by stage of life and level of activity. That makes the inclusion of an appropriate quantity of high quality protein in the diet critical for growth and development in children and the maintenance of good health in adults.

Protein can be sourced from plants and animals; important sources include cereals, meat, poultry, fish and dairy foods. These sources differ in the relative bioavailability of protein and, in particular, essential amino acids. In general the digestibility of proteins from vegetable sources is lower than for those of animal origin, being around 78-85% as opposed to 94-97% for meat, dairy and eggs. Animal sources of protein also generally have higher levels of the essential amino acids.

Proteomics, through the application of gel and non-gel approaches, offers a powerful new way to characterise the protein component of foods. Whereas genomics provides information on the total genome of the organism, proteomics reveals which proteins are actually expressed in each tissue type. Furthermore the application of proteomic techniques offers a way to investigate differences in the protein composition of different tissues within a specific animal or vegetable food type, as well as between different varieties of it. In addition it has the power to follow changes in the protein component of various tissues during growth, maturation and post-mortem or post-harvest, as well as downstream treatments such as cooking.

3. Food proteomics

This section overviews and summarises the application of classical proteomics in food science, broken down into major food protein groups.

3.1 Dairy

Proteomics has been successfully applied to the study of milk proteins by many research teams, and significant effort has gone into the characterisation of the milk proteome. Also, bioactive milk components are of enormous scientific and commercial interest. Proteomics approaches have been used to compare milk from different species, while proteomic evaluation of other dairy products such as cheese has been a specialist subject that has also received attention.

In general, the dynamic range of proteins in milk poses a challenge to proteomics technologies. This is because the proteins in milk tend to be dominated by the caseins, which make up some 80% of the total protein content. Even when these are removed, the minor components in whey are dominated by one or two proteins; in bovine milk these are α-lactalbumin and β-lactoglobulin. Dynamic range issues have been overcome thanks to improvements in mass spectrometer sensitivity, coupled with the application of depletion and/or fractionation techniques. Casein, for example, is easily removed by acid precipitation. Alternatively affinity purification has been successfully used to remove IgA, lactoferrin, α-lactalbumin and serum albumin from human colostrum, allowing the identification of 151 proteins, over half of which have not been previously identified in colostrum or in milk (Palmer et al., 2006).

Following fractionation, substantial attention has been given to the identification of minor (including bioactive) components in milk. In one study 2-dimensional electrophoretic (2-DE) proteomic methods were applied to bovine whey after it had been fractionated into acidic, basic and non-bound components by semi-coupled anion and cation exchange chromatography (Fong et al., 2008). Utilising this approach, a large number of minor whey proteins were identified, some of which had not been previously reported in milk; in particular the acidic fraction was found to have a group of osteopontin peptides. Other investigators have used proteomics approaches to compare milk from different species to evaluate their suitability as a substitute for human milk (D'Auria et al., 2005).

Discovery and characterisation of bioactive milk components is of enormous scientific and commercial interest. For example, the host defence proteins in milk and colostrum have the potential to add significant value to the dairy industry, and techniques have been put in place to fractionate and analyse them using proteomics (Smolenski et al., 2007; Stelwagen et al., 2009).

The milk fat globule membrane (MFGM) constitutes another important component of milk. MFGM composition is of interest because it is known to be rich in bioactive components and there is increasing evidence that there are significant health benefits associated with the consumption of MFGM. Vanderghem *et al.* (2008), investigating a simple and rapid approach for the extraction of the MFGM, evaluated a number of different detergents and found that the inclusion of 4% CHAPS resulted in the removal of the highest amount of skim milk proteins as evaluated by 2-DE. Affolter *et al.* (2010) profiled two fractions of the MFGM, a whey protein concentrate (WPC) and a buttermilk protein concentrate (BPC) using three different approaches. Using a LC-MS/MS shotgun proteomics approach, 244 proteins were identified in WPC and 133 in BPC, while label-free profiling was used for semi-quantitative profiling and the determination of protein fingerprints.

Water soluble extracts of Teleme cheeses prepared from ovine, caprine and bovine milk were separated by 2-D gel electrophoresis and analysed by MALDI-MS and by HPLC in conjunction with Edman degradation, MS and tandem MS (Pappa et al., 2008). The 2-DE gels tended to be dominated by the casein and whey proteins, while in the MS analysis of the RP-HPLC-separated peptide fractions a few predominant peptides tended to mask the minor components. Nevertheless, enough differences were observed to enable the source of milk to be identified. Species specific differences were also observed in the tandem MS of peptides originating from casein. The effect of variations in milk protein composition on cheese yield of chymosin-separated sweet whey and casein fractions was examined by 2-D electrophoresis in conjunction with MS and multivariate data analysis (Wedholm, 2008). Using this approach it was possible to identify a range of proteins which had a significant effect on the transfer of proteins from milk to cheese. Included among these was a C-terminal fragment of β-casein, as well as a combination of several other minor fragments of β-, αs1 and αs-2 caseins, whose individual effect was relatively low.

3.2 Meat

The field of meat proteomics has seen steady growth over the past five years. A large number of studies have been performed, with a few distinct categories emerging. The post mortem conversion of muscle into meat is a significant series of events connected with protein modification and breakdown, and has received substantial attention. This has been extended to changes occurring in cooked/cured products. Protein marker discovery for various meat quality attributes is another important area. With countless animal breeds all providing different muscle/meat parameters of interest, there is significant potential for exciting discoveries leading to improved quality/value attributes and product differentiation.

Bjarnadottir et al, (2010) examined the insoluble protein fraction of meat (*longissimus thoracis*) from eight male Norwegian Red dual-purpose cattle during the first 48 hour post-mortem period using 2-DE and MALDI-TOF MS/MS and noted significant changes in 35 proteins, divided into three different groups based on predicted function, specifically: metabolism, cellular defence/stress and structural. Of these, most of the metabolic enzymes involved energy metabolism in the cell, while the cellular defence/stress proteins related mainly to the regulation and stabilisation of myofibrillar proteins. Laville et al. (2009) examined proteomic changes during meat aging in tough and tender beef using 2-DE and found that a higher proportion of proteins from the inner and outer membrane of mitochondria were found in the tender group, suggesting that a more extensive degradation of this organelle may be related to the apoptotic process. In another study into relationships between the protein composition of myofibrillar beef muscle proteins and tenderness Zapata et al. (2009) analysed by 1-DE and nano-LC-MS/MS the myofibrillar muscle fraction of *longissimus dorsi* from 22 Angus cross steers that had linearly regressed to shear values. Six bands showed a significant relationship and were found to contain a wide variety of cellular pathways involving structural, metabolic, chaperone and developmental functions.

Bouley et al. (2003) applied 2-DE to the separation of proteins from semitendinous muscles from Belgian Blue bulls that were either double-muscled homozygotes or non double muscled homozygotes. With the aim of identifying markers for muscle hypertrophy they found 26 proteins whose expression varied between homozygote types and hence had potential as markers for this trait. Shibata et al. (2009), in contrast, were interested in how the type of feed affected the muscle proteome of Japanese Black Cattle, in particular the difference between grass and grain feeding. The cattle were individually housed in a barn and fed a combination of ad libitum and Italian ryegrass hay until 21 months of age, whereupon half were put onto pasture outdoors until slaughter at 27 months of age. From 2-DE gels, differences were apparent and involved 20 spots from the sarcoplasmic and 9 from the myofibrillar fraction, of which adenylate kinase 1 and myoglobin from the sarcoplasmic fraction and slow twitch myosin light chain 2 from the myofibrillar fraction were significantly higher in the grazing group. All of this was indicative of a change from slow-twitch tissues to fast twitch tissues when the cattle were grazed in the latter fattening period.

The application of proteomics to following post-mortem changes in porcine muscles has also been of interest. Using a proteomic approach, te Pas et al. (2009) identified several proteins associated with drip loss and shear force in Yorkshire and Duroc pigs but none for cooking loss. In a study of post-mortem changes in *longissimus dorsi* muscles in pigs, Choi et al. (2010) used a 2-DE approach to show that muscles exhibiting a higher degree of protein denaturation not only displayed lower muscle pH, paler surfaces, and higher degrees of fluid loss of exudation, but were also characterised by myofibrillar and metabolic protein degradation. As a result of this work it would appear that myosin, actin, the troponin T 4f isoform and glycogen phosphorylase fragments have potential for explaining variations in the degree of protein degradation and meat quality. Hwang et al. (2005) also studied post-mortem changes in *longissimus dorsi* muscles of Landrace pigs using a gel-based approach, comparing those that had been fasted for 18 hours before slaughter with those fed on the morning of slaughter. They noted semi-quantitative changes in 27 proteins with muscle ageing, including myosin light chain 1, desmin, troponin T, cofilin 2, F-actin capping protein β subunit, ATP synthase, carbonate dehydrase, triosephosphate isomerise, actin, peroxiredoxin 2, α-b crystalline and heat shock protein 27 kDa. Lametsch and Benedixen (2001) used 2-DE to follow post-mortem changes in porcine muscles from 4 to 48 hours, paying particular attention to those focusing between 5-200 kDa and pH 4-9, of which 15 were found to show noticeable changes with time. Morzel et al. (2004) used proteomics to monitor the effects of two different pre-slaughter handling techniques on meat aging, specifically mixing of animals from different pens and transport to the abattoir 12 hours before slaughter with no mixing of animals and immediate slaughter after transport. They found significant changes in protein composition with ageing and in animals that were not mixed and slaughtered immediately there was over-expression of mitochondrial ATPase and increased concentration of a protein thought to be a phosphorylated form of myosin light chain 2. Lametsch et al. (2003) also looked at post-mortem changes in pig muscle tissue in relation to tenderness and found significant changes in 103 protein spots, of which 27 of the most prominent were identified. They found significant correlations between shear force and three of the identified actin fragments, the myosin heavy chain, as well as the myosin light chain II and triose phosphate isomerase I. A 2-DE approach was used to study relationships between protein composition and various meat quality traits in the *longissimus dorsi* muscle from Landrace pigs at various times after slaughter (Hwang, 2004). He found that high lightness values (as measured by Hunter L*) and drip loss coincided with high proteolysis rates. Moreover, 12 proteins appeared to be related to L* values, including α-actin, myosin light chain 1, cofilin 2, troponin T and α-b crystalline chaperone proteins. In addition four proteins were related to drip values, these being troponin T, adenylate cyclise, ATP-dependent proteinase SP-22 and DJ1 protein.

Proteomics has also been applied to the more general area of protein content and meat quality in pork. Kwasiborski et al. (2008) related 2-DE spots of the soluble portion from pig *longissimus lumborum* tissue to known meat quality traits using multiple regression analysis and determined that one to two proteins could explain between 25 and 85% of the variability in quality in this tissue. Relationships between the water holding capacity of pork and protein content were examined by van de Wiel and Zhang (2008) using a 2-DE approach. They were able to identify up to eight proteins that appeared to be significantly related to water holding capacity; the most clearly related were creatine phosphokinase M-type (CPK), desmin and a transcription activator. Another proteomic study provided evidence for pig muscles that were darker in colour having more abundant mitochondrial enzymes of the respiratory chain, haemoglobin and chaperone or regulator proteins (Sayd et al., 2006). Lighter coloured meats, in contrast were found to have more glycolytic enzymes. The genotype of the pig was also found to have an effect on meat quality (Laville et al., 2009). They used 2-DE to compare the sarcoplasmic protein profiles of semi-membranous muscles from early post-mortem pigs with different HAL genotypes (RYR1 mutation 1841T/C) and from ANOVA analysis found that 18% of the total matched protein spots were influenced by genotype, while hierarchical clustering analysis identified a further 10% that were coregulated with these proteins. One of the genotypes was found to contain fewer proteins of the oxidative metabolic pathway, fewer antioxidants and more protein fragments. The effect of the positive influence of compensatory growth on meat tenderness was examined by Lametsch et al. (2006), who divided female pigs into two groups, one of which had free access to the diet, another of which were feed-restricted from day 28 to 80 and were then given free access to the diet. Forty-eight hours post-mortem, proteins affected by compensatory growth were all found to be full length proteins, a result that goes against previous hypotheses that compensatory growth results in increased post-mortem proteolysis.

There has been some interest in applying proteomic techniques to analysing cooked pork products, including ham. Di Luccia et al. (2005) followed the progressive loss by hydrolysis of myofibrillar proteins during the ripening of hams using 2-DE, even to the extent of identifying a novel form of actin in the 2D gel. Sentandreu et al. (2007) used MALDI-TOF MS and MS/MS to identify peptides generated during the processing of ham. Sequence homology analysis indicated that they originated from actin. The protein composition of tough and tender pig meat, as defined by its shear force after cooking, was examined by 2-DE (Laville et al., 2007). They found at least 14 spots that differed significantly in quantity between the two groups, in particular, proteins involved in lipid traffic and control of gene expression were overrepresented in the tender meats, along with proteins involved in protein folding and polymerization.

When it comes to poultry, proteomics has been applied to identifying proteins differentially expressed in Cornish and White Leghorn chickens (Jung et al., 2007). They investigated over 300 spots in their 2-DE gels and found one protein that was differentially expressed in *pectoralis* muscles and four in *peroneus longus* muscles, all of which are assumed to be associated with muscle development, growth stress and movement in chickens. Nakamura et al. (2010) used proteomics to examine the allergenicity of meat from transgenic chickens. Using an allergenome approach they were able to identify five IgE-binding proteins that from 2D-DIGE analysis were found not significantly changed between non-GM and GM chicken. Proteomics also has proved to be useful for detecting chicken in mixed meat preparations. Sentandreu et al (2010) developed a method for the extraction of myofibrillar proteins and the subsequent enrichment of target proteins using OFFGEL isoelectric focusing, which were then identified by LC-MS/MS. Quantitative detection of chicken meat was achieved using AQUA stable isotope peptides and it was possible to detect contaminating chicken down to levels of 0.5%.

3.3 Seafood

There are a wide variety of ways that proteomics can be applied to seafood including the identification of species, characterisation of post-mortem changes in the various species of fish, or the effect of additives during the processing of fish muscle (Martinez and Friis, 2004). Gebriel et al. (2010), interested in understanding the effects of environmental, nutritional, biological and industrial factors on fish meat quality, undertook a large scale proteomic approach to first characterise cod muscle tissue composition. Using 1-DE coupled with nanoflow LC and linear trap quadrupole MS they were able to identify 446 unique proteins in cod muscle.

Addis et al. (2010) were interested in how physiological conditions experienced by wild and farmed fish affected muscle proteome of gilthead sea bream. They applied 2-DE and mass spectrometry to systematically characterise the proteome of this sea bream along the production cycle in four offshore floating cage plants and two repopulation lagoons in different areas of Sardinia. From this they were able to conclude that the protein expression profile of muscle tissue is comparable between the farmed and those found in the wild and therefore that farming in offshore cages would be good for proper muscle development and enable the production of higher quality fish. Veiseth-Kent et al. (2010) were also interested in the effects of fish farming on muscle and blood plasma proteomes but this time in changes induced by crowding. They found that the proteins mainly affected were those involved in secondary and tertiary stress responses and thus provided insight into the mechanisms causing accelerated muscle pH decline and rigor mortis contraction in salmon living under crowded conditions.

The application of proteomics to the study of post-mortem changes of fish muscle has also been of interest. Kjaersgard and Jessen (2003) examined the alterations in cod (*Gadus morhua*) muscle proteins over eight days in storage on ice using 2-DE and noted significant increases in intensity for eight spots over the first two hours, while there were significant decreases in intensity for two other spots over the entire eight days.

3.4 Cereal

Given the importance of rice as a staple food, it is not surprising that significant attention has been devoted to its proteome. Komatsu and Tanaka (2004) described the construction of a rice proteome database based around 23 reference 2-DE maps ranging from the cataloguing of its individual proteins to the functional characterization of some its component proteins, including major proteins involved in growth and stress response. As a result of this work, a total of 13,129 proteins are contained within these maps, of which 5092 have been entered into the database. Xue et al (2010) applied proteomics to rice with the view to determining its potential for detecting the unintended effects of genetically modified crops. By examining the total seed protein expression of two strains of transgenic rice with 2-DE they found that some of the seed proteins in the two lines differed in their relative intensities after comparison with their respective control lines. Kim et al. (2009) examined the proteomes of two different cultivars of rice, one high-quality and the other low-quality, and identified 15 proteins that may have important roles in quality determination, regulation of protein stability of imparting disease resistance during grain filing and storage. Ferrari et al. (2009) studied the proteomic profile of rice bran with a view to understanding its functional properties. Through a combination of gel and gel-free approaches they were able to identify 43 proteins with functions ranging from signalling/regulation, enzymic activity, storage, transfer to structural. Kang et al., (2010) were interested in the responses of plants to stress and applied a proteomic approach to determine how rice leaves respond to various environmental factors.

Other cereals that have been examined using proteomics include barley, maize, wheat (Agrawal and Rakwal, 2006) and corn (Ricroch et al., 2011). In the case of wheat, studies have ranged from an examination of the proteome of polyploid wheat cultivars to determine the effect of genome interaction in protein expression (Islam et al., 2003), through to studying allergens in wheat. For instance, Akagawa et al. (2007) undertook a comprehensive characterisation of allergens in wheat using a 2-DE approach coupled with MALDI-TOF MS. As a result of this study they were able to identify nine subunits of low molecular weight glutenins as being the most predominant IgE-binding proteins. The 2-DE maps they were able to generate were also considered to have much potential for the diagnosis of wheat-allergic patients or the identification of wheat allergens in food. Proteomics has also been applied to identify stress-induced proteins in wheat lines that may have a special role in food science (Horváth-Szanics et al., 2006). Using separation on 2-DE gels coupled with MALDI MS they studied a set of drought-stressed wide-range herbicide resistant transgenic spring wheat lines and noted that a number of inhibitor-like proteins were dominant in the stressed transgenic lines.

Akagawa et al. (2007) undertook a comprehensive characterisation of allergens in wheat using a 2-DE approach coupled with MALDI-TOF MS. As a result of this study they were able to identify nine subunits of low molecular weight glutenins as being the most predominant IgE-binding proteins. The 2-DE maps they were able to generate were also considered to have much potential for the diagnosis of wheat-allergic patients or the identification of wheat allergens in food.

3.5 Other

Proteomic studies of eggs have ranged from simple exploratory investigations to locating allergenic proteins or more complex interactome studies. Mann *et al.* (2008), using high-throughput mass spectrometry based techniques to search for bioactive compounds in eggs, identified 119 proteins in egg yolk, 78 in egg white and a further 528 proteins in the decalcified egg shell organic matrix. These included some 39 phosphoproteins from the egg shell soluble matrix, 22 of which had not been previously identified as phosphoproteins. D'Alessandro et al. (2010) took this one stage further by taking data from the literature and then regrouping and elaborating them for network and pathway analysis with the view to developing a unified view of the proteomes in egg, specifically those of egg white and yolk. As a result they were able to highlight roles for proteins in cell development or proliferation, cell-to-cell interactions and haematological system development in the egg yolk. The egg white proteins were found to be mainly in pathways involved with cell migration. Lee and Kim (2009) compared immunochemical methods such as ELISA, PCR and a proteomic approach involving MALDI-TOF and LC-tandem quadrupole-TOF MS for their ability to detect some egg allergens. Of these, ELISA proved to be very sensitive and specific; the proteomic approach was not able to detect some egg allergens such as ovomucoid because of its non-denaturing properties with urea and trypsin, while PCR proved unable to distinguish between eggs and chicken meat because it was tissue-non-specific.

Proteomics has also been applied to determine the extent of natural variation in the ripening of tomatoes and ultimately whether it would be possible to set up criteria for ripening at each stage (Kok et al., 2008).

Lei et al. (2011) developed a legume specific protein database incorporating sequences from seven different legume species which, when applied, resulted in a 54% increase in the average protein score and a 50% increase in the average number of matched peptides. When using MALDI-TOF MS data they found their success rate in identifying proteins increased from 19% with the NCBI nr database to 34% when they used the LegProt database.

Legumes are considered an important crop for sustainable agriculture but are relatively poor model systems for genetics and proteomics research, largely because of limited availability of sequence information. This is also the case for soybean as the sequencing of its genome is not yet finished (Komatsu and Ahsan, 2009). Nevertheless, despite the limited information available on soybean protein sequences Krishnan and Nelson (2011) were able to establish that the soybean cultivars with protein contents greater than 45% of the dry seed weight had a significantly higher content of seed storage proteins than the standard soybean cultivar (Williams 82); the largest difference in higher protein quality was within the 11S storage globulins. Proteomic studies have even been used to back up claims of the health benefits of soy (Erickson, 2005). In a recent study, an isoflavone from soybean has been found to prevent the regulation of proteins that induce apoptosis, cells which are the prime targets of arteriosclerotic stressors such as oxidised low density lipoproteins (Fuchs et al., 2005).

4. Protein modification

Most difficulties encountered in the proteomic evaluation of foods are due to the complexity of food protein modification. Even with advanced proteomic technologies, locating and tracking numerous modifications of proteins and peptides at trace levels adducts can remain a challenge. The need for accurate evaluation of a wide range of low abundance modifications is a differentiating factor associated with the proteomics of *ex vivo* proteins, such as food proteins. This section provides an overview of non-biological (environmental or process-induced) post-translation modifications that affect the performance and quality of food proteins.

Extracellular protein modification, particularly oxidative modification, is implicated in damage and degenerative processes within a diverse range of biological systems, including natural fibres, skin, eyes, pharmaceuticals and notably also in foods and ingredients (O'Sullivan and Kerry, 2009; Østdal et al., 2008). Protein and peptide modification plays a key role in all food quality and value attributes. *Ex vivo* modifications such as amino acid side-chain oxidation, protein-protein cross-linking and backbone cleavage can cause dramatic changes in product properties including shelf life, nutritional value, digestibility, functionality, health benefits and consumer appeal (Kerwin and Remmele, 2007). For instance, heat and/or alkali treatment of foods, feeds or pure proteins is very common in processing and manufacturing steps for a wide range of foods and ingredients, including dairy, meat, cereal and seafood products. However, such treatments can lead to the formation of a series of xenobiotics, including lysinoalanine and ornithoalanine, crosslinked modifications which have been implicated in nutritional damage and adverse health effects (Fay and Brevard, 2005; Gliguem and Birlouez-Aragon, 2005; Rerat et al., 2002; Silvestre et al., 2006).

Protein modification at the primary level is caused by a range of molecular mechanisms, but in food proteins, redox damage and Maillard chemistries are the major causes of modifications.

Reactive oxygen species (ROS) can be generated through exposure to heat and light, in particular, leading to oxidative attack (Dyer et al., 2006a; Grosvenor et al., 2010a; b). Two ROS in particular are implicated in protein oxidative degradation: singlet oxygen and the hydroxyl radical. Singlet oxygen is a highly reactive electrophilic species that reacts with proteins to form hydroperoxides, which decompose to form a variety of secondary oxidation products (Min and Boff, 2002). Where lipids are also present in the food system, initial oxidation to form lipid hydroperoxides can then result in subsequent secondary protein oxidative damage. Such lipid-mediated oxidation is brought about both through the triggering of radical chains and via the by-products of free radical lipid oxidation (Trautinger, 2007). On the other hand, hydroxyl radicals are highly reactive free radical species that can form from the initial generation of hydrogen peroxide and superoxide (Lee et al., 2004). Hydroxyl radicals are capable of reacting with any amino acid residue through α-hydrogen abstraction or direct residue side chain attack. In particular, they have an affinity for electron rich molecular structures, such as aromatic rings and sulphur-based moieties (Nukuna et al., 2001).

The Maillard reaction is a non-enzymatic process extremely important in the preparation of many food types. It results from initial chemical reaction between an amino acid and a reducing sugar and usually involves heat (Chichester, 1986). Maillard reaction products (MRPs) have a profound influence on food flavour and colour, but also affect all other key food properties. It is known that protein-bound MRPs are present at relatively high levels in common bakery, milk and cooked meat products. However, as any given modification is present in low relative abundance, traditional analytical techniques have had difficulty studying these changes at the molecular level. Proteomic approaches offer a highly sensitive and specific approach to evaluating Maillard products in foods.

The next sub-sections examine in more details the types of protein redox damage generated, followed by an overview of Maillard chemistries.

4.1 Side-chain redox chemistry

For whole proteins, model studies have indicated that protein backbone α- or β- positions are not the primary site of attack for reactive oxygen species, but that residue side chains are often the first point of ROS attack (Goshe et al., 2000). The protein residues most sensitive to oxidative modification are the aromatic and sulfur-containing amino acids (Berlett and Stadtman, 1997). Model studies on free amino acid oxidation have found that tryptophan, tyrosine, histidine, phenylalanine, and cysteine are the most susceptible to oxidation in solution. This correlates well with observations in whole proteins under oxidative conditions (Asquith et al., 1971; Asquith and Rivett, 1969; Boreen et al., 2008; Davies and Truscott, 2001; Schäfer et al., 1997).

Tryptophan and tyrosine oxidation results in a wide range of modified products, some of which can be used as oxidative markers (Asquith and Rivett, 1971; Davies et al., 1999; Dean et al., 1997; Guedes et al., 2009; Maskos et al., 1992; Simat and Steinhart, 1998; Żegota et al., 2005). Many of these products absorb light in the visible range, such as the tryptophan-derived yellow chromophore hydroxykynurenine, therefore affecting the colour of the protein material.

Additionally, the sulfur-containing residues cysteine and methionine are highly susceptible to oxidative degradation. Within proteins, cysteines are present both as free thiols and within disulfide bonds (cystine). Since disulfide networks play an important role in the structural and mechanical properties of proteins, oxidative modification resulting in non-reversible cleavage of these crosslinks is a critical consideration with respect to food texture, moisture retention and rheology. It is also noteworthy that disulfide bonds can have a protective effect on other residues, due to the same ROS quenching properties that make them vulnerable to oxidation (Li et al., 1992). Methionine oxidation results ultimately in the generation of methionine sulfoxide, which although a reversible oxidative modification *in vivo*, generally accumulates in extracellular proteins (Garner and Spector, 1980).

For proteins with low relative amounts of tryptophan, such as collagen, oxidation of other non-aromatic residues, such as proline, becomes more prevalent. Proline is susceptible to hydroxylation to form hydroxyproline, but other degradation products include glutamic semialdehyde and pyroglutamic acid (Sionkowska and Kaminska, 1999; Stadtman and Levine, 2006a).

Another form of redox modification is side chain scission, with cleavage typically occurring at the β-position (Dean et al., 1997; Stadtman and Levine, 2006a). This scission results in generation of carbonyl containing moieties as side-chain derivatives. For example cleavage of alanine and valine side-chains leads to formation of formaldehyde and acetone respectively. Side chain scission can also result in the introduction of a carbon centred radical into the protein backbone itself (see also sub-section 4.2). Other non-residue specific modifications of note are deamidation and decarboxylation. Deamidation can occur at asparagine and glutamine residues, as well as on peptide and protein *N*-termini. Decarboxylation particularly affects terminal residues (Stadtman and Berlett, 1991).

4.2 Carbonylation and protein backbone cleavage

The term carbonylation refers to the introduction of carbonyl groups into proteins, either on side-chains or on the protein backbone itself, and is a form of modification that has long been associated with protein oxidative damage (Holt and Milligan, 1977; Stadtman, 2006). Large relative quantities of carbonyl groups can be formed within proteins under oxidative conditions, and carbonylation therefore represents the most abundant kind of oxidative damage at the molecular level (Stadtman and Levine, 2006a). Carbonyl moieties are often formed on residue side chains, notably proline, threonine, arginine and lysine (Scaloni, 2006).

Protein backbone cleavage can occur when ROS extract α-hydrogen atoms, leading to the formation of carbon-centred radicals (Dalle-Donne et al., 2006). These radicals can subsequently react with molecular oxygen to form peroxy radicals, then peroxides and alkoxyl derivatives. Protein and peptide bond cleavage occurs via two different pathways; the diamide and α-amidation pathways (Stadtman and Levine, 2006a). The diamide pathways produces peptide fragments with diamide and isocyanate moieties, while the α-amidation pathway leads to peptide fragments with amide and α-ketoacyl moieties.

4.3 Protein crosslinking

Exposure of proteins to heat, alkali and/or oxidative conditions can lead to significant protein-protein inter and intramolecular crosslinking (Lee et al., 2004; Stadtman and Levine, 2006b). Residues with a propensity to form crosslinks include lysine, arginine, tyrosine, serine, cysteine and histidine (Taylor and Wang, 2007).

Lysine can form a range of crosslinks with other residues (Silvestre et al., 2006). Since lysine is an essential amino acid for nutrition, process-induced lysine crosslinking is of concern in the food industry. The best characterised crosslinked modification product is lysinoalanine. Lysinoalanine is formed from initial dehydration of cysteine or serine to form dehydroalanine, which subsequently reacts with proximal lysine residues.

Another key crosslinked modification in proteins is the formation of lanthionine from cystine, a crosslinked derivative which cannot be reductively cleaved (Bessems et al., 1987; Earland and Raven, 1961). Lanthionine formation is strongly associated with heat and alkali treatment of proteins. For proteins under oxidative conditions, the crosslink dityrosine is formed from two tyrosine residues after ROS attack to form tyrosyl radicals, followed by radical-radical combination (Dyer et al., 2006a).

4.4 Maillard chemistry

Particularly in the case of thermal modification, Maillard chemistries can result in a cascade of complex modifications that can profoundly affect colour, flavour, digestibility and nutritional value in foods (Gerrard, 2002). The Maillard reaction is a non-enzymatic process extremely important in the preparation of many food types. It results from initial chemical reaction between an amino acid and a reducing sugar and usually involves heat. The sugar carbonyl group reacts with the nucleophilic amino moiety of a free amino acid or protein residue to initially form an N-substitute glycosylamine. This unstable compound subsequently undergoes Amadori re-arrangement and a complex mixture of resultant products can form, many as yet poorly understood. However, these products are responsible for a wide range of flavours and odours in foods, with each food type having distinct patterns of Maillard reaction products (MRP).

Application of proteomics in the area of Maillard chemistry was first performed only within the last decade, and has subsequently been increasingly applied to study protein modification introduced during food processing (Cotham et al., 2003; Gerrard, 2006). Challenges still remain in the areas of quantification and development of procedures with increased sensitivity to locate novel modifications. Research correlating processing parameters to effect on nutritional damage to proteins at the molecular level due to Maillard chemistry has been limited to date.

The effect of MRP on the nutritional value of food is of critical concern and is a topic of considerable interest to the food industry. Processing, packaging and storage conditions can have a profound influence on the nutritional quality and health effects of food products (Ames, 2009; Seiquer et al., 2006). Maillard reactions can lead to a decrease in bio-availability of several essential amino acids (Hewedi et al., 1994; Shin et al., 2003), decreased protein digestibility and formation of some undesirable compounds (Taylor et al., 2003; Uribarri and Tuttle, 2006). The risks and benefits associated with consumption of MRPs remains a controversial issue (Lee and Shibamoto, 2002). Profiling and tracking nutritional quality of proteinaceous foods, such as meat or dairy, throughout processing and storage is of particular importance, due to the continuous development of MRP products, even at reduced temperatures. The enhanced formation of such products associated with oxidation from vitamins, such as ascorbic acid, frequently used as preservatives, is also a concern. MRP formation encompasses a complex array of reactions, starting with the glycation of proteins and progressing to form sugar-derived protein adducts and cross-links also known as advanced glycation end products (AGE). Impaired nutritional value can occur due to changes in protein integrity and function through protein crosslinking mediated by AGE. On the other hand, increased antioxidant activity in food systems can also be attributed to these reactions.

The profile and relative abundance of all these modification products vary between specific proteins and have a significant effect on all critical food quality traits, as discussed in the following sub-section.

4.5 Effects of protein modification on foods

Progressive generation of modifications at the protein primary level results in changes to secondary, tertiary and higher orders of protein structure through denaturation, protein breakdown and conformational re-ordering, leading to observable changes in the holistic properties of the food (Dean et al., 1997). An overview of the effects on food proteins is here presented based on food type.

4.5.1 Dairy

The main dairy protein types are caseins, albumins and globulins (Farrell et al., 2004). These proteins impart many of the important nutritional and functional characteristics to dairy food products. Degradation of milk and dairy products represents a significant area of concern for the dairy industry, due to the development of off-flavours, impaired nutritional quality and function. These problems can appear very rapidly under the right conditions, and create consumer complaints (Havemose et al., 2004; Jung et al., 1998; Mestdagh et al., 2005; Østdal et al., 2008).

Exposure to heat and/or light, in particular, can result in significant degradation in dairy products. For instance, flavour changes become noticeable rapidly in milk exposed to sunlight (Mestdagh et al., 2005). Development of off-flavours in milk and cheeses on light exposure is photosensitised by riboflavin (Becker et al., 2005). Such photo-oxidation increases protein carbonyl levels and leads to the formation of numerous tryptophan, tyrosine, histidine, and methionine oxidative products, as well as crosslink formation, notably dityrosine (Dalsgaard et al., 2007). Oxidation alters the structure of all the major milk proteins, with caseins particularly vulnerable to modification (Dalsgaard et al., 2008). Changes in protein folding, and polymerisation have been noted particularly for lactalbumin and lactoglobulin (Dalsgaard et al., 2007). The hydrolysis of dairy proteins is also affected by oxidation. Impaired accessibility of chymosin to oxidised dairy proteins has been observed, resulting in altered peptide profiles, including lowered total amounts of free N-termini (Dalsgaard and Larsen, 2009). Other notable modifications for dairy products are lysine-derived protein-protein crosslinks. In particular, lysinoalanine is formed from alkali or heat treatment of dairy proteins, and is therefore typically present in all common dairy products and ingredients at varying levels.

4.5.2 Meat and seafood

Flavour, colour, texture and nutritional value comprise the most essential quality attributes for all varieties of meats, seafoods and their derivative products, with a direct correlation to value and quality (Kerry and Ledward, 2009). These attributes are directly linked to the protein content. Meat typically consists of 10-30% proteins, depending on the species and the cut (Paul and Southgate, 1985). Meat flavour is derived from its profile of proteins, peptides, carbohydrates and lipids (Spanier et al., 2004; Wood et al., 1999). For meat products, lipid oxidation must be considered together with protein modification. Aldehyde-derived secondary lipid oxidation products are associated with the formation of proteinaceous chromophores, and therefore influence colour (Sun et al., 2001). Proteolysis plays an essential role in the developing flavour of meat and seafood products, while protein and lipid oxidation products are inter-related, and influence specific flavour and odour. An important component of meat colour is myoglobin chemistry, along with other proteins such as haemoglobin and cytochrome C (Kerry and Ledward, 2009; Mancini and Hunt, 2005). The texture of meat is directly related to its protein matrix, and is affected by protein backbone cleavage (proteolysis of structural proteins) and crosslinking. In terms of nutritional value, proteins represent a key macronutrient in meat and seafood, while protein-protein and protein-lipid crosslinks, protein cleavage and amino acid damage are associated with the deterioration of nutritional value (Erickson, 1997; Love and Pearson, 1971). Resultant food structural modifications in turn affect properties such as moisture retention, solubility and digestibility (Xiong, 2000).

Meat modification is further influenced by conditions immediately prior to, and after, animal slaughter. Physiological reactions to stress may influence the rate and extent of post mortem pH decline, altering the modification profile and thereby impacting quality parameters such as tenderness, colour and water-holding capacity (Ferguson et al., 2001; Ferguson and Warner, 2008). Recent findings suggest that pre-slaughter stress may enhance oxidative reactions during ageing (Picard et al., 2010). Oxidative damage is implicated in meat quality deterioration, for example in high O_2 packaging during long-term chilled storage. Protein *ex vivo* modification in seafood products is typically less well studied than meat, however it is of equal concern with respect to the quality and value of the resultant food product (Baron et al., 2007; Kinoshita and Sato, 2007; Kjaersgard and Jessen, 2004).

4.5.3 Grains and cereals

Degradative modification also significantly influences the quality of high protein plant-derived foods, notably grains and cereals. Oxidation of these proteins also influences the formation of off-flavours and the nutritive value (Heinio et al., 2002). Plant proteins are increasingly being used as ingredients and additives and ingredients in processed food, and therefore modifications induced in these proteins during processing and storage and an important consideration for the food as a whole.

5. Redox proteomics

The emerging proteomic sub-discipline of redox proteomics is based on the study of key reductive and oxidative chemistries occurring within proteins (Dalle-Donne et al., 2006). In complex biological systems exposed to varied processing and environmental conditions, such as is the case for foods, mapping the vast potential array of protein modifications is particularly challenging. Traditionally, holistic evaluation methodologies have been employed for assessing food proteins, including evaluation of protein extractability and determination of total carbonyl content. However, the advent of advanced proteomic tools and approaches means that mapping and tracking molecular level changes directly within the proteins is now possible. Redox proteomics adopts and customises techniques from more classical shotgun and differential proteomic approaches to provide an unprecedented overview of these molecular changes. Some of the underpinning steps in such an approach are outlined here.

5.1 Characterisation of modifications

The first critical step in mapping food protein modifications is characterisation of the modifications. This can be very challenging as modification patterns can be highly complex, and any given modification, at a specific site in a specific protein, is likely to be present in very low relative abundance. In addition, the induction of protein-protein crosslink networks throughout the substrate further complicates peptide extraction and identification. As a further note, it is critical that any extraction procedure selected for subsequent proteomic analysis does not induce any further protein modification. The protein sample is generally enzymatically digested and analysed by one- or multi-dimensional LC-ESI-MS/MS or LC-MALDI-MS/MS, in a typical MudPIT-type approach (Thomas et al., 2006). Detailed MS/MS evaluation allows specific peptides to be selected and fragmented to provide structural information on modifications, and modification pathways from the key affected amino acid residues can be constructed. Bioinformatic analysis allows the location of modifications within proteins. For food systems where extensive crosslinking inhibits enzymatic digestion, a combined chemical/enzyme approach can be employed. For instance, initial chemical digestion with 2-nitro 5-thiocyanobenzoic acid (NTCB), which cleaves next to cysteine residues, followed by more standard tryptic digestion, can be effective at digesting crosslinked proteins to produce peptides suitable to mass spectrometric evaluation (Koehn et al., 2009).

5.2 Profiling and tracking

Profiling protein modifications allows construction of degradation pathways, while tracking individual or collective modifications provides precise molecular evaluation of processing and/or environmentally-induced changes, as well as validation of any amelioration strategy (Davies et al., 1999; Gracanin et al., 2009).

Both redox and Maillard chemistries generate a range of products in susceptible protein regions, and it is the type and abundance profile of these multiple products that results in the overall food properties. Changes in relative abundance in mass spectrometric analysis can be used to track the progressive loss of native peptides and proteins of parent peptides, and the corresponding formation of modification products within those proteins and peptides (Grosvenor et al., 2009). This can be used to evaluate differing protein processing or treatment parameters, as wells as monitor variation under a range of environmental conditions. However, the effectiveness of MS ion abundance data to track modification is limited by factors such as variation between analyses.

A related approach to evaluating molecular level is the characterisation and tracking of marker peptides. An initial analysis of the protein substrate is performed to identify peptides which fulfil selected criteria, such as reproducible extraction and a unique mass within the MS profile of the sample (Grosvenor et al., 2009). Individual marker peptides are then monitored and tracked. Another emerging approach to profile protein modifications is to utilise a scoring system, where different modifications are assigned a score and the range of modifications observed at a certain MS/MS identification threshold contribute to the overall score (Dyer et al., 2010). The data can be broken down to evaluate selected modification types, and scores between samples allow effective comparison and contrast.

A recent, novel approach to tracking protein modifications is to utilise stable isotope tags, such as iTRAQ (Wiese et al., 2007; Wu et al., 2006). Peptides from protein samples subjected to a range of conditions, such as differing hydrothermal processing times, are each tagged with a different iTRAQ label after enzymatic digestion and subsequently combined into one sample. MS/MS analysis of individual peptides generates reporter ions which allow direct tracking of the relative abundance of specific native or modified peptides across the conditions. Proof-of-concept for this approach has been demonstrated in model peptides, but its utility in more complex samples, such as whole protein foods, is yet to be fully explored (Grosvenor et al., 2010a; b; 2011).

5.3 Mapping carbonylation

As discussed earlier in this chapter, holistic evaluation of protein carbonyl content is a relatively good indicator of overall oxidative degradation (Thomas et al., 2006). However, to locate where such carbonylation is occurring, more advanced proteomic strategies are required.

Characterisation and mapping of carbonylation sites within the proteome can be achieved by affinity purification coupled to LC-MS/MS. Typically, in this approach reaction of biocytin hydrazide to tag and trap the carbonyl groups results in hydrazone conjugate formation (Thomas et al., 2006). Streptavidin immobilised on agarose is then used to select out the carbonylated proteins or peptides, and then evaluation of the proteins is performed using a standard MudPIT approach. Mapping of carbonylation within food proteins has been very limited to date, but promises to further advance our understanding of food protein molecular changes under oxidative stress.

5.4 Crosslink characterisation

Protein-protein crosslinks are very difficult to characterise and locate proteomically. Specific crosslinks are typically present in very low abundance within the protein substrate. In addition, protein digestion techniques must leave crosslinked peptides intact during extraction. To add to the challenge, software for identifying and sequencing crosslinked peptides are highly limited in their utility. However, more recently new approaches are emerging which may progressively make comprehensive elucidation of crosslinks in complex food protein mixtures possible.

Many crosslink evaluation strategies developed to date are based initially on partial digestion of the protein substrate, often followed by isotopic labelling (Back et al., 2002; Chen et al., 1999). The most common isotopic labelling approach is based around the action of trypsin, which adds an oxygen atom to the C-terminus at its point of protein cleavage (Mirza et al., 2008). Since crosslinked peptides have at least two C-termini, tryptic digestion in ^{18}O isotopically enriched water results in two or more ^{18}O atoms being incorporated where a crosslink is present. A parallel digestion is performed in regular water. Subsequent proteomic analysis of the resultant peptides can then select for crosslinked peptides through identification of a characteristic mass shift of 8 amu relative to the unlabelled peptide in the case of a single crosslink (Back et al., 2002).

6. Redox proteomics case study

6.1 Introduction

In this case study we present the concept and practical application of an advanced damage scoring framework to characterise the oxidative and non-oxidative damage occurring in UVB-irradiated protein fractions and in commercially available dairy products.

As established in the previous section, redox proteomics is an essential component of the molecular-level characterisation of food. In particular, it can provide strong evidence for protein damage due to food processing and packaging, incurred during retail display or during preparation. Since nearly all food product quality parameters correlate directly or indirectly with protein quality and function, molecular-level characterisation of protein modification and damage is of crucial importance for the food industry. Understanding of protein damage at the amino acid residue level and correlation with processing parameters will ultimately allow process improvement and mitigation/repair of protein damage.

We have previously used a damage scoring system for the characterisation of UVB-induced photo-oxidation in wool (Dyer 2010). This system allowed us to assign an absolute numerical value to the degree of damage in the samples under investigation.

In this case study, we describe a significantly refined version of our damage scoring system, allowing us to perform an advanced evaluation of redox and non-redox protein modifications, with precise weighting and thresholding functionality. This is achieved in a state-of-the-art redox proteomics context, based on LC-MS/MS data, ensuring representative coverage across all key proteins in a sample.

First, in a proof of concept approach, we show how damage scoring helps to quantify and characterise the damage in two model proteins as induced by a redox event. Then, we demonstrate the application of our damage scoring system to commercially available dairy products, sourced from a local supermarket.

6.1.1 Proof of concept – lactoferrin and β-lactoglobulin

Our previous work has allowed us to develop an advanced understanding of the photochemistry underpinning oxidative modification of proteins subjected to UVB irradiation (Dyer et al., 2006a; Dyer et al., 2006b; Dyer et al., 2009; Dyer et al., 2010; Grosvenor et al., 2010a). Therefore, UVB irradiation was chosen as the method to reliably induce oxidative modifications in dairy proteins. The aim of this proof of concept preamble was to establish that our damage scoring system is able to detect redox proteomic differences between control and treated samples.

Bovine lactoferrin and β-lactoglobulin were chosen as the model proteins. Lactoglobulin is the major protein in whey and is extremely well characterised, including its potential post-translational modifications. Lactoferrin is a high value milk protein that encodes a number of antimicrobial peptides, however its potential modifications are less well characterised (Arseneault et al., 2010; Murdock et al., 2010; Stelwagen et al., 2009).

The proteins, at a concentration of 2.4 mg/ml in PBS, were subjected to UVB light for 40 minutes, while controls were left untreated. This procedure was expected to induce a significant amount of oxidative modifications compared to the control.

After reduction, alkylation and trypsin digestion of the proteins, the peptide samples were analysed using nanoLC-MS/MS. This allowed the acquisition of MS/MS information for a large number of peptides. This MS/MS data was then utilised for peptide/protein identification and characterisation of modifications. Specifically, data files were converted to peak lists, imported in the ProteinScape data warehousing software, and serial Mascot searching of the data using specific sets of variable modifications was initiated. A number of specific searches including a selection of variable modifications were conducted. Each search focused on a specific set of modifications, e.g. non-oxidative modifications such as methylation, deamidation, ubiquitylation, or on progressive sets of oxidative modifications such as 1st, 2nd, 3rd and 4th stage oxidation events of tryptophan. Searching was performed in the NCBInr database, querying approximately 40,000 *Bos taurus* sequences. A peptide homology threshold FDR of 6% was obtained. The various search results were then compiled per sample and exported to Excel. A collection of Visual Basic scripts was developed to remove redundancy, count, weight, score and sum the different damage modifications. Essentially, each targeted modification is given a factor reflecting its severity, based on its rank in the modification pathway (Dyer et al., 2010). The number of occurrences of each modification (multiplied by the modification factor) is then weighted against the number of observed susceptible residues. The sum of weighted scores constitutes the damage score. In a scenario-based approach, the trade-off between selectivity and sensitivity as determined by the peptide score acceptance threshold was explored. A score threshold of 35 was chosen as the best compromise. Figure 1 shows the damage scores obtained for the control and irradiated lactoferrin and β-lactoglobulin samples.

The current version of our scoring model provides a non-oxidative damage score (NODS) comprising 6 modifications, and an oxidative damage score (ODS) comprising 12 modifications. Together they make up the total damage score (TDS).

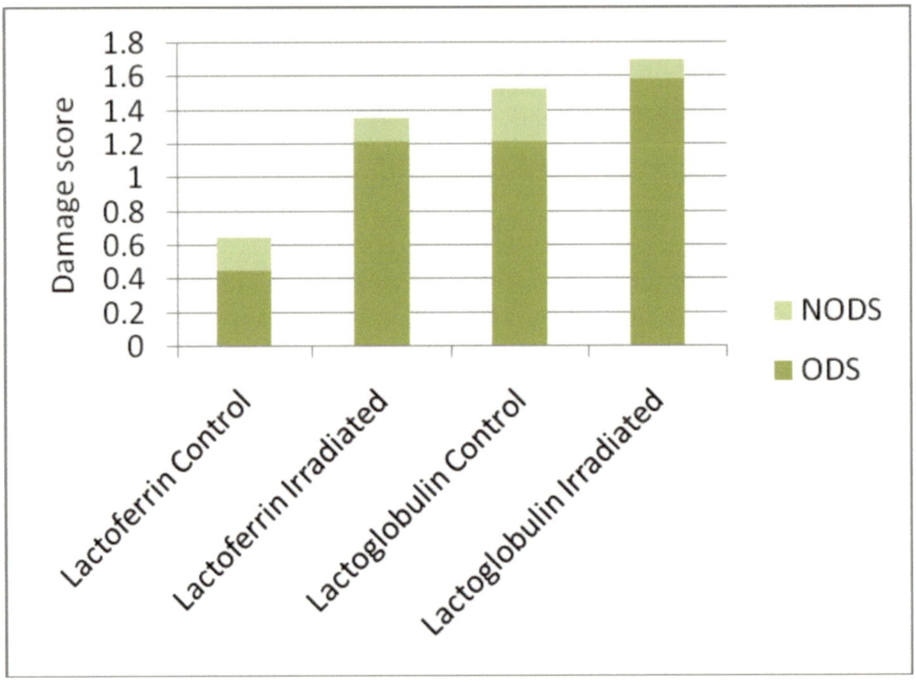

Fig. 1. Damage scores obtained for the control and irradiated lactoferrin and β-lactoglobulin samples. NODS, non-oxidative damage score; ODS, oxidative damage score.

Lactoferrin starts off with a relatively low TDS, made up of approximately 70% ODS and 30% NODS. After UV irradiation, the lactoferrin TDS more than doubles, with oxidative damage representing approximately 90% of the TDS.

Lactoglobulin has a much higher base damage level compared to lactoferrin - close to the damage level of irradiated lactoferrin. The reason for this is unknown, but it is possible that the process used to produce lactoglobulin includes a step that introduces more damage, compared to the process for lactoferrin. This finding highlights an additional strength of this scoring system: unexpected or unsuspected protein damage is revealed. In lactoglobulin, oxidative damage is responsible for approximately 80% of the TDS in the control sample. After irradiation, the TDS increases by 11%, with the oxidative damage component seeing a 30% increase.

It is presently unclear why the NODS appears to decrease between the control and irradiated samples. The fact that this is observed both for lactoferrin and lactoglobulin suggests that it is a direct or indirect effect of the irradiation. It is possible that some non-oxidative modifications remain undetected in the irradiated samples because of severe damage (e.g. crosslinking). Replicate analyses, detailed analysis of the modifications involved, and comparison with other proteins may help elucidate this further.

Using our scoring system, we observed an approximate doubling of the damage score for lactoferrin, and a substantial increase for lactoglobulin after irradiation. The ODS, in particular, increased significantly in the irradiated samples - in accordance with the experimental design, which strongly favoured oxidative modification.

This provides unequivocal proof that our system is able to detect differences in damage between control and treated samples using proteomics data.

6.1.2 Application – Commercial dairy products

With proof of concept established, we sought to apply our scoring system to commercially available dairy products, and investigate how various processing treatments influenced the damage score compared to fresh farm-sourced milk, expected to return the lowest damage score.

Nine commercially available dairy products were sourced from a local supermarket. Milk sourced directly from a local farm was included in our analysis as an expected low damage control.

The following samples were included:

- Farm sourced milk
- Standard fat milk
- Standard fat milk powder
- Standard fat milk UHT
- Skim milk UHT
- Skim milk powder
- Evaporated milk
- Low fat evaporated milk
- Lactose free UHT
- Skim calcium enriched UHT

Because of challenges associated with the protein concentration dynamic range in dairy, whey fractions were obtained from all samples in order to eliminate casein.

These fractions were loaded on anion exchange cartridges, and the unretained fraction was selected for further analysis. This fraction, among many other proteins, contains lactoferrin.

Samples were analysed as described in the previous section. Briefly, samples were reduced, alkylated, trypsin digested and analysed using LC-MS/MS. Data was imported in ProteinScape, repeatedly searched using specific sets of variable modifications, and combined search results were exported to Excel for scoring. As before, a score threshold of 35 was chosen. Figure 2 shows the damage scores obtained for the dairy samples.

Farm sourced milk and standard fat milk have very similar damage levels, although the NODS for farm sourced milk is slightly higher. In general, the levels of non-oxidative damage are in the same range in all samples, whereas the ODS differs substantially between samples, and gives rise to dramatic differences in TDS.

The TDS in standard fat milk powder is increased compared to standard fat milk, likely reflecting the additional processing step the former has undergone. UHT treatment dramatically increases the TDS for this product family, giving the highest damage score in this test. The other UHT-treated products also return some of the highest damage scores, underscoring the potential of UHT to induce hydrothermal damage, measured primarily as oxidative side chain modifications.

When heat-treated standard fat products are compared with heat-treated low fat products, substantially lower overall damage is recorded in the latter. Both standard fat milk UHT compared to skim milk UHT, and evaporated milk compared to low fat evaporated milk support this observation. Observations in a range of proteinaceous systems indicate that lipid oxidation has a synergistic effect with protein oxidation during damaging processing treatments or environmental insult. Consequently, products with lower lipid levels would be expected in many damaging conditions to sustain lower oxidative damage. Intriguingly however, this trend is not observed in powder products, with skim milk powder returning a

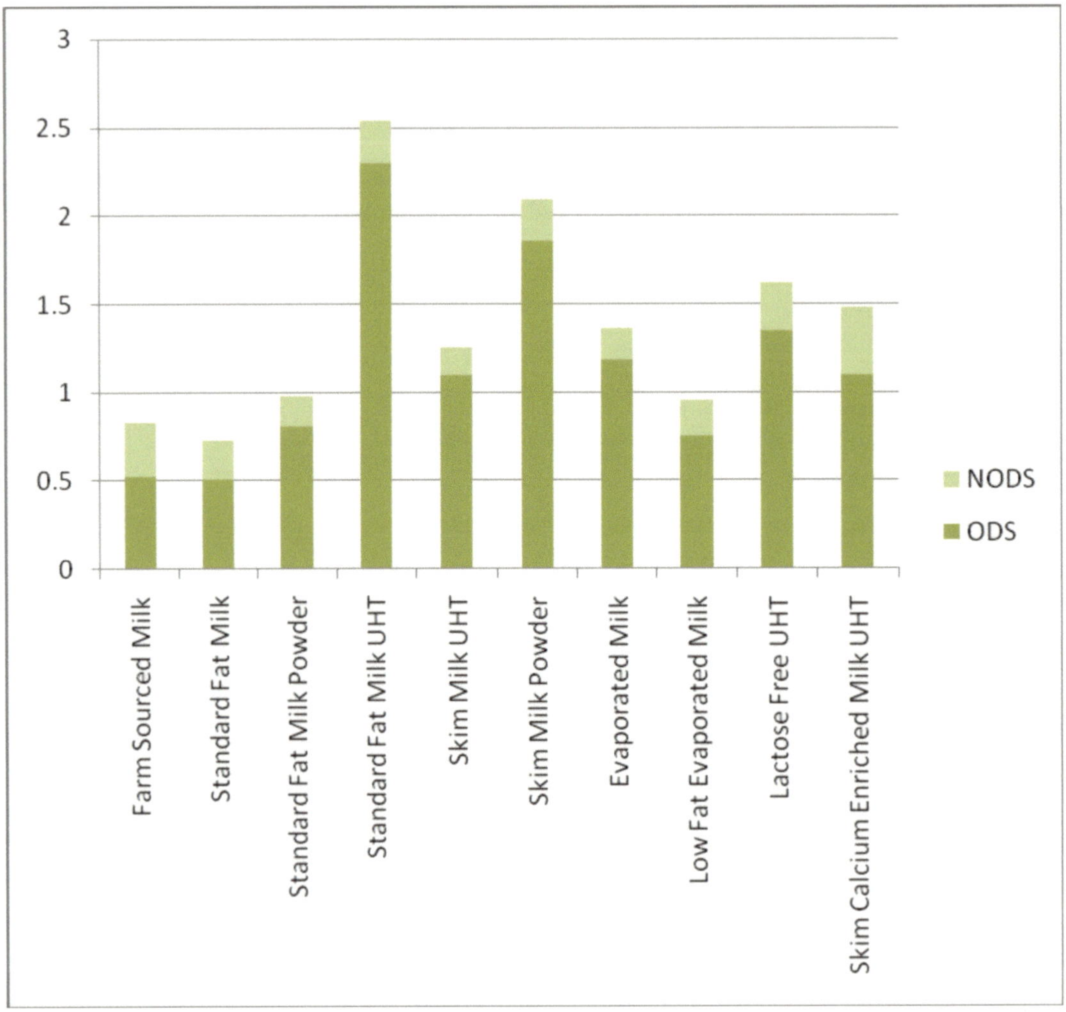

Fig. 2. Damage scores obtained for nine off-the-shelf dairy products and one farm-sourced milk sample. NODS, non-oxidative damage score; ODS, oxidative damage score.

very high TDS compared to standard fat milk powder. The molecular damage processes induced during skimming and during food protein spray drying processes therefore require further investigation.

6.2 Redox proteomics conclusions

In a proof of concept study, we showed that it is possible to detect molecular-level damage in a redox proteomics context. Specifically, we investigated control and UVB irradiated lactoferrin and lactoglobulin. This revealed substantial differences in the amount of oxidative modification and hence damage score between control and treated samples.

We then applied our damage scoring system to commercially available dairy products. This represented a significant increase in complexity, with analyses scaled up from individual proteins in the proof of concept, to the proteome level in the commercial samples. Using LC-

MS/MS approaches and automated data analysis routines, we were able to successfully acquire and process data, allowing damage scoring at the sample proteome level.
Scores returned for the dairy samples revealed that differently treated products can harbour substantial amounts of molecular-level damage. Farm-sourced and standard fat milk had the lowest levels of damage, while samples that had undergone processing treatments such as UHT or spray drying, showed elevated damage scores.
The importance of detailed protein damage scoring at the molecular level in food systems, with direct correlation and implications to holistic product quality parameters, cannot be overstated. Indeed, the scoring system we have developed is equally applicable to other protein foods such as meat, seafood, cereal or soy. Global trends indicate growing consumer awareness of correlations between excessive processing and decreased product quality. Consumers also increasingly demand pure, untreated, clean and healthy foods. Product quality parameters such as flavour, odour, eating quality, nutritional value and digestibility are directly correlated with protein modification and molecular level damage. Therefore, understanding and evaluating protein quality and function is of crucial importance for the food industry. It will allow food producers to track and evaluate improvements to processing parameters, will allow substantiation of marketing claims to increasingly discerning and knowledgeable consumers, and will ultimately provide the world with healthier, more nutritious food.

7. Future directions

The application and modification of proteomic approaches to analyse the complexity of food protein modification is anticipated to become increasingly important in the area of general food science, quality assurance and product differentiation. In addition, we anticipate further application of these approaches to understanding food protein modification with respect to subsequent digestive processes, and to actually tracking the molecular modification and truncation of proteins during human consumption and digestion. Until now, this area has been limited by the lack of sensitive and specific technologies to track molecular level damage mechanisms. However, development of the enabling proteomic technologies summarised in this chapter indicates that the goal of mapping the complex process of food protein digestion is attainable. This will facilitate the correlation of modification and digestion profiles with food quality traits, particularly nutritional and health effects.
Another key area of future development is in the development of advanced food quality and function. Future step-change advances in meat and dairy biotechnologies, such as advanced phenotyping for quality traits and predictive selection for resistance to specific oxidative stresses, will require integrated evaluation at the molecular level. In addition, to truly understand and control these phenotypes and product attributes, it is critical that proteomics is linked to lipidomics, as protein and lipid modification are correlated. These approaches have the potential to open up considerable new areas, enabling study of the networks involving the functional interactions and pathways of food proteins both *in vivo* and *ex vivo*.

8. References

Addis M.F., Cappuccinelli R., Tedde V., Pagnozzi D., Porcu M.C., Bonaglini E., Roggio T., Uzzau S. 2010. Proteomic analysis of muscle tissue from gilthead sea bream (Sparus aurata, L.) farmed in offshore floating cages. Aquaculture 309, 245-252.

Affolter M., Grass L., Vanrobaeys F., Casado B., Kussmann M. 2010. Qualitative and quantitative profiling of the bovine milk fat globule membrane proteome. J Proteomics 73, 1079-1088.

Agrawal G.K., Rakwal R. 2006. Rice proteomics: A cornerstone for cereal food crop proteomes. Mass Spec Rev 25, 1-53.

Akagawa M., Handoyo T., Ishii T., Kumazawa S., Morita N., Suyama K. 2007. Proteomic analysis of wheat flour allergens. J Agric Food Chem 55, 6863-6870.

Ames J.M. 2009. Dietary Maillard reaction products: implications for human health and disease. Czech Journal of Food Sciences 27 (Special Issue), S66-S69.

Arseneault M., Bédard S., Boulet-Audet M., Pézolet M. 2010. Study of the interaction of lactoferricin B with phospholipid monolayers and bilayers. Langmuir 26, 3468-3478.

Asquith R.S., Hirst L., Rivett D.E. 1971. Effects of ultraviolet radiation as related to the yellowing of wool. Applied Polymer Symposium 18, 333-335.

Asquith R.S., Rivett D.E. 1969. The photolysis of tyrosine and its possible relationship to the yellowing of wool. Text Res J 39, 633-637.

Asquith R.S., Rivett D.E. 1971. Studies on the photooxidation of tryptophan. Biochim Biophys Acta 252, 111-116.

Back J.W., Notenboom V., De Koning L.J., Muijsers A.O., Sixma T.K., De Koster C.G., De Jong L. 2002. Identification of cross-linked peptides for protein interaction studies using mass spectrometry and 18O labeling. Anal Chem 74, 4417-4422.

Baron C.P., Kjærsgård I.V.H., Jessen F., Jacobsen C. 2007. Protein and lipid oxidation during frozen storage of rainbow trout (*Oncorhynchus mykiss*). J Agric Food Chem 55, 8118-8125.

Becker E.M., Cardoso D.R., Skibsted L.H. 2005. Deactivation of riboflavin triplet-excited state by phenolic antioxidants: mechanism behind protective effects in photooxidation of milk-based beverages. Eu Food Res Tech 221, 382-386.

Berlett B.S., Stadtman E.R. 1997. Protein oxidation in aging, disease, and oxidative stress. J Biol Chem 272, 20313-20316.

Bessems G.J., Rennen H.J., Hoenders H.J. 1987. Lanthionine, a protein cross-link in cataractous human lenses. Exp Eye Res 44, 691-695.

Bjarnadóttir S.G., Hollung K., Færgestad E.M., Veiseth-Kent E. 2010. Proteome changes in bovine longissimus thoracis muscle during the first 48 h postmortem: Shifts in energy status and myofibrillar stability. J Agric Food Chem 58, 7408-7414.

Boreen A.L., Edhlund B.L., Cotner J.B., McNeill K.M. 2008. Indirect photodegradation of dissolved free amino acids: The contribution of singlet oxygen and the differential reactivity of DOM from various sources. Environ Sci Technol 42, 5492-5498.

Bouley J., Chambon C., Picard B. 2003. Proteome analysis applied to the study of muscle development and sensorial qualities of bovine meat. Sciences des Ailments 23, 75-78.

Chen X., Chen Y.H., Anderson V.E. 1999. Protein cross-links: Universal isolation and characterization by isotopic derivatization and electrospray ionization mass spectrometry. Anal Biochem 273, 192-203.

Chichester C.O. 1986. Advances in food research (advances in food and nutrition research). Academic Press. Boston.

Choi Y.M., Lee S.H., Choe J.H., Rhee M.S., Lee S.K., Joo S.T., Kim B.C. 2010. Protein solubility is related to myosin isoforms, muscle fiber types, meat quality traits, and postmortem protein changes in porcine longissimus dorsi muscle. 127, 183-191.

Cotham W.E., Hinton D.J., Metz T.O., Brock J.W., Thorpe S.R., Baynes J.W., Ames J.M., Tanaka S. 2003. Mass spectrometric analysis of glucose-modified ribonuclease. Biochem Soc Trans 31, 1426-1427.

D'Alessandro A., Righetti P.G., Fasoli E., Zolla L. 2010. The egg white and yolk interactomes as gleaned from extensive proteomic data. J Proteomics 73, 1028-1042.

D'Auria E., Agostoni C., Giovannini M., Riva E., Zetterström R., Fortin R., Franco Greppi G., Bonizzi L., Roncada P. 2005. Proteomic evaluation of milk from different mammalian species as a substitute for breast milk. Acta Pædiatrica 94, 1708-1713.

Dalle-Donne I., Scaloni A., Butterfield D.A., editors. 2006. Redox Proteomics - From protein modifications to cellular dysfunctions and diseases. John Wiley and Sons, Hoboken.

Dalsgaard T.K., Heegaard C.W., Larsen L.B. 2008. Plasmin digestion of photooxidized milk proteins. J Dairy Sci 91, 2175-2183.

Dalsgaard T.K., Larsen L.B. 2009. Effect of photo-oxidation of major milk proteins on protein structure and hydrolysis by chymosin. Int Dairy J 19, 362-371.

Dalsgaard T.K., Otzen D., Nielsen J., Larsen L.B. 2007. Changes in structures of milk proteins upon photo-oxidation. J Agric Food Chem 55, 10968-10976.

Davies M.J., Fu S., Wang H., Dean R.T. 1999. Stable markers of oxidant damage to proteins and their application in the study of human disease. Free Rad Biol Med 27, 1151-1163.

Davies M.J., Truscott R.J.W. 2001. Photo-oxidation of proteins and its role in cataractogenesis. J Photochem Photobiol B 63, 114-125.

Dean R.T., Fu S., Stocker R., Davies M.J. 1997. Biochemistry and pathology of radical-mediated protein oxidation. Biochem J 324, 1-18.

Di Luccia A., Picariello G., Cacace G., Scaloni A., Faccia M., Liuzzi V., Alviti G., Musso S.S. 2005. Proteomic analysis of water soluble and myofibrillar protein changes occurring in dry-cured hams. Meat Sci 69, 479-491.

Dyer J.M., Bringans S., Bryson W.G. 2006a. Characterisation of photo-oxidation products within photoyellowed wool proteins: tryptophan and tyrosine derived chromophores. Photochem Photobiol Sci 5, 698-706.

Dyer J.M., Bringans S.D., Bryson W.G. 2006b. Determination of photo-oxidation products within photoyellowed bleached wool proteins. Photochem Photobiol 82, 551-557.

Dyer J.M., Clerens S., Cornellison C.D., Murphy C.J., Maurdev G., Millington K.R. 2009. Photoproducts formed in the photoyellowing of collagen in the presence of a fluorescent whitening agent. Photochem Photobiol 85, 1314-1321.

Dyer J.M., Plowman J., Krsinic G., Deb-Choudhury S., Koehn H., Millington K., Clerens S. 2010. Proteomic evaluation and location of UVB-induced photomodification in wool. Photochem Photobiol B Biol 98, 118-127.

Earland C., Raven D.J. 1961. Lanthionine formation in keratin. Nature 191, 384-384.

Erickson B.E. 2005. Proteomics data back up soy health claims. J Proteome Res 4, 219.

Erickson M.C. 1997. Lipid oxidation: Flavor and nutritional quality deterioration in frozen foods. In: Erickson M.P., Hung Y.C., eds). Quality in Frozen Food. Springer, New York, pp 141-173.

Farrell H.M., Jr., Jimenez-Flores R., Bleck G.T., Brown E.M., Butler J.E., Creamer L.K., Hicks C.L., Hollar C.M., Ng-Kwai-Hang K.F., Swaisgood H.E. 2004. Nomenclature of the proteins of cows' milk - Sixth revision. J Dairy Sci 87, 1641-1674.

Fay L.B., Brevard H. 2005. Contribution of mass spectrometry to the study of the Maillard reaction in food. Mass Spec Rev 24, 487-507.

Ferguson D.M., Bruce H.L., Thompson J.M., Egan A.F., Perry D., Shorthose W.R. 2001. Factors affecting beef palatability — farmgate to chilled carcass. Aust J Exp Agr 41, 879-891.

Ferguson D.M., Warner R.D. 2008. Have we underestimated the impact of pre-slaughter stress on meat quality in ruminants. Meat Sci 80, 12-19.

Ferrari F., Fumagalli M., Profumo A., Viglio S., Sala A., Dolcini L., Temporini C., Nicolis S., Merli D., Corana F., Casado B., Iadarola P. 2009. Deciphering the proteomic profile of rice (Oryza sativa) bran: A pilot study. Electrophoresis 30, 4083-4094.

Fong B.Y., Norris C.S., Palmano K.P. 2008. Fractionation of bovine whey proteins and characterisation by proteomic techniques. Int Dairy J 18, 23-46.

Fuchs D., Erhard P., Turner R., Rimbach G., Daniel H., Wenzel U. 2005. Genistein reverses changes of the proteome induced by oxidized-LDL in EA.hy 926 human endothelial cells. J Proteome Res 4, 369-376.

Garner M.H., Spector A. 1980. Selective oxidation of cysteine and methionine in normal and senile cataractous lenses. Proc Nat Acad Sci USA 77, 1274-1277.

Gebriel M., Uleberg K.E., Larssen E., Hjelle B.A., Sivertsvik M., Møller S.G. 2010. Cod (Gadus morhua) muscle proteome cataloging using 1D-PAGE protein separation, nano-liquid chromatography peptide fractionation, and linear trap quadrupole (LTQ) mass spectrometry. Journal of Agriculture & Food Chemistry 58, 12307-12312.

Gerrard J.A. 2002. New aspects of an AGEing Chemistry — recent developments concerning the Maillard reaction. ChemInform 33, 243-243.

Gerrard J.A. 2006. The Maillard reaction in food: Progress made, challenges ahead. Trends Food Sci Technol 17, 324-330.

Gliguem H., Birlouez-Aragon I. 2005. Effects of sterilization, packaging, and storage on vitamin C degradation, protein denaturation, and glycation in fortified milks. J Dairy Sci 88, 891-899.

Goshe M.B., Chen Y.H., Anderson V.E. 2000. Identification of the sites of hydroxyl radical reaction with peptides by hydrogen/deuterium exchange: prevalence of reactions with the side chains. Biochem 39, 1761-1770.

Gracanin M., Hawkins C.L., Pattison D.I., Davies M.J. 2009. Singlet-oxygen-mediated amino acid and protein oxidation: Formation of tryptophan peroxides and decomposition products. Free Rad Biol Med 47, 92-102.

Grosvenor A.J., Morton J.D., Dyer J.M. 2009. Profiling of residue-level photo-oxidative damage in peptides. Peptides In preparation.

Grosvenor A.J., Morton J.D., Dyer J.M. 2010a. Isobaric labelling approach to the tracking and relative quantitation of peptide damage at the primary structural level. J Agric Food Chem 58, 12672-12677.

Grosvenor A.J., Morton J.D., Dyer J.M. 2010b. Profiling of residue-level photo-oxidative damage in peptides. Amino Acids 39, 285-296.

Grosvenor A.J., Morton J.D., Dyer J.M. 2011. Proteomic characterisation of hydrothermal redox damage. J Sci Food Agric, Early View.

Guedes S., Vitorino R., Domingues R., Amado F., Domingues P. 2009. Oxidation of bovine serum albumin: identification of oxidation products and structural modifications. Rap Comm Mass Spec 23, 2307-2315.

Havemose M.S., Weisbjerg M.R., Bredie W.L.P., Nielsen J.H. 2004. Influence of feeding different types of roughage on the oxidative stability of milk. Int Dairy J 14, 563-570.

Heinio R.-L., Lehtinen P., Oksman-Caldentey K.-M., Poutanen K. 2002. Differences between sensory profiles and development of rancidity during long-term storage of native and processed oat. Cereal Chem 79, 367.

Hewedi M.M., Kiesner C., Meissner K., Hartkopf J., Erbersdobler H.J. 1994. Effects of UHT heating of milk in an experimental plant on several indicators of heat treatment. J Dairy Res 61, 305-309.

Holt L.A., Milligan B. 1977. The formation of carbonyl groups during irradiation of wool and its relevance to photoyellowing. Text Res J 47, 620-624.

Horváth-Szanics E., Szabó Z., Janáky T., Pauk J., Hajós G. 2006. Proteomics as an emergent tool for identification of stress-induced proteins in control and genetically modified wheat lines. Chromatographia 63, S143-S147.

Hwang I. 2004. Proteomics approach in meat science: a model study for Hunter L* value and drip loss. Food Science and Biotechnology 13, 208-214.

Hwang I.H., Park B.Y., Kim J.H., Cho S.H., Lee J.M. 2005. Assessment of postmortem proteolysis by gel-based proteome analysis and its relationship to meat quality traits in pig longissimus. Meat Sci 69, 79-91.

Islam N., Tsujimoto H., Hirano H. 2003. Proteome analysis of diploid, tetraploid and hexaploid wheat: towards understanding genome interaction in protein expression. Proteomics 3, 549-557.

Jung K.C., Jung W.Y., Lee Y.J., Yu S.L., Choi K.D., Jang B.G., Jeon J.T., Lee J.H. 2007. Comparisons of chicken muscles between layer and broiler breeds using proteomics. AJAS 20, 307-312.

Jung M.Y., Yoon S.H., Lee H.O., Min D.B. 1998. Singlet oxygen and ascorbic acid effects on dimethyl disulfide and off-flavor in skim milk exposed to light. J Food Sci 63, 408-412.

Kang S., Chen S., Dai S. 2010. Proteomics characteristics of rice leaves in response to environmental factors. Frontiers in Biology 5, 246-254.

Kerry J.P., Ledward D., editors. 2009. Improving the sensory and nutritional quality of fresh meat. Woodhead Publishing Ltd, Cambridge, UK.

Kerwin B.A., Remmele R.L.J. 2007. Protect from light: Photodegradation and protein biologics. J Pharm Sci 96, 1468-1479.

Kim Y.J., Choi S.H., Park B.S., Song J.T., Kim M.C., Koh H.J., Seo H.S. 2009. Proteomic analysis of the rice seed for quality improvement. Plant Breeding 128, 541-550.

Kinoshita Y., Sato T. 2007. Proteomic studies on protein oxidation in bonito (*Katsuwonus pelamis*) muscle. Food Science and Technology Research 13, 133-138.

Kjaersgard I.V.H., Jessen F. 2004. Two-dimensional gel electrophoresis detection of protein oxidation in fresh and tainted rainbow trout muscle. J Agric Food Chem 52, 7101-7107.

Kjærsgård I.V.H., Jessen F. 2003. Proteome analysis elucidating post-mortem changes in cod (Gadus morhua) muscle proteins. J Agric Food Chem 51, 3985-3991.

Koehn H., Clerens S., Deb-Choudhury S., Morton J., Dyer J.M., Plowman J.E. 2009. Higher sequence coverage and improved confidence in the identification of cysteine-rich proteins from the wool cuticle using combined chemical and enzymatic digestion J Proteomics 73, 323-330.

Kok E.J., Lehesranta S.J., van Dijk J.P., Helsdingen J.R., Dijksma W.T.P., Van Hoef A.M.A., Koistinen K.M., Karenlampi S.O., Kuiper H.A., Keijer J. 2008. Changes in gene and protein expression during tomato ripening - consequences for the safety assessment of new crop plant varieties. Food Science and Technology International 14, 503-518.

Komatsu S., Ahsan N. 2009. Soybean proteomics and its application to functional analysis. J Proteomics 72, 325-336.

Komatsu S., Tanaka N. 2004. Rice proteome analysis: A step toward functional analysis of the rice genome. Proteomics 5, 938-949.

Krishnan H.B., Nelson R.L. 2011. Proteomic analysis of high protein soybean (Glycine max) accessions demonstrates the contribution of novel glycinin subunits. J Agric Food Chem 59, 2432-2439.

Kwasiborski A., Sayd T., Chambon C., Santé-Lhoutellier V., Rocha D., Terlouw C. 2008. Pig *Longissimus lumborum* proteome: Part II: Relationships between protein content and meat quality. Meat Sci 80, 982-996.

Lametsch R., Bendixen E. 2001. Proteome analysis applied to meat science: Characterizing post mortem changes in porcine muscle. J Agric Food Chem 49, 4531-4537.

Lametsch R., Karlsson A., Rosenvold K., Andersen H.J., Roepstorff P., Bendixen E. 2003. Postmortem proteome changes of porcine muscle related to tenderness. J Agric Food Chem 51, 6992-6997.

Lametsch R., Kristensen L., Larsen M.R., Therkildsen M., Oksbjerg N., Ertbjerg P. 2006. Changes in the muscle proteome after compensatory growth in pigs. J Anim Sci 84, 918-924.

Laville E., Sayd T., Morzel M., Blinet S., Chambon C., Lepetit J., Renand G., Hocquette H.F. 2009. Proteome changes during meat aging in tough and tender beef suggest the importance of apoptosis and protein solubility for beef aging and tenderization. J Agric Food Chem 57, 10755-10764.

Laville E., Sayd T., Terlouw C., Chambon C., Damon M., Larzul C., Leroy P., Glénisson J., Chérel P. 2007. Comparison of sarcoplasmic proteomes between two groups of pig muscles selected for shear force of cooked meat. J Agric Food Chem 55, 5834-5841.

Lee J.-Y., Kim C.J. 2009. Determination of allergenic egg proteins in food by protein-, mass spectrometry-, and DNA-based methods. Journal of AOAC International 93, 462-477.

Lee J., Koo N., Min D.B. 2004. Reactive oxygen species, aging, and antioxidative nutraceuticals. Comp Rev Food Sci Food Safety 3, 21-33.

Lee K.G., Shibamoto T. 2002. Toxicology and antioxidant activities of non-enzymatic browning reaction products: Review. Food Reviews International 18, 151-175.

Lei Z., Dai X., Watson B.S., Zhao P.X., Sumner L.W. 2011. A legume specific protein database (LegProt) improves the number of identified peptides, confidence scores and overall protein identification success rates for legume proteomics. Photochem 72, 1020-1027.

Li Z., Bruce A., Galley W.C. 1992. Temperature dependence of the disulfide perturbation to the triplet state of tryptophan. Biophys J 61, 1364-1371.

Love J.D., Pearson A.M. 1971. Lipid oxidation in meat and meat products - a review. J Am Oil Chem Soc 48, 547-549.

Mancini R.A., Hunt M.C. 2005. Current research in meat color. Meat Sci 71, 100-121.

Mann K., Olsen J.V., Maček B., Gnad F., Mann M. 2008. Identification of new chicken egg proteins by mass spectrometry-based proteomic analysis. World's Poultry Science Journal 64, 209-218.

Martinez I., Friis T.J. 2004. Application of proteome analysis to seafood authentication. Proteomics 4, 347-354.

Maskos J., Rush J.D., Koppenol W.H. 1992. The hydroxylation of tryptophan. Arch Biochem Biophys 296, 514-520.

Mestdagh F., De Meulenaer B., De Clippeleer J., Devlieghere F., Huyghebaert A. 2005. Protective influence of several packaging materials on light oxidation of milk. J Dairy Sci 88, 499-510.

Min D.B., Boff J.M. 2002. Chemistry and reaction of singlet oxygen in foods. Comp Rev Food Sci Food Safety 1, 58-72.

Mirza S.P., Greene A.S., Olivier M. 2008. ^{18}O labeling over a coffee break: A rapid strategy for quantitative proteomics. J Proteome Res 7, 3042-3048.

Morzel M., Chambon C., Hamelin M., Sante-Lhoutellier V., Sayd T., Monin G. 2004. Proteome changes during pork meat ageing following use of two different pre-slaughter handling procedures. Meat Sci 67, 689-696.

Murdock C., Chikindas M.L., Matthews K.R. 2010. The pepsin hydrolysate of bovine lactoferrin causes a collapse of the membrane potential in Escherichia coli O157:H7. Probiotics and Antimicrobial Proteins 2, 112-119.

Nakamura R., Nakamura R., Nakano M., Arisawa K., Ezaki R., Horiuchi H., Teshima R. 2010. Allergenicity study of EGFP-transgenic chicken meat by serological and 2D-DIGE analysis. Food and Chemical Toxicology 48, 1302-1310.

Nukuna B.N., Goshe M.B., Anderson V.E. 2001. Sites of hydroxyl radical reaction with amino acids identified by 2H NMR detection of induced 1H/2H exchange. J Am Chem Soc 123, 1208-1214.

O'Sullivan M.G., Kerry J.P. 2009. Sensory and quality properties of packaged meat. In: Kerry J.P., Ledward D., eds). Improving the Sensory and Nutritional Quality of Fresh Meat. Woodhead Publishing Limited. Cambridge. pp 595-597.

Østdal H., Weisbjerg M.R., Skibsted L.H., Nielsen J.H. 2008. Protection against photooxidation of milk by high urate content. Milchwissenschaft 63, 119-122.

Palmer D.J., Kelly V.C., Smit A.M., Kuy S., Knight C.G., Cooper G.J. 2006. Human colostrum: Identification of minor proteins in the aqueous phase by proteomics. Proteomics 6, 2208-2216.

Pappa E.C., Robertson J.A., Rigby N.M., Mellon F., Kandarakis I., Mills E.N.C. 2008. Application of proteomic techniques to protein and peptide profiling of Teleme cheese made from different types of milk. Int Dairy J 18, 605-614.

Paul A.A., Southgate D.A.T. 1985. McCance and Widdowson's the composition of foods. Elsevier. Amsterdam.

Picard B., Berri C., Lafaucheur L., Molette C., Sayd T., Terlouw C. 2010. Skeletal muscle proteomics in livestock production. Briefings Funct Genom Proteom 9, 259-278.

Rerat A., Calmes R., Vaissade P., Finot P.-A. 2002. Nutritional and metabolic consequences of the early Maillard reaction of heat treated milk in the pig. Significance for man. Eur J Nutr 41, 1-11.

Ricroch A.E., Bergé J.B., Kuntz M. 2011. Evaluation of genetically engineered crops using transcriptomic, proteomic, and metabolomic profiling techniques. Plant Physiol 155, 1752-1761.

Sayd T., Morzel M., Chambon C., Franck M., Figwer P., Larzul C., Le Roy P., Monin G., Cherel P., Laville E. 2006. Proteome analysis of the sarcoplasmic fraction of pig Semimembranosus muscle: Implications on meat color development. J Agric Food Chem 54, 2732-2737.

Scaloni A. 2006. Mass spectrometry approaches for the molecular characterisation of oxidatively/nitrosatively modified proteins. In: Dalle-Donne I., Scaloni A., Desiderio D.M., Nibbering N.M., eds). Redox Proteomics, John Wiley and Sons, Hoboken, pp 59-122.

Schäfer K., Goddinger D., Höcker H. 1997. Photodegradation of tryptophan in wool. J Soc Dyers & Colourists 133, 350-355.

Seiquer I., Díaz-Alguacil J., Delgado-Andrade C., López-Frías M., Muñoz Hoyos A., Galdó G., Navarro M.P. 2006. Diets rich in Maillard reaction products affect protein digestibility in adolescent males aged 11–14 y. The American Journal of Clinical Nutrition 83, 1082-1088.

Sentandreu M.A., Armenteros M.N., Calvete J.J., Ouali A., Aristoy M.-C.N., Toldraä F. 2007. Proteomic identification of actin-derived oligopeptides in dry-cured ham. J Agric Food Chem 55, 3613-3619.

Sentandreu M.A., Fraser P.D., Halket J., Patel R., Bramley P.M. 2010. A proteomic-based approach for detection of chicken in meat mixes. J Proteome Res 9, 3374-3383.

Shibata M., Matsumoto K., Oe M., Ohnishi-Kameyama M., Ojima K., Nakajima I., Muroya S., Chikuni K. 2009. Differential expression of the skeletal muscle proteome in grazed cattle. J Anim Sci 87, 2700-2708.

Shin S.H., Park H., Park D. 2003. Influence of different oligosaccharides and inulin on heterocyclic aromatic amine formation and overall metagenicity in fried ground beef patties. J Agric Food Chem 51, 6726-6730.

Silvestre D., Ferrer E., Gaya J., Jareno E., Miranda M., Muriach M., Romero F.J. 2006. Available lysine content in human milk: stability during manipulation prior to ingestion. Biofactors 26, 71-79.

Simat T.J., Steinhart H. 1998. Oxidation of free tryptophan and tryptophan residues in peptides and proteins. J Agric Food Chem 46, 490-498.

Sionkowska A., Kaminska A. 1999. Thermal helix-coil transition in UV irradiated collagen from rat tail tendon. Int J Biol Macromol 24, 337-340.

Smolenski G., Haines S., Kwan F.Y.S., Bond J., Farr V., Davis S.R., Stelwagen K., Wheeler T.T. 2007. Characterisation of host defence proteins in milk using a proteomic approach. J Proteome Res 6, 207-215.

Spanier A.M., Flores M., Toldra F., Aristoy M.C., Bett K.L., P. B., Bland J.M. 2004. Meat flavor: contribution of proteins and peptides to the flavor of beef. In: Shahidi F., Spanier A.M., Ho C.-T., Braggins T., eds). Quality of fresh and processed foods. Kluwer Academic/Plenum Publishers. New York. pp 33-49.

Stadtman E.R. 2006. Protein oxidation and aging. Free Rad Res 40, 1250-1258.

Stadtman E.R., Berlett B.S. 1991. Fenton chemistry. Amino acid oxidation. J Biol Chem 266, 17201-17211.

Stadtman E.R., Levine R.L. 2006a. Chemical modification of proteins by reactive oxygen species. In: Dalle-Donne I., Scaloni A., Butterfield D.A., eds). Redox Proteomics - From Protein Modifications to Cellular Dysfunctions and Diseases. John Wiley and Sons, Inc. Hoboken. pp 3-23.

Stadtman E.R., Levine R.L. 2006b. Chemical modification of proteins by reactive oxygen species. In: Dalle-Donne I., Scaloni A., Butterfield D.A., eds). Redox proteomics - from protein modifications to cellular disfunctions and diseases. John Wiley and Sons. Hoboken. pp 3-23.

Stelwagen K., Carpenter E., Haigh B., Hodgkinson A., Wheeler T.T. 2009. Immune components of bovine colostrum and milk. J Anim Sci 87, 3-9.

Sun Q., Faustman C., Senecal A., Wilkinson A.L., Furr H. 2001. Aldehyde reactivity with 2-thiobarbituric acid and TBARS in freeze-dried beef during accelerated storage. Meat Sci 57, 55-60.

Taylor C.M., Wang W. 2007. Histidinoalanine: a crosslinking amino acid. Tetrahedron 63, 9033-9047.

Taylor J.L.S., Demyttenaere J.C.R., Abbaspour Tehrani K., Olave C.A., Regniers L., Verschaeve L., Maes A., Elgorashi E.E., van Staden J., De Kimpe N. 2003. Genotoxicity of melanoidin fractions derived from a standard glucose/glycine model. J Agric Food Chem 52, 318-323.

te Pas M.F.W., Jansen J., Broekman K.C.J.A., Reimert H., Heuven H.C.M. 2009. Postmortem proteome degradation profiles of longissimus muscle in Yorkshire and Duroc pigs and their relationship with pork quality traits. Meat Sci 83, 744-751.

Thomas S.N., Lu B.-W., Nikolskaya T., Nikolsky Y., Yang A.J. 2006. MudPIT (multidimensional protein identification technology) for identification of post-translational protein modifications in complex biological mixtures. In: Dalle-Donne I., Scaloni A., Butterfield D.A., eds). Redox Proteomics - From Protein Modifications to Cellular Dysfunctions and Diseases. John Wiley and Sons, Inc. Hoboken. pp 233-252.

Trautinger F. 2007. Damaged proteins: Repair or removal? In: Giacomoni P.U., (ed. Biophysical and physiological effects of solar radiation on human skin. RSC Publishing. Cambridge, UK. pp 311-319.

Uribarri J., Tuttle K.R. 2006. Advanced glycation end products and nephrotoxicity of high-protein diets. Clinical Journal of the American Society of Nephrology 1, 1293-1299.

Van de Wiel D.F.M., Zhang W. 2008. Identification of pork quality parameters by proteomics. Meat Sci 77, 46-53.

Vanderghem C., Blecker C., Danthine S., Deroanne C., Haubruge E., Guillonneau F., De Pauw E., Francis F. 2008. Proteome analysis of the bovine milk fat globule: Enhancement of membrane purification. Int Dairy J 18, 885-893.

Veiseth-Kent E., Grove H., Færgestad E.M., Fjæra S.O. 2010. Changes in muscle and blood plasma proteomes of Atlantic salmon (Salmo salar) induced by crowding. Aquaculture 309, 272-279.

Wedholm A. 2008. Variation in milk protein composition and its importance for the quality of cheese milk. Uppsala: Swedish University of Agricultural Sciences.

Wiese S., Reidegeld K.A., Meyer H.E., Warscheid B. 2007. Protein labeling by iTRAQ: A new tool for quantitative mass spectrometry in proteome research. Proteomics 7, 340-350.

Wood J.D., Enser M., Fisher A.V., Nute G.R., Richardson R.I., Sheard P.R. 1999. Manipulating meat quality and composition. Proc Nutr Sci 58, 363-370.

Wu W.W., Wang G., Baek S.J., Shen R.-F. 2006. Comparative study of three proteomic quantitative methods, DIGE, cICAT, and iTRAQ, using 2D gel- or LC-MALDI TOF/TOF. J Proteome Res 5, 651-658.

Xiong Y.L. 2000. Protein oxidation and implications for muscle food quality. In: Decker E., Faustman C., Lopez-Bote C.J., eds). Antioxidants in muscle foods - nutritional strategies to improve quality. Wiley Interscience, Hoboken, pp 85-112.

Xue K., Liu B., Yang J., Xue D. 2010. The integrated risk assessment of transgenic rice: A comparative proteomics approach. Electronic Journal of Environmental, Agricultural and Food Chemistry 9, 1693-1700.

Zapata I., Zerby H.N., Wick M. 2009. Functional proteomic analysis predicts beef tenderness and the tenderness differential. J Agric Food Chem 57, 4956-4963.

Żegota H., Kołodziejczyk K., Król M., Król B. 2005. o-Tyrosine hydroxylation by OH· radicals.2,3-DOPA and 2,5-DOPA formation in γ-irradiated aqueous solution. Radiat Phys Chem 72, 25-33.

2

Fruit Proteomics

Ariel Orellana and Ricardo Nilo
FONDAP Center for Genome Regulation,
Millennium Nucleus in Plant Cell Biotechnology,
Centro de Biotecnología Vegetal,
Facultad de Ciencias Biológicas, Universidad Andrés Bello
Chile

1. Introduction

Obesity has been recognized as a major threat to human health in the 21st century [Yun, 2010]. One of the central causes to such nutritional disturbance relies in the consumption boost of the so called "fast food", which is characterized by high levels of fat, salt and sugars [Rosenheck, 2008]. On the opposite side of the spectrum are the plant fruits, which are characterized by high levels of relevant nutrients such as phenolic compounds, vitamins and essential minerals [Prasanna et al., 2007]. Besides its direct positive effects on human health, fruit intake has also been associated with the prevention of age-related neurodegeneration and cognitive decline [Spencer, 2010]. Some relevant aspects that may help fruit become an alternative to the ingestion of "fast food" are its easiness of consumption, which helps in its fast intake, and attractive organoleptic characteristics, aspects that have driven the "fast food" adoption by the society. However, fruit are very perishable and most of the species have a distinct seasonal producing pattern, making access throughout the year a difficult task and increasing their costs. Therefore, a great effort has been placed to understand the molecular mechanisms that could affect the pre- and post-harvest life of fruit, based on the hypothesis that this knowledge could improve the quality and accessibility of these goods to the society [Palma et al., 2011].

In the present chapter, the main proteomic approaches used to assess fruit development and ripening are described. Examples that will help the reader understand and recognize the advantages and drawbacks of each method, in order to decide the one that best suits their own objectives, are provided.

2. Fruit ontogeny

The ovule, being the female structure that develops into seed, is central for seed-bearing plant reproduction. During evolution, a specialized structure was generated to protect it, giving rise to the angiosperms, as opposed to the more ancient gymnosperms. This organ, termed carpel, encloses the seeds, being the fruit precursor [Scutt et al., 2006]. Carpels usually are located at the innermost whorl of the angiosperm flower, the so called gynoecium. Either individual carpels or syncarpic gynoecia (where both organs are fused together) are divided into tissues which perform distinct roles in reproduction, such as the ovary, which accommodates the ovules and in which fertilisation takes place [Ferrandiz et

al., 2010]. Upon ovule fertilisation, the carpel tissues undergo a series of developmental changes that leads to the formation of the fruit, which not only protects and supports the developing seeds, but also contributes to its later dissemination [Scutt et al., 2006].
When an ovary develops into a fruit, the ovary wall becomes the pericarp, the fruit wall which is composed of three layers with characteristics that are species dependent: the exocarp, mesocarp and endocarp. The first one is the outermost protective layer, also known as peel or skin. The mesocarp, located at the middle, holds the succulent edible part of fruits such as peaches and mangoes, among others. The third inner layer is the endocarp [Levetin & McMahon, 2008].
Fruit can be broadly classified as dry or fleshy. In the former, the pericarp may be hard and woody or thin and papery [Levetin & McMahon, 2008]. Regarding the latter, virtually all parts of the total inflorescence structure could be, depending on the species, developed into fruit flesh, a bulky, succulent parenchymatous tissue that accumulates water and many organic compounds [Coombe, 1976]. Any of these diverse tissues could be the subject of study, making a universal protocol for its evaluation a difficult task to fulfil. Additionally, a variety of fruits are characterized by having large variations in interfering metabolites that occur during their development, mainly during the process of ripening [Martínez-Esteso et al., 2011; Palma et al., 2011]. This situation imposes further hurdles to the analysis of the samples, with one protocol suited for a particular developmental stage not necessarily the most appropriate for another.
Summary: Fruit ontogeny is quite complex, which entails difficulties in establishing a unique protocol for the proteomic analysis of their derived tissues. An empirical evaluation is almost certainly necessary for fruit that has not been tested before, even though certain guidelines can be followed on the basis of the previous work in the field.

3. Two dimensional gel electrophoresis

The process that leads to the successful completion of the two dimensional gel electrophoresis (2-DE), meaning a SDS-PAGE gel derived image with well resolved spots representing a proteome fraction of the fruit tissue under evaluation is comprised by five main steps: Protein extraction, isoelectrical focusing, equilibration, SDS-PAGE and protein visualization [Rabilloud & Lelong, 2011]. Since the nature of the fruit tissues is so diverse and particular, each of the above mentioned steps may have to be improved in order to achieve a proper final result, often through an empiric evaluation. However some general guidelines can be given as well as a rational basis to refine these steps.

3.1 Protein extraction from fruit tissues

Plant cells are characterized by the presence of extensive amounts of water, a crucial feature to maintain cell turgor, which helps the cell to accomplish several physiological processes. In terms of fruit post-harvest life, the turgor is directly involved in the organ integrity. However, this characteristic represents an important drawback for the protein extraction, since the amount of protein present per cell mass is very low due to this massive presence of water inside the fruit cells [Saravanan & Rose, 2004]. The presence of a cell wall also poses a difficulty for protein recovery, due to the nonspecific sticking of proteins to this polysaccharide matrix [Rose et al., 2004]. In addition, unlike other plant tissues, fruit tissues display a high content of proteases and metabolites such as phenolics, organic acids, lipids, pigments and polysaccharides, which interfere with protein extraction and gel image analysis [Carpentier et al., 2005; Wang et al., 2008]. The presence of such contaminants may result in horizontal and vertical streaking as well as smearing, with the consequent reduction in the number of distinctly resolved protein spots on 2-DE gels [Saravanan & Rose, 2004]. Therefore, for a fruit-based proteomics analysis, the protein extraction method is a critical issue to address.

Of foremost importance is the avoidance of protein modifications during the extraction steps in order to diminish the probability of generating artefacts, such as false spots unrepresentative of the sample under analysis, which can lead to misleading conclusions. These modifications may be generated by chemical alterations of the proteins [Righetti, 2006] or by biological compounds such as proteases [Rabilloud & Lelong, 2011].

3.1.1 Tissue disruption

Even though tissue disruption could be considered the simplest step, the efficiency of the entire process relies heavily on this step [Giavalisco et al., 2003]. Based on our extensive experience on this topic and in the literature, by far the most used and efficient method to render proteins available for extraction is the liquid nitrogen assisted mortar/pestle method of tissue grinding (Table 1). Most of the times the finer the powder the higher the protein yield, therefore the use of auxiliary materials to improve the final grinding result, such as quartz sand, or equipment such as stainless steel blenders, may be advisable when dealing with hard, fibrous tissues such as non-ripe firm fruit [Giavalisco et al., 2003; Vincent et al., 2006]. Few authors report the use of sonication or homogenizers to assist fruit tissue disruption/sample homogenization [Lee et al., 2006; Di Carli et al., 2011]. Other methods have been proposed to accomplish similar and more reproducible plant tissue disintegration techniques such as acoustic related technologies [Giavalisco et al., 2003; Toorchi et al., 2008]. However, these methods require access to specialized equipment.

3.1.2 Sample homogenization

Upon proper cell disruption, the recovered tissue is homogenized. The main objectives of this step are to capture and separate the proteins from other metabolites that may interfere with the subsequent proteome characterization. The direct recovery of the proteins from the disrupted tissues by solubilizing the samples in an IEF lysis buffer has proved to be inadequate for these kinds of samples [Wang et al., 2003; Carpentier et al., 2005]. Therefore, alternative and more labour intensive procedures must be used. At least two methods have been widely used to perform this task and are extensively described in the literature. Tissue homogenization in an aqueous buffer followed by protein extraction with phenol or protein precipitation with trichloroacetic acid [Wang et al., 2008]. Importantly, both render proteins amenable for mass spectrometric analysis [Sheoran et. al, 2009].

In the phenol based method, an aqueous buffer is added to the pulverized tissue, followed by protein extraction with this solvent [Hurkman & Tanaka, 1986]. The nature of this buffer may differ greatly among protocols, but is usually composed of reducing and chelating agents which helps cope with polyphenols, metalloproteases and polyphenol oxidases, salts that promote protein extraction, and protease inhibitors dissolved in high pH buffer (Table 1). Polyvinylpolypyrrolidone (PVPP) has also been used to adsorb polyphenols, even though its action is restricted to those molecules in non-ionized states, such as in low pH environments [Carpentier et al., 2005]. The use of SDS and sample heating has been reported [Hurkman & Tanaka, 1986; Hu et al., 2011], albeit the surfactant should be removed prior isoelectric focusing (IEF) in order to avoid its interference on this step [Molloy, 2000; Görg et al., 2004]. A recent report, where mesocarp proteins from *Prunus persica* fruit were evaluated, suggests that the direct phenol extraction of freeze-dried tissue, followed by the addition of an aqueous buffer, could improve both the protein yield as well as the number of detectable spots on 2-DE gels [Prinsi et al., 2011]. Thus, variations of the method have been performed, although the most used version is the one described by Hurkman and Tanaka (Table 1) [Hurkman & Tanaka, 1986].

Regarding the second method, several versions have been generated, most employing the addition of trichloroacetic acid (TCA) and acetone to a sample extracts to achieve protein precipitation. This step is followed by resolubilization in an appropriate IEF buffer (see below). Variations are mainly focused in the solubilisation of the pulverized tissue in an aqueous buffer, similar to the one used in the phenol based method, prior to the addition of TCA/acetone [Saravanan & Rose, 2004]. A combination of TCA/acetone washes followed by phenol-based protein extraction proved to be successful in dealing with plant samples rich in lipids and pigments, such as mature grape berry clusters [Wang et al., 2003; Vincent et al., 2006]. These interfering compounds are the main contaminants of the phenol-based protocol, since they do not partition in the buffer phase during the first steps of this procedure [Carpentier et al., 2005].

Direct comparisons of these methods, using tissues such as tomato pericarp and grape berry, indicates that the phenol based procedure outperforms the TCA/acetone precipitation method both in terms of protein yield and qualitative characteristics of the 2-DE gels (Table 1)[Saravanan & Rose, 2004; Carpentier et al., 2005]. These differences may arise from dissimilar capacities of both protocols to nullify the proteases activity, and in difficulties in resolubilizing the proteins precipitated by the TCA/acetone protocol [Carpentier et al., 2005]. Since the latter is still the method of choice for many researchers, alternative methodologies to overcome this problem have been evaluated (see below).

More elaborated pre-treatments have been used for extraction of proteins from highly recalcitrant tissues, such as grape berry pericarp [Martínez-Esteso et al., 2011]. Mesocarp were homogenized at 4°C in extraction buffer containing 50 mM Na_2HPO_2 pH 7.0, 1 mM EDTA, 0.1 M PVPP, 1 mM $Na_2O_5S_2$, 10 mM ascorbic acid, and a cocktail of protease inhibitors. The homogenate was filtered through eight layers of cotton gauze and the filtrate was centrifuged. The resulting pellet was washed once in a buffer containing 50 mM Na_2HPO_2 pH 7.0, 1 mM EDTA, 0.1 M NaCl, 10 mM ascorbic acid, and recovered by centrifugation. Afterwards the pellet was cleaned with ethyl acetate:ethanol 1:2 (v/v), followed by TCA and acetone, as described by Wang and others [Wang et al., 2003; Martínez-Esteso et al., 2011]. Another alternative cited in the literature, with a similar performance to TCA/acetone, was used to extract protein from coffee seeds, tissues rich in polyphenols. Samples were milled with liquid nitrogen and extracted in a solution containing 0.1 M acetic acid, 3 M urea and 0.01% CTAB. Extracts were then centrifuged and supernatants were precipitated in an anhydrous solution of acetone and methanol. The samples were stored at low temperature and then centrifuged. The resulting pellet was resuspended in an appropriate IEF buffer [Gil-Agusti et al., 2005].

Summary: An efficient tissue disruption using liquid nitrogen assisted mortar and pestle followed by phenol-based extraction of the proteins has proven to be the best option to achieve a reproducible and adequate amount of proteins that can be used in the subsequent electrophoretic separation. If the samples are especially rich in lipids and pigments, an initial wash with organic solvents, such as TCA and acetone, prior to protein extraction with phenol, is recommended.

Species	Tissue	Tissue disruption			Homogenization buffer		Protein extraction		Resuspension/solubilization buffer											References
		N₂ assisted grinding		Other	Tris pH 8, reducing agent, protease inhibitor, EDTA, salt	Tris pH 8, SDS, salt, incubated at high °T	Aqueous/phenol two phase[1]	TCA/acetone, reducing agent	Chaotrope			Detergent				Reducing agent		Ampholytes		
			PVPP						Urea		Thiourea	CHAPS	ASB-14	SB3-10	OTHER	DTT	OTHER	pH 3-10	pH 4/5-7/8	
									7M	5M	2M									
Capsicum annuum	Placental tissue	X		X[2]					>[3]			X								Lee et al., 2006
Citrus reticulata	Juice sacs	X					?		>			X				X			X	Yun et al., 2010
Elaeagnus umbellata	Mesocarp	X			X		X		>		X	X			X[4]	X		X		Wu et al., 2011
Fragaria x ananassa	Whole fruit	X			X		X		>			X				X		X		Hjernø et al., 2006
Fragaria x ananassa	Whole fruit[5]	X		X[6]		X		X	X		X	X	X			X		X		Zheng et al., 2007
Fragaria x ananassa	Whole fruit	X	X	X[6]	X		X		X		X	X				X		X		Zheng et al., 2007
Fragaria x ananassa	Accrescent receptacle	X			X		X		>			X				X[7]		?[7]	?[7]	Bianco et al., 2009
Malus domestica	Pseudocarp	X			X		X		>			X				X		X		Guarino et al., 2007
Malus domestica	Peel	X		X[6]		X		X	X		X	X	X			X		X		Zheng et al., 2007
Malus domestica	Peel	X	X	X[6]	X		X		X		X	X				X		X		Zheng et al., 2007
Malus domestica	Pericarp[8]	X		X[6]	X			X		X	X	X		X		X	X[9]	X	X	Song et al., 2006
Malus domestica	Pericarp[8]	X		X[6]		X		X	X		X	X	X			X		X	X	Song et al., 2006
Musa spp	Meristem cultures	X						X	X		X	X				X		?	?	Carpentier et al., 2005
Musa spp	Meristem cultures	X			X		X		X		X	X				X		?	?	Carpentier et al., 2005
Musa spp	Mesocarp	X		X[6]	X			X		X	X	X		X		X	X[9]	X	X	Song et al., 2006
Musa spp	Mesocarp	X		X[6]		X		X	X		X	X	X			X		X	X	Song et al., 2006
Musa spp	Meristematic tissue	X			X		X		X		X	X				X		?	?	Carpentier et al., 2007
Persea americana	Exocarp	X	X					X	X		X	X				X				Barraclough et al., 2004
Prunus avium	Mesocarp	X			X[10]		X		X		X	X				X		X	X	Chan et al., 2008
Prunus persica	Mesocarp	X	X		X[11]		X[12]		X		X	X								Borsani et al., 2009
Prunus persica	Mesocarp	X			X[10]			X	X		X	X			X[13]	X		X	X	Chan et al., 2007
Prunus persica	Mesocarp	X			X		X		X		X	X	X			X	X[9]	X	X	Nilo et al., 2010
Prunus persica	Mesocarp	X			X		X[14]		X		X	X			X[13]	X		X		Prinsi et al., 2011
Prunus persica	Mesocarp				X		X[14]		X		X	X			X[13]	X		X		Prinsi et al., 2011
Prunus persica	Mesocarp	X	X			X	X		X		X	X				X			X	Hu et al., 2011
Prunus persica	Endocarp	X	X			X	X		X		X	X				X			X	Hu et al., 2011
Pyrus communis	Flesh[15]	X			X[16]		X		X		X	X				X				Pedreschi et al., 2007
Pyrus communis	Flesh	X			X		X		X		X	X								Pedreschi et al., 2009
Solanum lycopersicum	Pericarp	X	X	X[17]	X		X		X		X	X				X			X	Saravanan & Rose, 2004
Solanum lycopersicum	Pericarp	X	X	X[17]				X	X		X	X				X			X	Saravanan & Rose, 2004
Solanum lycopersicum	Pericarp	X	X	X[17]	X			X	X		X	X				X			X	Saravanan & Rose, 2004
Solanum lycopersicum	Fruit	X			X		X		>			X			X[4]	X		X		Rocco et al., 2006
Solanum lycopersicum	Pericarp	X			X		X		>			X			X[4]	X		X		Faurobert et al., 2007
Vitis vinifera	Berries[18]	X	X	X				X[19]	X		X	X			X[4]	X		X		Giribaldi et al., 2007
Vitis vinifera	Berries and stem	X		X[20]				X	>			X	X				X[21]	X	X	Vincent et al., 2006
Vitis vinifera	Berries and stem			X[20]				X[22]	X		X	X			X[4]	X		X	X	Vincent et al., 2006
Vitis vinifera	Berries and stem			X[20]	X		X			X	X	X				X		X	X	Vincent et al., 2006
Vitis vinifera	Pericarp	X	X	X[23]	X		X	X	X		X	X								Martínez-Esteso et al., 2011
Vitis vinifera	Mesocarp	X	X	X[24]	X		X	X	X		X	X								Martínez-Esteso et al., 2011
Vitis vinifera	Mesocarp			X[25]				X[22]	X		X	X			X[4]	X		X	X	Sarry et al., 2004
Vitis vinifera	Exocarp	X	X		X		X		X		X	X				X		?	?	Deytieux et al., 2007
Arachis hypogaea	Peanut pegs	X					X[26]	X	X[26]											Zhang et al., 2011

Table 1. Sample preparation

1. Aqueous/phenol two phase protein recovery followed by cold NH4-acetate dissolved in methanol precipitation and washes with the same solution and acetone.
2. Sonication.
3. > - higher values than the annotated were used.
4. Triton X-100.
5. Protein extracts obtained from this protocol were further purified using a 2-D Clean-Up Kit.
6. Stainless steel blender was used prior to mortar and pestle.
7. According to the manufacturer, samples cannot be solubilised in a buffer with any primary amines, such as ampholites and DTT, if they are to be labelled with DIGE CyDyes (Chakravarti et al., 2005).
8. Pericarp was reported, even though the succulent tissue from this fruit is denominated pseudocarp.
9. TCEP.
10. Triton X-100 was added.
11. SDS was added.
12. Final wash used cold ethanol.
13. NP-40.
14. The mixing sequence was inverted, see text.
15. Tissue samples were taken from the equatorial region excluding the skin and core.
16. Variant II is reported.
17. The mixture was homogenized at low temperature using a polytron PT 10/35 with an SM standard generator.
18. Berries were cut, deseeded and pulverized with a steel roll-on mechanical grinder half filled with liquid nitrogen.
19. Frozen powder was vortexed in Tris-HCl (pH 7.5) containing 2 M thiourea, 7 M urea, 2% Triton X-100, 1% DTT and 2% PVPP previous to TCA/acetone wash.
20. Stainless steel blender plus dry ice was used prior to mortar and pestle under liquid nitrogen.
21. HED.
22. Washed twice with ethanol.
23. Frozen pericarp were directly washed with ethyl acetate:ethanol at -20 °C with periodic vortexing, and the pellet recovered by centrifugation.
24. Mesocarp were homogenized in 50 mM Na2HPO2 pH 7.0, 1 mM EDTA, 0.1 M NaCl, PVPP, 1 mM Na2O5S2, 10 mM ascorbic acid, and a cocktail of protease inhibitors, filtered, centrifuged, and washed in 50 mM Na2HPO2 pH 7.0, 1 mM EDTA, 0.1 M NaCl, 10 mM ascorbic acid, and recovered by centrifugation.
25. Raw material was crushed in TCA/acetone.
26. Samples were washed with TCA/acetone, precipitated and finally with phenol plus DTT. For more details refer to Zhang et al., 2011.

3.2 Isoelectrical focusing

The initial step in the process of two dimensional gel electrophoresis first described by O'Farrell [O'Farrell, 1975] is based on the protein separation due to their intrinsic charge, in a process called isoelectric focusing. Even though this procedure is of foremost importance for the correct completion of the two-dimensional gel electrophoresis, many publications that deal with fruit tissues rely on protocols developed for animal tissues. Therefore, the results are far from optimal and reflected in gels of poor quality and a low number of spots displayed, greatly undermining the capacity of this approach. A more exhaustive approach requires the appropriate selection of isoelectric focusing buffer used for the resuspension of samples. This result in a consistent way to improve the protein profiles detected in 2-DE gels.

Prior to IEF, proteins should be completely solubilised, disaggregated, denatured and reduced in order to resolve as many of the molecules as possible [Shaw and Riederer, 2003]. Under these conditions proteins are loaded onto an immobilized pH gradient strip and subject to increasingly higher field strengths, until they reach their isoelectric point (pI). However, at this moment, when their net charge is closest to zero, they have a tendency to aggregate and precipitate [Rabilloud & Lelong, 2011]. In order to overcome these constrains, methodological procedures have been optimized and a series of chemical reagents tested, leading to continuous improvements in IEF.

3.2.1 IEF solubilisation buffer

One of the main focuses to improve IEF has been the evaluation and introduction of novel chaotropes, detergents and reducing agents that could help in sample solubilisation. The presence of chaotropes, compounds that disrupt non-covalent interactions between the molecules present in the sample, are essential to render proteins disaggregated and denatured [Rabilloud et al., 1997; England & Haran, 2011]. However, the exposition of the hydrophobic patches, normally buried inside these molecules, to a hydrophilic environment increases the already strong tendency of proteins to precipitate [Molloy, 2000; Rabilloud & Lelong, 2011]. In order to avoid this phenomenon, surfactants are added to the solubilisation buffer. Due to their amphipathic nature, these molecules help in the protein dispersion both through the stabilization of the proteins hydrophobic patches as well as by interacting with ionic and hydrogen bonds of the molecules in solution. The disruption of intramolecular and intermolecular disulfide bonds for complete protein unfolding and linearity is also mandatory, not only at this stage, but also for proper molecular weight based separation in the SDS-PAGE gels [Molloy, 2000]. This can be accomplished with the use of reducing agents.

Two different chaotropes, both of which do not display a net electric charge in solution over the pH range used for IEF, are the most used in at this stage: urea and thiourea [Shaw and Riederer, 2003; Rabilloud, 2009]. The capacity of the latter to improve the protein solubilisation has prompted its wide use (Table 1). However, certain constraints to the composition of the IEF buffer have been imposed by its presence, since thiourea is only soluble in a water-based buffer when high concentrations of urea are added. In turn, the most efficient surfactants already tested are not compatible with these urea concentrations, limiting therefore the amount of thiourea that can be used to solubilize proteins [Rabilloud et al., 1997].

Among the detergents, the most frequently used is the 3-[(3-cholamidopropyl) dimethylammonio] propane sulfonate (CHAPS), a sulfobetaine-type switterionic surfactant. Its compatibility with high urea concentrations commonly used in 2-DE and superior efficiency compared to nonionic detergents have driven its use [Rabilloud et al., 1997; Molloy, 2000]. Other alternatives include amidosulfobetaine-14 (ASB14), Sulfobetaine 3-10 (SB 3-10), 4-n-Octylbenzoylamido-propyl-dimethylammoniosulfobetaine (C8Φ) and 3-(4-heptyl) phenyl 3-hydroxypropyl dimethylammonio propane sulfonate (C7BzO) [Molloy, 2000; Maserti et al., 2007]. In terms of disulfide reducing agents, thiol-reducing agents and phosphines have gained widespread use in 2-DE, being dithiothreitol (DTT) the most often used. Since DTT is charged, especially at alkaline pH, during IEF it will migrate out of the gel, with a concomitant loss of solubility for some proteins and 2-DE horizontal streaking [Herbert, 1999; Molloy, 2000]. Therefore, its use in combination with other reducers or its substitution by compounds such as tributyl phosphine (TBP), Tris (2-carboxyethyl) phosphine hydrochloride (TCEP-HCP) and hydroxyethyl disulfide (HED) is advisable [Méchin et al., 2003; Sarma et al., 2008; Acín et al., 2009; Zhang et al., 2011]. Another advantage of using phosphines is the possibility of shortening the length of the equilibration step, therefore diminishing the loss of proteins at this point [Zuo & Speicher, 2000]. This can be accomplished by performing the reducing and alkylating procedures at the same time, since the phosphines such as TBP do not react with alkylating agents such as acrylamide and 2-vinylpyridine [Molloy, 2000].

Salt ions help stabilize proteins; therefore their absence may lead to protein precipitation. One way to overcome this situation is to add ampholytes to the IEF solution. These molecules enhance solubility of individual proteins as they approach their pI. They also buffer changes in conductivity, scavenge cyanate derived from urea, prevent interactions between hydrophobic proteins and IEF matrix and assist nucleic acids precipitation during centrifugation [Shaw and Riederer, 2003; Khoudoli et al., 2004; Gorg et al., 2009; Rabilloud & Lelong, 2011].

As expected, improvements in the composition of the IEF solubilisation buffer should help overcome some of the problems mentioned earlier in this chapter. For instance, the use of a reducing and an alkylating agent, TBP and 2-vinylpyridine, dissolved in a strong chaotrope such as guanidine hydrochloride to resuspend a dry fruit (e.g. peanut pegs) protein pellet obtained after TCA/acetone washes and phenol-based precipitation, have improved the spot number and resolution on 2-DE gels [Zhang et al., 2011]. Advances in solubilisation of acetone precipitated plant proteins have also been achieved by incremental changes in the concentration of Tris-base in the resuspension buffer, with a maximum effect obtained at 200 mM Tris-base. This result was probably due to the reduction in the protein-protein associations existing at this salt concentration, enhancing their release into the solution [Cho et al. 2010]. It is important to mention that a final dilution of the high salt IEF buffer was performed, in order to avoid a possible Joule heating during the focusing process [Wu et al., 2010; Rabilloud & Lelong, 2011].

Also interesting is the powerful result achieved with maize endosperms proteins when 2% of the surfactant SB 3-10, which is not compatible with high concentrations of urea, was combined with urea 5M, thiourea 2M, CHAPS 2%, DTT 20 mM, TCEP 5 mM, and two carrier ampholites (designated R2D2 by the authors). Compared to the more classical mixture of urea 7M, thiourea 2M, CHAPS 4%, DTT 25 mM and ampholytes, protein solubilisation and spot resolution were clearly enhanced [Méchin et al., 2003]. A similar improvement was observed when mesocarp derived *P. persica* 2-DE protein patterns were compared among samples resuspended in the R2D2 buffer and the T8 buffer evaluated by Méchin and co-authors (Nilo et al., 2011 - submitted).

3.2.2 Sample application

The sample application protocol has also demonstrated its relevance in improving the final 2-DE protein pattern. The now widely used immobilized pH gradients are supplied as a dehydrated gel matrix with plastic backing. Therefore, they have to be rehydrated before the IEF run, by "sample in-gel rehydration" or without the protein samples present in the rehydration solution by cup-loading or by paper-bridge loading. There are advantages and

disadvantages for each technique, mainly when working with hydrophobic or very high molecular weight proteins. Nonetheless, in some cases the use of one or the other method may be crucial, e.g. very alkaline proteins should be loaded by cup-loading, even though paper-bridge has been reported as a good alternative especially when samples are scarce and a broad pI range is to be assessed [Kane et al., 2006; Gorg et al., 2009]. The sample in-gel rehydration can be performed by a passive or active IPG strip rehydration. The latter option improves the entry of higher molecular weight proteins into the gel matrix [Gorg et al., 2009].
Recently, a novel strategy for sample loading, called G-electrode-loading method (GELM), has been introduced [Koga, 2008; Koga and Minohata, 2011]. This method allows a higher amount of protein to be loaded and therefore available for IEF. However, its performance has not been tested thus far with fruit derived proteins.

3.2.3 IEF running program improvement

The quality of the IEF is fundamental to achieve high quality 2-DE gels. However, most of the times an empirical assessment of the IEF program is performed until satisfactory results are achieved. This process can be time consuming and even be detrimental to the equipment being used, since the high heat generated by a sample that has not been properly desalted can burn the IEF machine plastic support where the samples are applied. For instance, salt interference is highly detrimental for 2-DE reproducibility, with concentrations lower than 10 mM recommended [Heppelmann et al., 2007]. Salt ions may affect IEF by slowing down its progression due to increased conductivity; producing artefacts and inducing protein modifications. Unfortunately, mandatory salt removal procedures will lead to sample loss and can result in the generation of a technical bias [Wu et al., 2010]. Therefore careful and reproducible procedures have to be implemented to deal with this kind of contamination.
One of the symptoms of salt contamination is the generation of a low voltage during the initial focusing of the IPG strip, which leads to suboptimal focusing. The presence of protein gaps and of streaking at the end of second dimension gels are also indicators of this problem [Heppelmann et al., 2007]. An estimation of salt conductivity, through the use of instruments such as portable conductivity meters, could help to confirm the presence or absence of salts as the source of these problems [Wu et al., 2010]. Additionally, it has been reported that IPGs washes, even when the focusing process has already commenced, could help to get remove salts and help to achieve adequate 2-DE results [Heppelmann et al., 2007].
In order to evaluate and compare results from different IEF runs the Volt hour (Vh) values should be recorded. The Vh reflects the total supplied energy to the system and should be optimized to produce the lowest value. The amount should be sufficient to reach a steady-state IEF, appropriate for protein focusing. This will depend on the sample, but also on the pH gradient, the IEF gel size and the amount of protein loaded (Table 2) [Gorg et al., 2009]. One way to avoid the cumbersome empirical evaluation of the IEF program for each new sample would be the use of a recently published algorithm, designed to predict the total Vh required for proper protein focusing during IEF [Wu et al., 2010].
Summary: IEF quality is fundamental in achieving high quality 2-DE gels. Besides the appropriate selection of IEF resuspension buffer components, some of which are almost standard nowadays (e.g. urea, thiourea, CHAPS, DTT), a careful evaluation of the sample application procedures and program settings required for reproducible IEF are crucial. High salt concentrations in the sample must be avoided. It is highly advisable that all of these points have been evaluated and optimized prior to running highly expensive experiments with scarce samples.

Species	Tissue	IEF Program			pI range	IEF gel size (cm)	# spots analyzed [1]	Image analysis	Spot visualization	References
		Pre-focusing step		Final kVh						
		Passive	Active							
Capsicum annuum	Placental tissue	?[2]	?	?	4-7; 4.5-5.5; 5.5-6.7; 6-9 L	?	1200; 600; 550; 200	Melanie IV	Coomassie	Lee et al., 2006
Citrus reticulata	Juice sacs	X		80	4-7 L	17	489	PDQuest	Coomassie	Yun et al., 2010
Elaeagnus umbellata	Mesocarp	?	?	?	4-7 L	?	1030	PDQuest	Silver	Wu et al., 2011
Fragaria x ananassa	Whole fruit	X		73	3-10 NL	24	1000	DeCyder	DIGE	Hjernø et al., 2006
Fragaria x ananassa	Whole fruit	X		30	3-11 L	18	956	PDQuest	SYPRO Ruby	Zheng et al., 2007
Fragaria x ananassa	Whole fruit	X		30	3-11 L	18	1368	PDQuest	SYPRO Ruby	Zheng et al., 2007
Fragaria x ananassa	Accrescent receptacle	X		27	3-10 (?)	18	622	Image Master 2D Platinum	DIGE	Bianco et al., 2009
Malus domestica	Pseudocarp	X		52	4-7 L	18	470	PDQuest	Coomassie	Guarino et al., 2007
Malus domestica	Peel	X		30	3-11 L	18	849	PDQuest	SYPRO Ruby	Zheng et al., 2007
Malus domestica	Peel	X		30	3-11 L	18	1422	PDQuest	SYPRO Ruby	Zheng et al., 2007
Malus domestica	Pericarp	X		30	3-11 NL	11	500	PDQuest	Silver	Song et al., 2006
Malus domestica	Pericarp	X		30	3-11 NL	11	500	PDQuest	Silver	Song et al., 2006
Musa spp	Meristem cultures	X		60	3-10 (?)	24	1348	Image Master 2D Platinum	Silver	Carpentier et al., 2005
Musa spp	Meristem cultures	X		60	3-10 (?)	24	1500	Image Master 2D Platinum	Silver	Carpentier et al., 2005
Musa spp	Mesocarp	X		30	3-11 NL	11	394	PDQuest	Silver	Song et al., 2006
Musa spp	Mesocarp	X		30	3-11 NL	11	394	PDQuest	Silver	Song et al., 2006
Musa spp	Meristematic tissue	X		55	4-7 L	24	1657	Image Master 2D Platinum	Coomassie	Carpentier et al., 2007
Persea americana	Exocarp			140	3-10 NL	18	?	ImageMaster 2D Elite software	SYPRO Ruby	Barraclough et al., 2004
Prunus avium	Mesocarp			9	3-10 (?)[3]	13	600	Image Master 2D Elite software	Coomassie	Chan et al., 2008
Prunus persica	Mesocarp		X	68	4-7 L	17	600	Image Master 2D Platinum	DIGE	Borsani et al., 2009
Prunus persica	Mesocarp			7	3-10 (?)[3]	13	?	Image Master 2D Elite software	Coomassie	Chan et al., 2007
Prunus persica	Mesocarp		X	70	3-10 NL	17	242	Delta 2D	DIGE	Nilo et al., 2010
Prunus persica	Mesocarp	X		90	3-10 NL	24	1128	Image Master 2D Platinum	Coomassie	Prinsi et al., 2011
Prunus persica	Mesocarp	X		90	3-10 NL	24	516	Image Master 2D Platinum	Coomassie	Prinsi et al., 2011
Prunus persica	Mesocarp		X	65	5-8 L	24	601	PDQuest	Coomassie	Hu et al., 2011
Prunus persica	Endocarp		X	65	5-8 L	24	714	PDQuest	Coomassie	Hu et al., 2011
Pyrus communis	Flesh	X		24	5-8 L[4]	24	800	Image Master 2D Platinum	Silver	Pedreschi et al., 2007
Pyrus communis	Flesh		X	91	4-7 L	24	?	Progenesis	DIGE	Pedreschi et al., 2009

Species	Tissue	IEF Program			pI range	IEF gel size (cm)	# spots analyzed [1]	Image analysis	Spot visualization	References
		Pre-focusing step		Final kVh						
		Passive	Active							
Solanum lycopersicum	Pericarp	X		100	4-7 L	17	679[5]	Progenesis	Coomassie	Saravanan & Rose, 2004
Solanum lycopersicum	Pericarp	X		100	4-7 L	17	679[5]	Progenesis	Coomassie	Saravanan & Rose, 2004
Solanum lycopersicum	Pericarp	X		100	4-7 L	17	679[5]	Progenesis	Coomassie	Saravanan & Rose, 2004
Solanum lycopersicum	Fruit	X		72	4-7 L	24	638	PDQuest	Coomassie	Rocco et al., 2006
Solanum lycopersicum	Pericarp	X		>64	4-7 L	24	1730	Image Master 2D Platinum	Silver	Faurobert et al., 2007
Vitis vinifera	Berries		X	105	3-10 NL	18	792	Image Master 2D Platinum	Coomassie	Giribaldi et al., 2007
Vitis vinifera	Berries and stem		X	85	3-10 NL	17	326	PDQuest	Coomassie	Vincent et al., 2006
Vitis vinifera	Berries and stem		X	85	3-10 NL	17	844	PDQuest	Coomassie	Vincent et al., 2006
Vitis vinifera	Berries and stem		X	85	3-10 NL	17	942	PDQuest	Coomassie	Vincent et al., 2006
Vitis vinifera	Pericarp	X		56	3-10 NL	18	921	Progenesis	DIGE	Martínez-Esteso et al., 2011
Vitis vinifera	Mesocarp	X		56	3-10 NL	18	804	Progenesis	DIGE	Martínez-Esteso et al., 2011
Vitis vinifera	Mesocarp		X	120	3-10 NL	?	270	PDQuest	Coomassie	Sarry et al., 2004
Vitis vinifera	Exocarp		X	64	3-10 NL	?	700	Image Master 2D Platinum	Coomassie	Deytieux et al., 2007
Arachis hypogaea	Peanut pegs	X		>80	3-10 ?	11	?	Dymension III	Silver	Zhang et al., 2011

Table 2. 2-DE Conditions

1. Maximum number evaluated.
2. Not determined.
3. Gels were polymerized in glass tubes: The IEF gel solution contained 10% NP-40, 30% w/v acrylamide, 9.5 M urea, 10% ammonium persulfate, and an equal mixture of 2% carrier ampholytes pH 3.5-10 and 5-8.
4. Other pI ranges were also reported.
5. A clear indication of the differences in the number of spots detected is not delivered.

3.3 Equilibration and SDS-page

After IEF, focused protein samples must be negatively charged with SDS to ensure exclusive molecular weight based separation during the second dimension. In parallel, proteins must be reduced and alkylated, a pre-requisite for keeping proteins unfolded during the SDS-PAGE step. This objective is accomplished in two main steps. First, the proteins are reduced by the action of DTT, and subsequently they are alkylated in the presence of iodoacetamide. Even though Gorg and colleagues have set the proper conditions for IPGs equilibration [Gorg et al., 2009], improvements can be achieved by speeding-up the process. These would allow a reduction in the levels of proteins lost during this step [Zuo & Speicher, 2000]. One way to achieve this task is to use vast excess of a high specific low molecular mass disulphide, which blocks the cysteines thiols [Olsson et al., 2002; Rabilloud, 2010]. Another option is to reduce and alkylate cysteine residue thiol groups prior to the IEF step, by using reagents such as TBP and 2-vinylpyridine [Zhang et al., 2011].

Regarding SDS-PAGE, some alternatives that may allow the strengthening of the fragile acrylamide-bis-acrylamide based matrix have been identified. However, their use has been

restricted due to problems associated with mass spectrometry (MS) incompatibility or by negatively affecting the electrophoresis itself. This has precluded the generation of large gels, which would have a much better resolution since this parameter is dependent on the surface area of the gel [Rabilloud, 2010]. An alternative to these large gels is to improve the area occupied by the spots during the second dimension in the conventional gels. This can be achieved through the generation of acrylamide gradients, which can encompass diverse ranges. Due to the difficulty in achieving reproducible gradient home-cast gels, their use is not widespread, with adoption by few authors whose work is summarized in Table 1 and 2 [Lee et al., 2006; Song et al., 2006; Nilo et al., 2010]. Finally, the 2-DE reproducibility heavily relies on this part of the process, with most of the noise and technical bias being generated at this stage [Choe and Lee, 2003; Lilley and Dupree, 2006]. Therefore, extreme care must be taken in order to avoid technical derived artefacts.
Summary: Equilibration is a well-defined and very important step of 2-DE, even though some improvements in the process can still be accomplished. Regarding SDS-PAGE, an increment in the gel resolution can be achieved through the use of acrylamide gradients. Due to the fact that SDS-PAGE is not a steady-state separation technique, an additional effort must be employed in order achieve highly standardized running conditions.

3.4 Protein visualization

After completion of SDS-PAGE, several alternatives are available for the detection of the protein spots present in this matrix. Some of them, such as Coomassie Brillant Blue (CBB) and its variant, colloidal Coomassie Brilliant Blue, as well as silver staining, are readily accessible and cost effective. Their use enables the detection of proteins in the sub-microgram range. However, silver has quite a poor linear dynamic range and proteins excised from gels stained by this means can be problematic to identify by MS [Patton, 2000].
Alternatives to these methods which are more sensitive (detection limit in the picogram range) as well as more reliable for protein quantitation, due to their linear dynamic range of at least three orders of magnitude, are the fluorescent dyes. Among the most sensitive are the Deep Purple (DP) and SYPRO Ruby (SR). Additionally, some of these fluorescent stains allow the detection of post-translational protein modifications on 2-DE gels, such as glycoproteins and phosphoproteins [Patton, 2000; Rieder 2008, Gauci et al., 2011].
Other factors that must be considered when choosing the visualization method are the inter-protein variability, ease of use, compatibility with subsequent MS analyses, among others [Gauci et al., 2011].

3.4.1 Difference gel electrophoresis – DIGE

Difference gel electrophoresis (DIGE) is a powerful tool for proteomics analysis. It provides the user with an internal standard control on each gel run, therefore strengthening the process of image comparison, which leads to more statistically robust results [Unlü et al., 1997; Lilley and Dupree, 2006]. Additionally, in the same way as other fluorophores, like DP and SR, allow the detection of protein amounts below the nanogram threshold [Patton, 2000; Gauci et al., 2011].
However, the use of this technology imposes several restrictions that may hamper its use when working with fruit samples. First, the protein concentration recommended by the manufacturers is of 5-10 mg/ml, which is not easy to achieve from fruit samples. Second, the sample pH needs to be adjusted to between 8.0 and 9.0. Fruit are characterized by

having highly acid components, which makes this adjustment a difficult task. In fact, since pH is adjusted with NaOH or Tris-HCl, it may lead to an increase of salts present in the sample, and therefore poor IEF [Tannu & Hemby, 2006; Wu et al., 2010]. Third, since the cye-dye labelling process must be performed at low temperatures, the solubility of compounds present at high molarity, such as thiourea and urea [Wahl et al., 2006], and also of detergents such as SB 3-10, will decrease. Since these molecules have proved very important to keep the proteins soluble for IEF, this situation may be detrimental in obtaining consistent 2-DE gels.

Summary: The use of fluorescent dyes to detect the presence of protein spots on 2-DE gels is advantageous both in terms of sensitivity, specificity and linearity. However, the requirement of high-cost equipment to excite and detect the fluorescence emitted by these molecules imposes some restrictions to their broad use. Some hurdles must also be addressed, concerning the use of the DIGE technology, to fully exploit its advantages for fruit proteome characterizations.

4. Literature evaluation – Fruit proteomics

A comprehensive search of the literature lead to the identification of 30 publications, produced in the last seven years, where 2-DE gels with fruit protein samples had been evaluated. Over 40% of the studies were performed by using well established plant fruit models such as tomato (*Solanum lycopersicum*), grape (*Vitis vinifera*) and peach (*Prunus persica*) (Figure 1A). The economical relevance of these species is also clear, with grapes been the most cultivated fruit plant throughout the world [Alexander & Grierson, 2002; Shulaev et al., 2008; Giribaldi & Giuffrida, 2010].

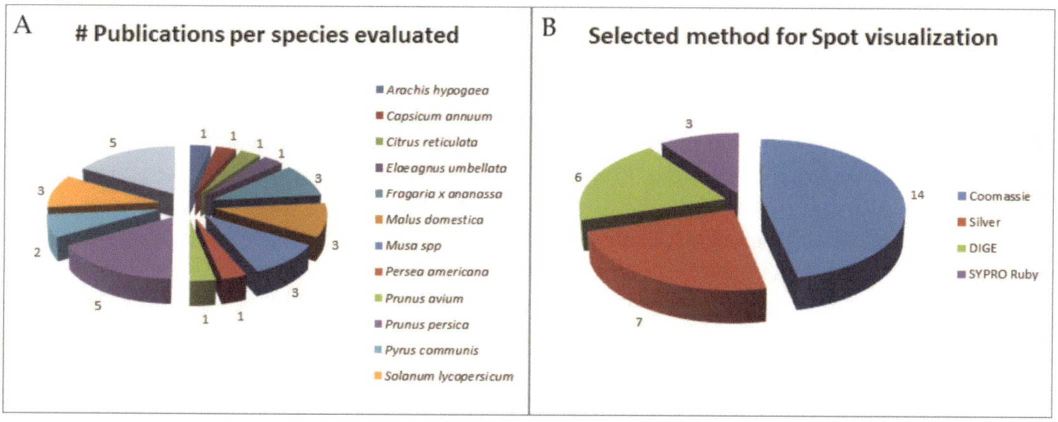

Fig. 1. General evaluation of the data displayed on Tables 1 and 2. The pie charge on **A** illustrates the species that have been assessed through a proteomic 2-DE gel based approach. Most of these evaluations have been performed using Coomassie based procedures (**B**).

One of the main goals of 2-DE is to maximise the numbers of detectable spots [Khoudoli et al., 2004]. Therefore, this criterion could be used to evaluate some of the parameters collected from the fruit proteomic literature (Tables 1 and 2), and used to discriminate which method would be the most relevant in order to achieve high quality 2-DE gels. However, there are a series of variables that may influence this parameter, as mentioned earlier. For instance, it has been reported that the number of spots detected in a gel are largely dependent on the software package used [Stessl et al., 2009; Dowsey et al., 2010]. The progress in these programs as well as the report in the literature of other quality parameters associated with spot resolution, such as intensity and circularity, could help to improve this kind of evaluations.

Contrary to what one would expect from the previous statement, by far the most used procedure to detect spots on 2-DE gels is the least sensitive, that being coomassie staining (Table 2 and Figure 1B). This is probably due to the simplicity of the protocol and associated the low costs. However, Carpentier et al. was able to reach the highest level of spot detection in the literature (Table 2) by using the most sensitive version of this staining protocol, colloidal coomassie, with 24 cm gels and loading as much as 400 micrograms of protein per gel [Carpentier et al., 2007].

Another striking point is the broad inclusion of thiourea, CHAPS and also of Triton X-100 in the IEF buffer. The use of more powerful surfactants is less popular, possibly due to lack of information regarding the benefits of their use. A similar phenomenon can be observed regarding the gel size, which is still mainly limited to 17-18 cm (Table 2). Regarding the method of IPG sample in-gel rehydration, the passive mode was preferred over the active for most of the researchers.

5. Concluding remarks

Despite the enormous relevance of fruit for human nutrition and its usefulness as a powerful biological model to understand processes of great scientific interest, to date fruit from very few species have been assessed through the use of 2-DE technology. As described in this chapter, this may be due to the intrinsic complexity of the fruit samples, which hampers the adequate development of the 2-DE generating process if a minimal set of precautions are not followed. Fortunately, several of the cited publications have reached outstanding results, which foster the use of this powerful proteomic tool to dissect the fruit associated phenomena under evaluation. Regarding the protein extraction method, the use of phenol-based approaches has proved to be superior compared to the other alternatives published. The development of alternative non-toxic compounds, with similar efficiency to extract proteins, but less prone to solubilize phenols and lipids, would be of great importance.

It is interesting to stress that there are no discernable trends in the use of protein solubilisation cocktails (Table 1). Few publications have addressed this point using a systematic assessment, probably due to the enormous number of factors that would have to be confronted. In other systems, the Taguchi method, a statistical tool that allows the evaluation of a limited number of experiments that generates the most information, has been used to achieve this goal [Khoudoli et al., 2004; Rao et al., 2008]. Using solely animal tissues, Khoudoli and colleagues were able to improve 2-DE gel aspects such as resolution and reproducibility [Khoudoli et al., 2004]. To date no similar studies have been performed with fruit tissues, even though similar enhancements were achieved by others with maize endosperms when similar guidelines were followed, namely combinations of zwitterionic detergents and optimization of the concentration of carrier ampholytes [Méchin et al., 2003]. In parallel, the development of an algorithm to improve the IEF running protocol by estimating the optimal amount of Vh required for protein focusing [Wu et al., 2010], will also be of invaluable interest for those that are beginning to work with scarce, complex fruit derived samples.

6. References

Acín P, Rayó J, Guerrero A, Quero C. Improved resolution in the acidic and basic region of 2-DE of insect antennae proteins using hydroxyethyl disulfide. Electrophoresis. 2009 Aug;30(15):2613-6.

Alexander L, Grierson D. Ethylene biosynthesis and action in tomato: a model for climacteric fruit ripening. J Exp Bot. 2002 Oct;53(377):2039-55.

Barraclough D., Obenland, DM, Laing W, Carrol T. A general method for two-dimensional protein electrophoresis of fruit samples. Postharvest Biology and Technology. 2004; 32:175–181.

Bianco L, Lopez L, Scalone AG, Di Carli M, Desiderio A, Benvenuto E, Perrotta G. Strawberry proteome characterization and its regulation during fruit ripening and in different genotypes. J Proteomics. 2009 May 2;72(4):586-607.

Borsani J, Budde CO, Porrini L, Lauxmann MA, Lombardo VA, Murray R, Andreo CS, Drincovich MF, Lara MV. Carbon metabolism of peach fruit after harvest: changes in enzymes involved in organic acid and sugar level modifications. J Exp Bot. 2009;60(6):1823-37.

Carpentier SC, Witters E, Laukens K, Deckers P, Swennen R, Panis B. Preparation of protein extracts from recalcitrant plant tissues: an evaluation of different methods for two-dimensional gel electrophoresis analysis. Proteomics. 2005 Jul; 5(10): 2497-507.

Carpentier SC, Witters E, Laukens K, Van Onckelen H, Swennen R, Panis B. Banana (Musa spp.) as a model to study the meristem proteome: acclimation to osmotic stress. Proteomics. 2007 Jan;7(1):92-105.

Chakravarti B, Gallagher SR, Chakravarti DN. Difference gel electrophoresis (DIGE) using CyDye DIGE fluor minimal dyes. Curr Protoc Mol Biol. 2005 Feb;Chapter 10:Unit 10.23.

Choe LH, Lee KH. Quantitative and qualitative measure of intralaboratory two-dimensional protein gel reproducibility and the effects of sample preparation, sample load, and image analysis. Electrophoresis. 2003 Oct;24(19-20):3500-7.

Chan Z, Qin G, Xu X, Li B, Tian S. Proteome approach to characterize proteins induced by antagonist yeast and salicylic acid in peach fruit. J Proteome Res. 2007 May;6(5):1677-88.

Chan Z, Wang Q, Xu X, Meng X, Qin G, Li B, Tian S. Functions of defense-related proteins and dehydrogenases in resistance response induced by salicylic acid in sweet cherry fruits at different maturity stages. Proteomics. 2008 Nov;8(22):4791-807.

Cho JH, Cho MH, Hwang H, Bhoo SH, Hahn TR. Improvement of plant protein solubilization and 2-DE gel resolution through optimization of the concentration of Tris in the solubilization buffer. Mol Cells. 2010 Jun;29(6):611-6.

Coombe B. The Development of Fleshy Fruits. Annu Rev Plant Phys. 1976 Jun;27:207-228.

Deytieux C, Geny L, Lapaillerie D, Claverol S, Bonneu M, Donèche B. Proteome analysis of grape skins during ripening. J Exp Bot. 2007;58(7):1851-62.

Di Carli M, Zamboni A, Pè ME, Pezzotti M, Lilley KS, Benvenuto E, Desiderio A. Two-dimensional differential in gel electrophoresis (2D-DIGE) analysis of grape berry proteome during postharvest withering. J Proteome Res. 2011 Feb 4;10(2):429-46.

Dowsey AW, English JA, Lisacek F, Morris JS, Yang GZ, Dunn MJ. Image analysis tools and emerging algorithms for expression proteomics. Proteomics. 2010 Dec;10(23):4226-57.

England JL, Haran G. Role of solvation effects in protein denaturation: from thermodynamics to single molecules and back. Annu Rev Phys Chem. 2011 May;62:257-77.

Faurobert M, Mihr C, Bertin N, Pawlowski T, Negroni L, Sommerer N, Causse M. Major proteome variations associated with cherry tomato pericarp development and ripening. Plant Physiol. 2007 Mar;143(3):1327-46.

Ferrandiz C, Fourquin C, Prunet N, Scutt CP, Sundberg E, Trehin C, Vialette-Guiraud ACM. Advances in Botanical Research. 2010;55:1-73.

Gauci VJ, Wright EP, Coorssen JR. Quantitative proteomics: assessing the spectrum of in-gel protein detection methods. J Chem Biol. 2011 Jan;4(1):3-29.

Giavalisco P, Nordhoff E, Lehrach H, Gobom J, Klose J. Extraction of proteins from plant tissues for two-dimensional electrophoresis analysis. Electrophoresis. 2003 Jan;24(1-2):207-16.

Gil-Agusti MT, Campostrini N, Zolla L, Ciambella C, Invernizzi C, Righetti PG. Two-dimensional mapping as a tool for classification of green coffee bean species. Proteomics. 2005 Feb;5(3):710-8.

Giribaldi M, Giuffrida MG. Heard it through the grapevine: proteomic perspective on grape and wine. J Proteomics. 2010 Aug 5;73(9):1647-55.

Giribaldi M, Perugini I, Sauvage FX, Schubert A. Analysis of protein changes during grape berry ripening by 2-DE and MALDI-TOF. Proteomics. 2007 Sep;7(17):3154-70.

Görg A, Weiss W, Dunn MJ. Current two-dimensional electrophoresis technology for proteomics. Proteomics. 2004 Dec;4(12):3665-85.

Görg A, Drews O, Lück C, Weiland F, Weiss W. 2-DE with IPGs. Electrophoresis. 2009 Jun;30 Suppl 1:S122-32.

Herbert B. Advances in protein solubilisation for two-dimensional electrophoresis. Electrophoresis. 1999 Apr-May;20(4-5):660-3.

Guarino C, Arena S, De Simone L, D'Ambrosio C, Santoro S, Rocco M, Scaloni A, Marra M. Proteomic analysis of the major soluble components in Annurca apple flesh. Mol Nutr Food Res. 2007 Feb;51(2):255-62.

Heppelmann CJ, Benson LM, Bergen HR 3rd. A simple method to remove contaminating salt from IPG strips prior to IEF. Electrophoresis. 2007 Nov;28(21):3988-91.

Hjernø K, Alm R, Canbäck B, Matthiesen R, Trajkovski K, Björk L, Roepstorff P, Emanuelsson C. Down-regulation of the strawberry Bet v 1-homologous allergen in concert with the flavonoid biosynthesis pathway in colorless strawberry mutant. Proteomics. 2006 Mar;6(5):1574-87.

Hu H, Liu Y, Shi GL, Liu YP, Wu RJ, Yang AZ, Wang YM, Hua BG, Wang YN. Proteomic analysis of peach endocarp and mesocarp during early fruit development. Physiol Plant. 2011 Aug;142(4):390-406.

Hurkman WJ, Tanaka CK. Solubilization of plant membrane proteins for analysis by two-dimensional gel electrophoresis. Plant Physiol. 1986 Jul;81(3):802-6.

Kane LA, Yung CK, Agnetti G, Neverova I, Van Eyk JE. Optimization of paper bridge loading for 2-DE analysis in the basic pH region: application to the mitochondrial subproteome. Proteomics. 2006 Nov;6(21):5683-7.

Khoudoli GA, Porter IM, Blow JJ, Swedlow JR. Optimisation of the two-dimensional gel electrophoresis protocol using the Taguchi approach. Proteome Sci. 2004 Sep 9;2(1):6.

Koga K. G-electrode-loading method for isoelectric focusing, enabling separation of low-abundance and high-molecular-mass proteins. Anal Biochem. 2008 Nov 1;382(1):23-8.

Koga K, Minohata T. An approach for identification of phosphoproteins using the G-electrode-loading method in two-dimensional gel electrophoresis. Proteomics. 2011 Apr;11(8):1545-9.

Lee JM, Kim S, Lee JY, Yoo EY, Cho MC, Cho MR, Kim BD, Bahk YY. A differentially expressed proteomic analysis in placental tissues in relation to pungency during the pepper fruit development. Proteomics. 2006 Oct;6(19):5248-59.

Levetin E, McMahon K. Plants and society. 2008. 5th ed, McGraw-Hill, Dubuque, Iowa.

Lilley KS, Dupree P. Methods of quantitative proteomics and their application to plant organelle characterization. J Exp Bot. 2006;57(7):1493-9.

Martínez-Esteso MJ, Sellés-Marchart S, Lijavetzky D, Pedreño MA, Bru-Martínez R. A DIGE-based quantitative proteomic analysis of grape berry flesh development and ripening reveals key events in sugar and organic acid metabolism. J Exp Bot. 2011 May;62(8):2521-69.

Maserti BE, Della Croce CM, Luro F, Morillon R, Cini M, Caltavuturo L. A general method for the extraction of citrus leaf proteins and separation by 2D electrophoresis: a follow up. J Chromatogr B Analyt Technol Biomed Life Sci. 2007 Apr 15;849(1-2):351-6.

Méchin V, Consoli L, Le Guilloux M, Damerval C. An efficient solubilization buffer for plant proteins focused in immobilized pH gradients. Proteomics. 2003 Jul;3(7):1299-302.

Molloy MP. Two-dimensional electrophoresis of membrane proteins using immobilized pH gradients. Anal Biochem. 2000 Apr 10;280(1):1-10.

Nilo R, Saffie C, Lilley K, Baeza-Yates R, Cambiazo V, Campos-Vargas R, González M, Meisel LA, Retamales J, Silva H, Orellana A. Proteomic analysis of peach fruit mesocarp softening and chilling injury using difference gel electrophoresis (DIGE). BMC Genomics. 2010 Jan 18;11:43.

O'Farrell PH. High resolution two-dimensional electrophoresis of proteins. J Biol Chem. 1975 May 25;250(10):4007-21.

Olsson I, Larsson K, Palmgren R, Bjellqvist B. Organic disulfides as a means to generate streak-free two-dimensional maps with narrow range basic immobilized pH gradient strips as first dimension. Proteomics. 2002 Nov;2(11):1630-2.

Palma JM, Corpas FJ, Del Río LA. Proteomics as an approach to the understanding of the molecular physiology of fruit development and ripening. J Proteomics. 2011 Apr 16.

Patton WF. A thousand points of light: the application of fluorescence detection technologies to two-dimensional gel electrophoresis and proteomics. Electrophoresis. 2000 Apr;21(6):1123-44.

Pedreschi R, Vanstreels E, Carpentier S, Hertog M, Lammertyn J, Robben J, Noben JP, Swennen R, Vanderleyden J, Nicolaï BM. Proteomic analysis of core breakdown disorder in Conference pears (Pyrus communis L.). Proteomics. 2007 Jun;7(12):2083-99.

Pedreschi R, Hertog M, Robben J, Lilley KS, Karp NA, Baggerman G, Vanderleyden J, Nicolaï B. Gel-based proteomics approach to the study of metabolic changes in pear tissue during storage. J Agric Food Chem. 2009 Aug 12;57(15):6997-7004.

Prasanna V, Prabha TN, Tharanathan RN. Fruit ripening phenomena--an overview. Crit Rev Food Sci Nutr. 2007;47(1):1-19.

Prinsi B, Negri AS, Fedeli C, Morgutti S, Negrini N, Cocucci M, Espen L. Peach fruit ripening: A proteomic comparative analysis of the mesocarp of two cultivars with different flesh firmness at two ripening stages. Phytochemistry. 2011 Jul;72(10):1251-62.

Rabilloud T, Adessi C, Giraudel A, Lunardi J. Improvement of the solubilization of proteins in two-dimensional electrophoresis with immobilized pH gradients. Electrophoresis. 1997 Mar-Apr;18(3-4):307-16.

Rabilloud T. Detergents and chaotropes for protein solubilization before two-dimensional electrophoresis. Methods Mol Biol. 2009;528:259-67.

Rabilloud T. Variations on a theme: changes to electrophoretic separations that can make a difference. J Proteomics. 2010 Jun 16;73(8):1562-72.

Rabilloud T, Lelong C. Two-dimensional gel electrophoresis in proteomics: A tutorial. J Proteomics. 2011 Jun 12.

Rao RS, Kumar CG, Prakasham RS, Hobbs PJ. The Taguchi methodology as a statistical tool for biotechnological applications: a critical appraisal. Biotechnol J. 2008 Apr;3(4): 510-23.

Riederer BM. Non-covalent and covalent protein labeling in two-dimensional gel electrophoresis. J Proteomics. 2008 Jul 21;71(2):231-44.

Righetti PG. Real and imaginary artefacts in proteome analysis via two-dimensional maps. J Chromatogr B Analyt Technol Biomed Life Sci. 2006 Sep 1;841(1-2):14-22.

Rocco M, D'Ambrosio C, Arena S, Faurobert M, Scaloni A, Marra M. Proteomic analysis of tomato fruits from two ecotypes during ripening. Proteomics. 2006 Jul;6(13):3781-91.

Rose JK, Bashir S, Giovannoni JJ, Jahn MM, Saravanan RS. Tackling the plant proteome: practical approaches, hurdles and experimental tools. Plant J. 2004 Sep;39(5):715-33.

Rosenheck R. Fast food consumption and increased caloric intake: a systematic review of a trajectory towards weight gain and obesity risk. Obes Rev. 2008 Nov;9(6):535-47.

Saravanan RS, Rose JK. A critical evaluation of sample extraction techniques for enhanced proteomic analysis of recalcitrant plant tissues. Proteomics. 2004 Sep;4(9):2522-32.

Sarma AD, Oehrle NW, Emerich DW. Plant protein isolation and stabilization for enhanced resolution of two-dimensional polyacrylamide gel electrophoresis. Anal Biochem. 2008 Aug 15;379(2):192-5.

Sarry JE, Sommerer N, Sauvage FX, Bergoin A, Rossignol M, Albagnac G, Romieu C. Grape berry biochemistry revisited upon proteomic analysis of the mesocarp. Proteomics. 2004 Jan;4(1):201-15.

Scutt CP, Vinauger-Douard M, Fourquin C, Finet C, Dumas C. An evolutionary perspective on the regulation of carpel development. J Exp Bot. 2006;57(10):2143-52.
Shaw MM, Riederer BM. Sample preparation for two-dimensional gel electrophoresis. Proteomics. 2003 Aug;3(8):1408-17.
Sheoran IS, Ross ARS, Olson DJH, Sawhney VK. Compatibility of plant protein extraction methods with mass spectrometry for proteome analysis. Plant Science. 2009; 176: 99–104.
Shulaev V, Korban SS, Sosinski B, Abbott AG, Aldwinckle HS, Folta KM, Iezzoni A, Main D, Arús P, Dandekar AM, Lewers K, Brown SK, Davis TM, Gardiner SE, Potter D, Veilleux RE. Multiple models for Rosaceae genomics. Plant Physiol. 2008 Jul;147(3):985-1003.
Song J, Braun G, Bevis E, Doncaster K. A simple protocol for protein extraction of recalcitrant fruit tissues suitable for 2-DE and MS analysis. Electrophoresis. 2006 Aug;27(15):3144-51.
Spencer JP. The impact of fruit flavonoids on memory and cognition. Br J Nutr. 2010 Oct;104 Suppl 3:S40-7.
Toorchi M, Nouri MZ, Tsumura M, Komatsu S. Acoustic technology for high-performance disruption and extraction of plant proteins. J Proteome Res. 2008 Jul;7(7):3035-41.
Stessl M, Noe CR, Lachmann B. Influence of image-analysis software on quantitation of two-dimensional gel electrophoresis data. Electrophoresis. 2009 Jan;30(2):325-8.
Tannu NS, Hemby SE. Two-dimensional fluorescence difference gel electrophoresis for comparative proteomics profiling. Nat Protoc. 2006;1(4):1732-42.
Unlü M, Morgan ME, Minden JS. Difference gel electrophoresis: a single gel method for detecting changes in protein extracts. Electrophoresis. 1997 Oct;18(11):2071-7.
Vincent D, Wheatley MD, Cramer GR. Optimization of protein extraction and solubilization for mature grape berry clusters. Electrophoresis. 2006 May;27(9):1853-65.
Wang W, Scali M, Vignani R, Spadafora A, Sensi E, Mazzuca S, Cresti M. Protein extraction for two-dimensional electrophoresis from olive leaf, a plant tissue containing high levels of interfering compounds. Electrophoresis. 2003 Jul;24(14):2369-75.
Wahl M, Kirsch R, Brockel U, Trapp S, Bottlinger M. Caking of urea prills. Chem. Eng. Tech. 2006; 29:674-678.
Wang W, Tai F, Chen S. Optimizing protein extraction from plant tissues for enhanced proteomics analysis. J Sep Sci. 2008 Jun;31(11):2032-9.
Wu HC, Chen TN, Kao SH, Shui HA, Chen WJ, Lin HJ, Chen HM. Isoelectric focusing management: an investigation for salt interference and an algorithm for optimization. J Proteome Res. 2010 Nov 5;9(11):5542-56.
Wu MC, Hu HT, Yang L, Yang L. Proteomic analysis of up-accumulated proteins associated with fruit quality during autumn olive (Elaeagnus umbellata) fruit ripening. J Agric Food Chem. 2011 Jan 26;59(2):577-83.
Yun JW. Possible anti-obesity therapeutics from nature--a review. Phytochemistry. 2010 Oct;71(14-15):1625-41.
Zhang E, Chen X, Liang X. Resolubilization of TCA precipitated plant proteins for 2-D electrophoresis. Electrophoresis. 2011 Mar;32(6-7):696-8.

Zhang L, Yu Z, Jiang L, Jiang J, Luo H, Fu L. Effect of post-harvest heat treatment on proteome change of peach fruit during ripening. J Proteomics. 2011 Jun 10;74(7):1135-49.

Zheng Q, Song J, Doncaster K, Rowland E, Byers DM. Qualitative and quantitative evaluation of protein extraction protocols for apple and strawberry fruit suitable for two-dimensional electrophoresis and mass spectrometry analysis. J Agric Food Chem. 2007 Mar 7;55(5):1663-73.

Zuo X, Speicher DW. Quantitative evaluation of protein recoveries in two-dimensional electrophoresis with immobilized pH gradients. Electrophoresis. 2000 Aug;21(14):3035-47.

Understanding the Pathogenesis of Cytopathic and Noncytopathic Bovine Viral Diarrhea Virus Infection Using Proteomics

Mais Ammari[1], Fiona McCarthy[1,2], Bindu Nanduri[1,2], George Pinchuk[3] and Lesya Pinchuk[1]
[1]*Department of Basic Sciences, Mississippi State University, Mississippi State, MS*
[2]*Institute for Genomics, Biotechnology and Biocomputing, Mississippi State University, Mississippi State, MS*
[3]*Department of Sciences and Mathematics, Mississippi University for Women, Columbus, MS*
USA

1. Introduction

Bovine Viral Diarrhea Virus (BVDV) is a single-stranded RNA virus in the *Pestivirus* genus within the *Flaviviridae* family. BVDV infections are seen in all ages and breeds of cattle worldwide and have significant economic impact due to productive and reproductive losses (Houe 2003). Two antigenically distinct genotypes of BVDV exist, type 1 and 2 (Ridpath *et al.* 1994). BVDV of both genotypes may occur as cytopathic (cp) or noncytopathic (ncp) biotypes, classified according to whether or not they produce visible changes in cell culture. Data indicate that cp biotypes of BVDV can actually be created through internal deletion of RNA of ncp biotypes, or through RNA recombination between ncp biotypes (Howard *et al.* 1992). Of the two BVDV biotypes, infection of a fetus by ncp BVDV can result in persistently infected (PI) calf that sheds the virus throughout its life without developing clinical signs of infection. PI animals are the major disseminators of BVDV in the cattle population and have been the cause of severe acute outbreaks (Carman *et al.* 1998). However cp BVDV is associated predominantly with animals that develop mucosal disease (MD), which can be acute, resulting in death within a few days of onset, or chronic, persisting for weeks or months before the afflicted animal dies (Houe 1999).
The interaction of BVDV with its host has several unique features, most notably the capacity to infect its host either transiently or persistently (Liebler-Tenorio *et al.* 2002; Bendfeldt S 2007). Initially the virus binds to CD46, a complement receptor expressed on lymphoid cells, monocytes, macrophages and dendritic cells and serving as a "magnet" for several viral and bacterial pathogens (Cattaneo 2004). Upon entry, the virus replicates and spreads in the lymphatic system, impairing the immunity of the infected animal, particularly antigen presenting cells (APC) function and production of interferons (IFN). Cytopathic BVDV biotype but not ncp biotype (Schweizer & Peterhans 2001) is implicated in the induction of apoptosis (Zhang *et al.* 1996; Schweizer & Peterhans 1999; Grummer *et al.* 2002; Jordan *et al.*

2002), and the existence of the two antigenically related biotype 'pairs' makes BVDV an important model for virus-induced apoptosis. In addition, cp BVDV readily trigger IFN type-I whereas infection with ncp BVDV fails to induce IFN generation (Peterhans *et al.* 2003). Also, BVDV has been reported to modulate functions of immune cells after infection *in vitro* including increased production of nitric oxide from infected macrophages (Adler *et al.* 1994), decreased production of TNF-α from activated macrophages (Adler *et al.* 1994), inhibited phagocytosis of alveolar macrophages (Brewoo *et al.* 2007) and decreased T-stimulatory ability of monocytes (Glew *et al.* 2003). Phagocytosis and macropinocytosis antigen uptake mechanisms play a crucial role in the innate immune responses by clearing pathogens at sites of infection. The endothytic pathways are also important early steps in triggering the adaptive immune responses which require processing of bacterial and viral pathogens and presentation of their antigens to CD4+ and CD8+ T cells (Boyd *et al.* 2004).

Since BVDV viruses are able to affect virtually all organs and systems in the body, including the innate and the adaptive immune system, it is important to know the mechanistic framework for the viral-host interactions in the complex etiology of the disease. The availability of BVDV genome sequence makes it a suitable target for genome-wide analyses. However, the corresponding viral proteome, the alterations in host proteomes upon BVDV infection and the dynamic nature of the BVDV proteins remain largely unknown.

New developments in comparative and quantification proteomics technology, especially mass spectrometry (MS), enabled a more comprehensive characterization of virions, protein location, protein isoforms and post-translational modifications, as well as protein-protein interactions involved in virus-host dynamics. One example on the application of these new advances is in the detection of virion protein composition that helped in identifying the role of specific viral proteins during infection. Since BVDV is an enveloped virus, it has a considerable potential to incorporate both viral and host proteins into its membrane as well as into the envelope, and these can be present at low levels, making their detection difficult using traditional methods. In recent viral proteomics studies, several different MS approaches (e.g., matrix-assisted laser desorption ionization (MALDI)-time of flight (TOF) mass spectrometry and LC-MS/MS) were successfully used to analyze the composition of a variety of virions, leading to the identification of previously unknown components of viral particles. For example, two enveloped viruses, SARS and HIV-1 were analyzed by these techniques and the investigators were able to confirm and identify virion proteins (Maxwell & Frappier 2007). Accurate identification of BVDV virion proteins is possible using these methods.

In this review, we discuss the work that has been done to date using proteomic-related approaches to understand BVDV viral protein structure, viral protein-protein interactions and viral-host protein interactions. At the end of each section we refer to some examples of new proteomics approaches that have been successfully used for different viral studies that can be applied for studying BVDV to develop a better understanding of its pathogenesis. We also include the effects of BVDV viral infection on host cell proteome where we place particular emphasis on MS-based approaches, highlighting how these new approaches facilitated the understanding of BVDV pathogenesis at a genomic scale.

2. Structural proteomics

The availability of genome sequences coupled with advances in molecular and structural biology guided the development of structural proteomics. Availability of three-dimensional structure information elucidates protein function and identifies targets for the attenuation of virus's replication. Currently, the Protein Data Bank (PDB) (H.M. Berman 2000) contains eight structures for BVDV-encoded proteins, all of which were solved in the past seven years (Table 1). In contrast, the whole *Flaviviridae* family has 312 protein structures.

Protein	PDB accession number	Type of structural data	Release date	Reference
RdRp (BVDV CP7-R12 RdRp (residues 3189-3907))	2CJQ	X-RAY DIFFRACTION	2006-07-19	(Choi *et al.* 2006)
*NS5A (NMR structure of the in-plane membrane anchor domain [1-28] of the monotopic NS5A of BVDV)	2AJJ	SOLUTION NMR	2005-08-23	(Sapay *et al.* 2006)
*NS5A (NMR structure of the in-plane membrane anchor domain [1-28] of the monotopic NS5A from the BVDV)	2AJM	SOLUTION NMR	2005-08-23	(Sapay *et al.* 2006)
*NS5A (NMR structure of the in-plane membrane anchor domain [1-28] of the monotopic NS5A from the BVDV)	2AJN	SOLUTION NMR	2005-08-23	(Sapay *et al.* 2006)
*NS5A (NMR structure of the in-plane membrane anchor domain [1-28] of the monotopic NS5A from the BVDV)	2AJO	SOLUTION NMR	2005-08-23	(Sapay *et al.* 2006)
RdRp (Crystal structure of RdRp construct 1 (residues 71-679) from BVDV)	1S48	X-RAY DIFFRACTION	2004-04-06	(Choi *et al.* 2004)
RdRp (Crystal Structure of RdRp construct 1 (residues 71-679) from BVDV complexed with GTP)	1S49	X-RAY DIFFRACTION	2004-04-06	(Choi *et al.* 2004)
RdRp (Crystal Structure of RdRp construct 2 (Residues 79-679) from BVDV)	1S4F	X-RAY DIFFRACTION	2004-04-06	(Choi *et al.* 2004)

Table 1. BVDV proteins structure. *For details on the differences between the four structures see indicated reference

BVDV genome is a positive-sense ssRNA of approximately 12.6-kb in length (Meyers et al. 1997; Brett D. Lindenbach 2007). The BVDV genome comprises a 5′ and 3′ untranslated regions (UTR), which flank a single open reading frame (ORF) (Collett et al. 1988; Brett D. Lindenbach 2007). The BVDV genome is translated into a single polyprotein NH2-N^{pro}-C-

E^{rns}-E1-E2-P7-NS2-3-NS4A-NS4B-NS5A-NS5B-COOH (Collett *et al.* 1988; Meyers *et al.* 1997; Brett D. Lindenbach 2007) (Figure 1). Upon synthesis, by a combination of host and viral proteases, the BVDV polyprotein is processed into at least four structural (C, E^{rns}, E1, E2) and six non-structural (NS2, NS3, NS4A, NS4B, NS5A, NS5B) proteins required for viral assembly and replication.

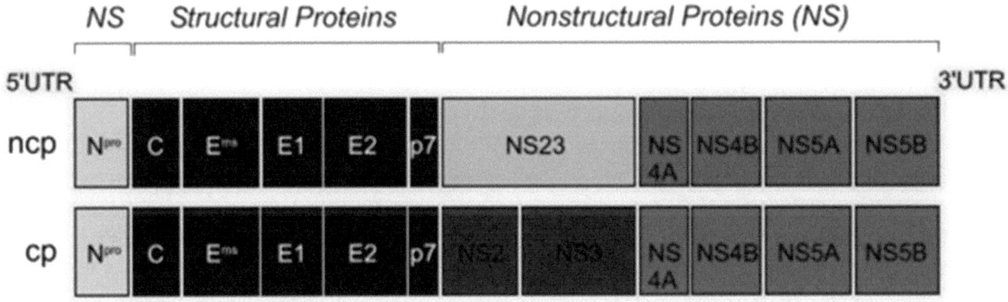

Fig. 1. BVDV genome structure

Among the NS proteins, NS5B has been shown to be the BVDV RNA-dependent RNA polymerase (RdRp) which is responsible for genomic replication as a part of larger, membrane associated, replicase complex (Brett D. Lindenbach 2007). Since BVDV uses a primer-independent (*de novo*) mechanism for RNA replication, its RNA polymerase requires GTP for initiating RNA synthesis (Brett D. Lindenbach 2007). 2.9 °A X-ray crystallography structural data of BVDV strain NADL RdRp revealed that it possesses a GTP N-terminal domain and identified the GTP-specific binding site required for *de novo* initiation (Choi *et al.* 2004). Comparison of the 2.6 °A X-ray crystal structure of BVDV CP7-R12, a BVDV CP7 polymerase recombinant with a single amino acid duplication of Asn438 (Choi *et al.* 2006), with the NADL BVDV polymerase showed that the alterations in the RdRp of the CP7-R12 derived mutant viruses could be allocated to a large fragment of the N-terminal domain indicating the role of this domain in the translocation of the template during catalysis. In addition, the study showed the formation of an unstable loop due to the insertion of an additional Asn438 in CP7-R12, which may account for the low replication activity of the mutant polymerase *in vivo* (Gallei *et al.* 2004).

BVDV NS5A function is not well determined. Three-dimensional NMR structure analyses of the membrane anchor (1-28) of NS5A from BVDV performed either in 50% TFE or in SDS micelles to mimic the membrane environment revealed that the N-terminal membrane anchor of NS5A includes a long amphipathic α-helix (aa 5-25). It interacts in-plane with the membrane interface and is divided into two portions separated by a flexible region centred around residue Gly19 (Sapay *et al.* 2006). The amphipathic α-helix exhibited a hydrophobic side buried in the membrane and a polar side accessible from the cytosol. These data were also confirmed and supported by molecular dynamic simulation at a water-dodecane interface. Despite the lack of amino acid sequence similarity, this amphipathic α-helix shows a common structural feature with that of the Hepatitis C virus (HCV). The phosphorylation state of NS5A is believed to be vital for HCV replication complex.

In addition, on the basis of sequence alignments of HCV and BVDV NS5A proteins, four cysteine residues involved in zinc binding were identified. BVDV secondary structure assignments of these were determined by computer prediction with the PSIPRED algorithm (Tellinghuisen *et al.* 2006).

Due mostly to solubility issues, generating structures using NMR and X-ray crystallography have been successful for only a small fraction of viral proteins. In contrast, the recent use of high-throughput approaches for multiple viral proteins structures and functions, for example, SARS virus enabled the investigators to elucidate the viral protein structure with high resolution.

3. Protein interactions

Physical interactions of viral proteins between each other and with their host proteins are important in allowing the pathogen to enter the host cell, manipulate host cellular processes, replicate and infect other cells. Identifying protein-protein interactions during the course of infection enables researchers to better understand the role of host-pathogen interactions and how these interactions affect infection, disease progression and the host immune response. Since interaction of intracellular infectious agents like BVDV with their host cells are mainly at the protein level, proteomics is the most suitable tool for investigating these interactions.

3.1 Virus protein-protein interactions

Physical association of viral proteins is important in the initial virus-host interaction and immune response against structural proteins. For example, by co- and sequential immuno precipitation a direct interaction of E^{rns} with E2 was reported very early after translation and showed to form a covalently linked heterodimer, which is later stabilized by disulfide bonds (Lazar *et al.* 2003). This interaction exists in both cp BVDV-infected MDBK cells and secreted virions.

Various cellular and viral processes are dependent on phosphorylation and dephosphorylation of specific proteins. Phosphorylation of multiple Flavivirus NS5 proteins correlates with subcellular localization and ability to associate with NS3. Studies of virion morphogenesis by immunoprecipitation show that the BVDV NS5A is phosphorylated by its associated serine/threonine kinase (Reed *et al.* 1998). Phosphorylation of NS5A may play a role in BVDV life cycle.

Analysis by radioimmunoprecipitation assays followed by SDS-PAGE under nonreducing conditions revealed that the envelope BVDV glycoproteins E1 and E2 interact through a disulfide bond to form a dimer (Weiland *et al.* 1990), which is thought to be a functional complex present on the surfaces of mature virion. In a follow up study, 30 min post-pulse, both E1 and E2 interacted independently and simultaneously with calnexin, an ER chaperone (Branza-Nichita *et al.* 2001). The inhibition of calnexin binding to the envelope proteins by α-glucosidase inhibitors resulted in the misflolding of those proteins and a decrease in the formation of E1-E2 heterodimers (Branza-Nichita *et al.* 2001).

Cp BVDV-infected cells were metabolically labeled and their proteins crosslinked with the cell-permeable and thiol-cleavable cross-linker DSP followed by immunoprecipitation with BVDV protein specific antiserum. Although not sufficient to establish direct protein-protein interactions, results indicate associations between NS3, NS4B, and NS5A (Qu *et al.* 2001).

3.2 Virus-host interacting partners

Identifying interactions between viral proteins and host proteins is important for understanding the mechanisms used by BVDV for successful replication and invasion of their host. Studies used the yeast-two hybrid (Y2H) screening system to screen individual BVDV proteins for host binding partners (Table 2). For example, using Y2H it was

Virus-host protein partner	Method	Reference
NS3-SphK1	Y2H	(Yamane et al. 2009a)
NS5A-α subunit of eEF1A	Y2H	(Johnson et al. 2001)
NS5A-NIBP	Y2H	(Zahoor et al. 2010)
NS2-Jiv	Co-precipitation	(Rinck et al. 2001)

Table 2. BVDV-host interacting proteins

demonstrated that NS3 from BVDV binds to and inhibits the catalytic activity of sphingosine kinase 1 (SphK1). NS3- SphK1 binding enhance BVDV replication and BVDV-induced apoptosis (Yamane *et al.* 2009a).

Using Y2H screening, the α subunit of bovine translation elongation factor 1A (eEF1A) was shown to interact with the NS5A polypeptide of BVDV (Johnson *et al.* 2001). This interaction was further analysed in a cell-free translation system and was found to be conserved among BVDV isolates of both genotypes and biotypes. Cell-free binding studies were done using a chimeric NS5A fused to glutathione S-transferase (GST–NS5A) expressed in bacteria. GST–NS5A bound specifically to both *in vitro* translated and mammalian cell expressed eEF1A. This interaction was suggested to play a role in the replication of BVDV (Johnson *et al.* 2001).

In addition, Y2H screening identified bovine NIK- and IKKβ-binding protein (NIBP), which is involved in protein trafficking and nuclear factor kappa B (NF-κB) signalling in cells (Zahoor *et al.* 2010). The interaction of NS5A with NIBP was confirmed both *in vitro* and *in vivo*. Supporting this data, confocal immunofluorescence results indicate that NS5A co-localized with NIBP on the endoplasmic reticulum in the cytoplasm of BVDV-infected cells. Moreover, the minimal residues of NIBP that interact with NS5A were mapped as aa 597–623. In addition, overexpression of NS5A inhibited NF-κB activation in HEK293 and LB9.K cells. The same study also showed that inhibition of endogenous NIBP by RNAi enhanced virus replication, indicating the importance of NIBP in BVDV pathogenesis. This is the first reported interaction between NIBP and a viral protein, suggesting a novel mechanism whereby viruses may subvert host-cell machinery for mediating trafficking and NF-κB signaling (Zahoor *et al.* 2010).

Efficient generation of NS2-3 cleavage product NS3 is required for the cytopathogenicity of the pestiviruses. Co-precipitation was used to identify the formation of a stable complex between Jiv, a member of the DnaJ-chaperone family, and BVDV nonstructural protein NS2 (Rinck *et al.* 2001). Jiv has the potential to induce in *trans* cleavage of NS2-3.

While Y2H data is commonly used for identifying protein-protein interaction, results of this technique are dependent on the library screened, the relative representation of each cDNA and the expression level of individual proteins. Also, immunoprecipitation approaches are dependent on the availability of high specificity and affinity antibodies. An alternative approach for identifying protein interaction is the use of tandem and column affinity followed by MS or sequencing analysis. These methods require a specific concentration and a highly purified target protein.

Also, the recent effort for the integration of experimentally established host-pathogen protein-protein interactions for several pathogens in public databases allows for cross-pathogen comparisons and the prediction of these interactions. Databases, for examples, VirHostNet, MINT and HPIDB, allow an investigator to search for homologous host-pathogen interactions and also to get a list of all host-pathogen protein-protein interactions available.

4. Virus-induced changes in the cellular proteome

Studying how viral infection or expression of specific viral proteins affects the expression of the host cell proteome provides insight into metabolic processes and critical regulatory events of the host cell. Although there have been multiple, comprehensive studies to profile viral induced changes in bovine cells at the transcriptional level in response to BVDV infection using microarrays (Werling *et al.* 2005; Maeda *et al.* 2009; Yamane *et al.* 2009b), these observed changes do not always correspond to changes at the protein level. Many viral proteins affect protein turnover without affecting the transcription rate of the protein. Thus, there will always be a need to determine changes at the protein level.

While both bovine and BVDV genome sequences are available, only a few studies have attempted a comprehensive survey of BVDV-induced host protein changes. These studies used MS-based approaches to identify BVDV-induced cellular proteome changes, focusing on host immune changes, which are assessed by comparing protein profiles before and after BVDV infection (Pinchuk *et al.* 2008; Lee *et al.* 2009; Ammari *et al.* 2010) (Figure 2). Proteome coverage (i.e. the proportion of the predicted proteome identified as expressed) is increased by the combined use of differential detergent fractionation (DDF) (McCarthy *et al.* 2005) of infected cells followed by multidimensional protein identification technology (MudPIT) for protein identification. DDF yields four electrophoretically distinct fractions (McCarthy *et al.* 2005). MudPIT was done using strong cation exchange (SCX) followed by reverse phase (RP) chromatography coupled directly in line with electrospray ionization (ESI) ion trap tandem mass spectrometry (2D-LC ESI MS/MS).

Analysis of proteins from uninfected bovine monocytes revealed proteins related to antigen pattern recognition, uptake and presentation to immunocompetent lymphocytes (Lee *et al.* 2006). DDF- MudPIT detected low-abundance cytosolic proteins, indicating the high sensitivity of this approach. Upon comparing proteins from BVDV infected and uninfected monocytes, label free protein quantification methods were used to utilize sampling parameters for estimating protein expression including the sum of cross correlation (ΣXcorr) of identified peptides to enhance the coverage of differentially expressed proteins between the two samples. This approach revealed alterations in the expression of proteins related to immune functions such as cell adhesion, apoptosis, antigen uptake, processing and presentation, acute phase response proteins and MHC class I- and II-related proteins (Lee *et al.* 2009).

In addition, proteomics showed the effects of BVDV biotypes infection on the expression of protein kinases and related proteins involved in the development of viral infection and oncogenic transformation of cells and proteins related to professional antigen presentation. We found that six protein kinases related to cell migration, anti-viral protection, sugar metabolism, and possibly the expression of the receptor for activated C kinase (RAC) were differentially expressed between the ncp and cp BVDV-infected monocytes (Pinchuk *et al.* 2008).

To link the observed differences in protein expression to their broader biological role, information regarding the functions of these gene products and how they interact was obtained using the Gene Ontology (GO) (Ashburner *et al.* 2000) and systems biology (Aggarwal & Lee 2003), respectively. GO information categorizes functional information for a gene product into three broad categories; identify molecular functions, biological processes and cellular components. A complementary approach, system biology, allows for the exploration and visualization of networks and pathways significantly represented in the proteomics datasets. Identification of altered host proteins by cp or ncp BVDV infection was based on rigorous statistical methods for peptide identification and control of false positive identification (Ammari *et al.* 2010). When followed by GO functional and pathway analysis, this study identified similarities and differences between BVDV biotypes and also showed, as expected, that cp BVDV had a more profound effect on infected monocytes than ncp BVDV. The top under- and over- represented GO functions are shown in figure 3. At Ingenuity Pathways Analysis (IPA; Ingenuity system, California) threshold of significance, 6 and 4 networks and 42 and 33 functions/diseases were significantly represented in the proteomes of ncp and cp BVDV-infected monocytes, respectively. The top ten functions/diseases and signalling pathways (ranked based on significance) are shown here in table 3 (for more details see reference (Ammari *et al.* 2010)). Interestingly, among 69 proteins that have been altered by both biotypes only two proteins, integrin alpha 2b (ITGA2B) and integrin beta 3 (ITGB3), were differentially altered by cp and ncp BVDV biotypes.

Since those studies focus on host immune-related proteins, another general analysis was done to identify the effect of BVDV infection on cells overall protein expression. This showed that the mitochondrial dysfunction pathway was the pathway most affected by cp BVDV, but not by ncp BVDV infection, indicating the induction of apoptosis by cp BVDV. BVDV biotypes differ in their effect on Na+-dependent phosphate transporter (SDPT). SDPT is a transmembrane transporter of inorganic phosphate and examples of cells responding to an increase in the inorganic phosphate by apoptosis are known (Di Marco *et al.* 2008). Its two most prevalent representatives are NaP_i1 and NaP_i2. The up-regulation of SDPT observed in our studies in cells infected by cp BVDV may indicate that this cytopathic virus uses the NaP_i to induce the apoptotic death of the cells that it infects. On the other hand, both cp and ncp BVDV biotypes down-regulated a different isoform of SDPT (NaP_i2b), suggesting that this other isoform of the transporter may not be directly involved in apoptosis. Also, our proteome analysis revealed that two proteins from the mitochondrial voltage-dependent anion channel (VDAC) family, as well as hexokinase (HK) were differentially regulated by cp and ncp BVDV. In addition, cp BVDV-infection contributed to oxidative stress by disturbance of cellular antioxidants system. These data are in line with previous findings about the involvement of these proteins in apoptosis (manuscript in preparation).

Overall, the use of MS-based methods provided a means to study BVDV pathogenesis in a more-high-throughput parallel fashion than individual immunopreciptation studies and this global scale of study increased our ability to assess immune-related protein interaction and cellular changes that are significantly compromised in monocytes infected with BVDV biotypes. Thus, they open new possibilities for discovery of previously unknown viral-host connections. Other proteomics techniques that have been used for other viruses include quantitative methods such as isotype-coded affinity tags (ICAT) and different gel electrophoresis (DIGE) (for a review on the use of proteomics in studying different viruses, see reference (Maxwell & Frappier 2007)).

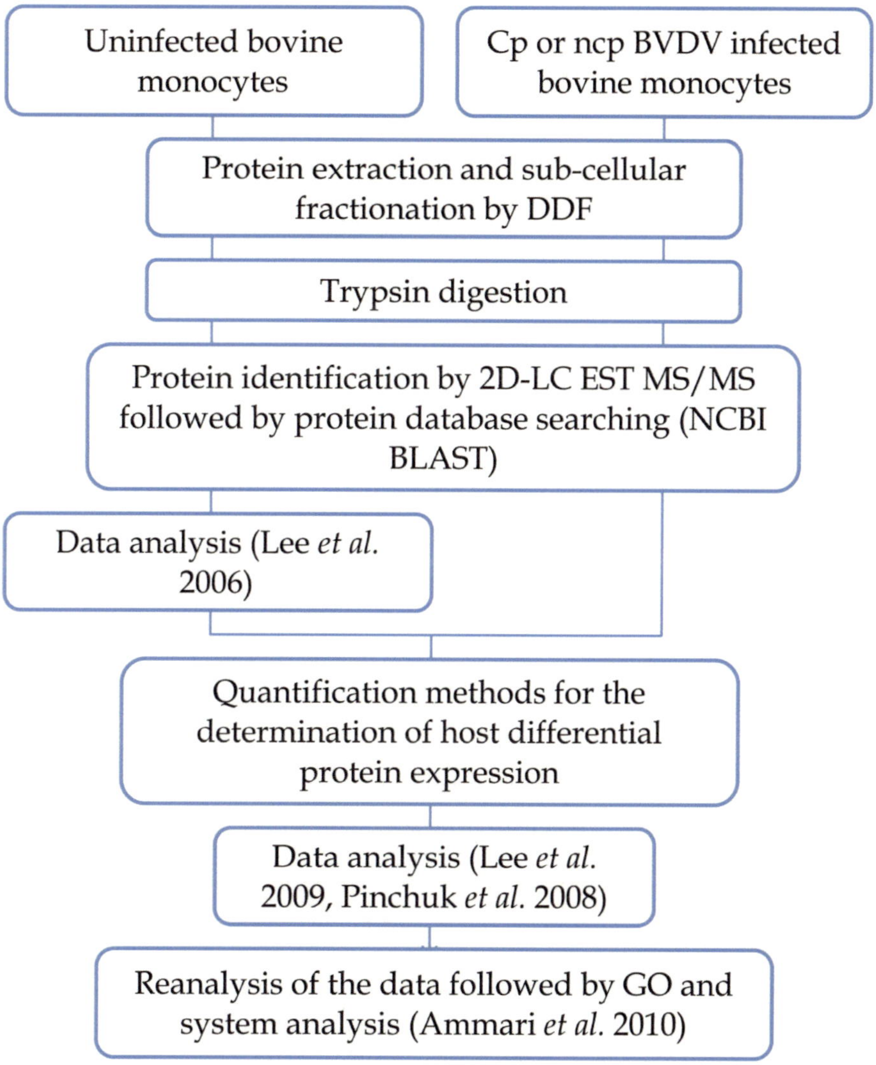

Fig. 2. Flow chart of MS-related work for identifying BVDV-induced changes in bovine monocytes

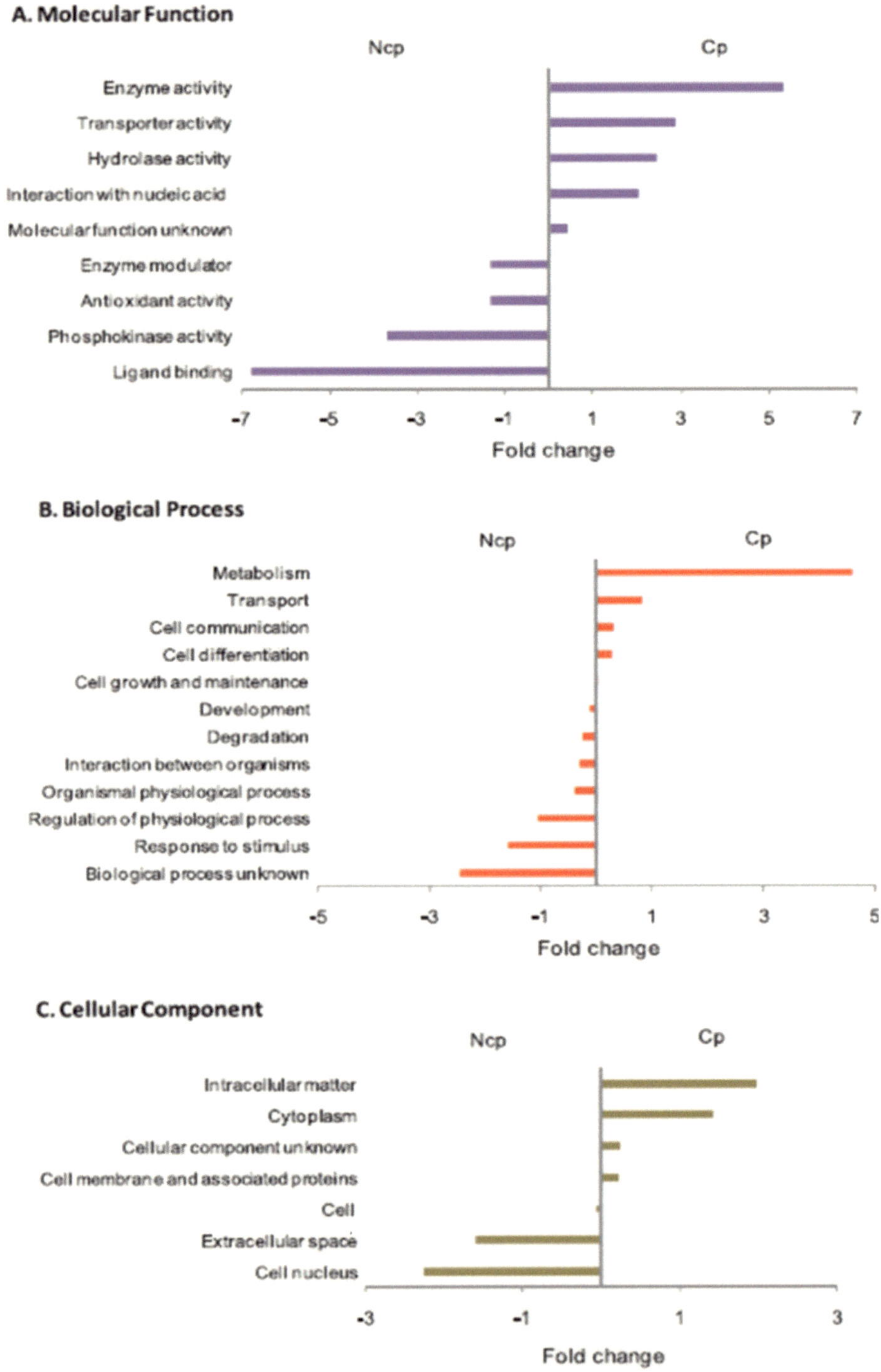

Fig. 3. GO functional analysis of host proteins differentially expressed in ncp or cp BVDV-infected monocytes (Ammari *et al*. 2010)

	Cp BVDV-infection	Ncp BVDV-infection
Networks	Four	Six
Pathways	Acute phase response signaling (12), Fcγ receptor mediated phagocytosis in macrophages and monocytes (8), **Actin cytoskeleton signaling (8)**, Antigen presentation pathway (3), B cell development (3), **RhoA signaling (5)**, Caveolae-mediated endocytosis signaling (4), **Clathrin-mediated endocytosis signaling (6)**, IL-10 signaling (3), **Interferon signaling (2)**	Acute phase response signaling (8), Caveolae-mediated endocytosis signaling (5), B cell development (3), IL-10 signaling (3), **Macropinocytosis signaling (3), Virus entry via endocytic pathway (3)**, Antigen presentation pathway (2), Fcγ receptor mediated phagocytosis in macrophages and monocytes (3), **Integrin signaling (4), Primary immunodeficiency signaling (2)**
Functions/ diseases	Cellular function and maintenance (7), cell-to-cell signaling and interaction (8), Inflammatory response (14), **Cellular assembly and organization (2), Protein synthesis (2), Cellular compromise (6), Cell morphology (3)**, Cellular development (4), Hematological system development and function (9), Immune cell trafficking (9)	Cellular movement (10), Immune cell trafficking (18), Inflammatory response (20), **Cell death (11)**, Hematological system development and function (21), Cell-to-cell signaling and interaction (10), Cellular function and maintenance (5), **Immunological disease (5), Antigen presentation (13), Inflammatory disease (5)**

Table 3. Top ten functions/diseases and pathways in BVDV- infected monocytes. Numbers between brackets indicates the number of altered proteins involved in the function/disease or pathway

5. Concluding remarks

By using proteomics it is possible now to separate very complex protein mixtures with high resolution, to extract the proteins of interest, to study them with MS and to identify them with high reliability. With the use of new proteomics approaches, much can be done in studying BVDV protein composition, structure, and interactions. We expect that the application of proteomic methods to study BVDV will provide valuable information about BVDV pathogenesis and reveal new insights about host-virus interactions, leading to better strategies to prevent or cure BVDV infection. However, one should keep in mind that proteomics should always be considered the starting point for functional studies rather than an end point that follows traditional methods.

6. References

Adler H, Frech B, Meier P, Jungi TW , Peterhans E (1994). Noncytopathic strains of bovine viral diarrhea virus prime bovine bone marrow-derived macrophages for enhanced generation of nitric oxide. *Biochemical and biophysical research communications*. 202, 1562-1568.

Aggarwal K , Lee HK (2003). Functional genomics and proteomics as a foundation for systems biology. *Briefings in functional genomics & proteomics*. 2, 175.

Ammari M, McCarthy FM, Nanduri B , Pinchuk LM (2010). Analysis of Bovine Viral Diarrhea Viruses-infected monocytes: identification of cytopathic and non-cytopathic biotype differences. *BMC Bioinformatics*. 11 Suppl 6, S9.

Ashburner M, Ball CA, Blake JA, Botstein D, Butler H, Cherry JM, Davis AP, Dolinski K, Dwight SS , Eppig JT (2000). Gene Ontology: tool for the unification of biology. *Nature genetics*. 25, 25.

Bendfeldt S RJ, Neill JD (2007). Activation of cell signaling pathways is dependant on the biotype of bovine viral diarrhea viruses type 2. *Virus Res*. Jun;126(1-2):96-105.

Boyd BL, Lee TM, Kruger EF , Pinchuk LM (2004). Cytopathic and non-cytopathic bovine viral diarrhoea virus biotypes affect fluid phase uptake and mannose receptor-mediated endocytosis in bovine monocytes. *Vet Immunol Immunopathol*. 102, 53-65.

Branza-Nichita N, Durantel D, Carrouee-Durantel S, Dwek RA , Zitzmann N (2001). Antiviral effect of N-butyldeoxynojirimycin against bovine viral diarrhea virus correlates with misfolding of E2 envelope proteins and impairment of their association into E1-E2 heterodimers. *J Virol*. 75, 3527-3536.

Brett D. Lindenbach H-JT, Charles M. Rice (2007). Flaviviridae: The Viruses and Their Replication. *Fields Virology*. 5th Edition.

Brewoo JN, Haase CJ, Sharp P , Schultz RD (2007). Leukocyte profile of cattle persistently infected with bovine viral diarrhea virus. *Veterinary immunology and immunopathology*. 115, 369-374.

Carman S, van Dreumel T, Ridpath J, Hazlett M, Alves D, Dubovi E, Tremblay R, Bolin S, Godkin A , Anderson N (1998). Severe acute bovine viral diarrhea in Ontario, 1993-1995. *J Vet Diagn Invest*. 10, 27-35.

Cattaneo R (2004). Four viruses, two bacteria, and one receptor: membrane cofactor protein (CD46) as pathogens' magnet. *J Virol*. 78, 4385-4388.

Choi KH, Gallei A, Becher P , Rossmann MG (2006). The structure of bovine viral diarrhea virus RNA-dependent RNA polymerase and its amino-terminal domain. *Structure*. 14, 1107-1113.

Choi KH, Groarke JM, Young DC, Kuhn RJ, Smith JL, Pevear DC , Rossmann MG (2004). The structure of the RNA-dependent RNA polymerase from bovine viral diarrhea virus establishes the role of GTP in de novo initiation. *Proc Natl Acad Sci U S A*. 101, 4425-4430.

Di Marco GS, Hausberg M, Hillebrand U, Rustemeyer P, Wittkowski W, Lang D , Pavenstaedt H (2008). Increased inorganic phosphate induces human endothelial cell apoptosis in vitro. *American Journal of Physiology-Renal Physiology*. 294, F1381.

Gallei A, Pankraz A, Thiel HJ , Becher P (2004). RNA recombination in vivo in the absence of viral replication. *J Virol*. 78, 6271-6281.

Glew EJ, Carr BV, Brackenbury LS, Hope JC, Charleston B , Howard CJ (2003). Differential effects of bovine viral diarrhoea virus on monocytes and dendritic cells. *J Gen Virol*. 84, 1771-1780.

Grummer B, Bendfeldt S, Wagner B , Greiser-Wilke I (2002). Induction of the intrinsic apoptotic pathway in cells infected with cytopathic bovine virus diarrhoea virus. *Virus Res*. 90, 143-153.

H.M. Berman JW, Z. Feng, G. Gilliland, T.N. Bhat, H. Weissig, I.N. Shindyalov, P.E. Bourne (2000). The Protein Data Bank Nucleic Acids Research. 28: 235-242.

Houe H (1999). Epidemiological features and economical importance of bovine virus diarrhoea virus (BVDV) infections. *Vet Microbiol.* 64, 89-107.

Houe H (2003). Economic impact of BVDV infection in dairies. *Biologicals.* 31, 137-143.

Howard CJ, Clarke MC, Sopp P , Brownlie J (1992). Immunity to bovine virus diarrhoea virus in calves: the role of different T-cell subpopulations analysed by specific depletion in vivo with monoclonal antibodies. *Vet Immunol Immunopathol.* 32, 303-314.

Johnson CM, Perez DR, French R, Merrick WC , Donis RO (2001). The NS5A protein of bovine viral diarrhoea virus interacts with the alpha subunit of translation elongation factor-1. *J Gen Virol.* 82, 2935-2943.

Jordan R, Wang L, Graczyk TM, Block TM , Romano PR (2002). Replication of a cytopathic strain of bovine viral diarrhea virus activates PERK and induces endoplasmic reticulum stress-mediated apoptosis of MDBK cells. *J Virol.* 76, 9588-9599.

Lazar C, Zitzmann N, Dwek RA , Branza-Nichita N (2003). The pestivirus E(rns) glycoprotein interacts with E2 in both infected cells and mature virions. *Virology.* 314, 696-705.

Lee SR, Nanduri B, Pharr GT, Stokes JV , Pinchuk LM (2009). Bovine viral diarrhea virus infection affects the expression of proteins related to professional antigen presentation in bovine monocytes. *Biochim Biophys Acta.* 1794, 14-22.

Lee SR, Pharr GT, Cooksey AM, McCarthy FM, Boyd BL , Pinchuk LM (2006). Differential detergent fractionation for non-electrophoretic bovine peripheral blood monocyte proteomics reveals proteins involved in professional antigen presentation. *Dev Comp Immunol.* 30, 1070-1083.

Liebler-Tenorio EM, Ridpath JF , Neill JD (2002). Distribution of viral antigen and development of lesions after experimental infection with highly virulent bovine viral diarrhea virus type 2 in calves. *American journal of veterinary research.* 63, 1575-1584.

Maeda K, Fujihara M , Harasawa R (2009). Bovine viral diarrhea virus 2 infection activates the unfolded protein response in MDBK cells, leading to apoptosis. *J Vet Med Sci.* 71, 801-805.

Maxwell KL , Frappier L (2007). Viral proteomics. *Microbiol Mol Biol Rev.* 71, 398-411.

McCarthy FM, Burgess SC, van den Berg BH, Koter MD , Pharr GT (2005). Differential detergent fractionation for non-electrophoretic eukaryote cell proteomics. *J Proteome Res.* 4, 316-324.

Peterhans E, Jungi TW , Schweizer M (2003). BVDV and innate immunity. *Biologicals.* 31, 107-112.

Pinchuk GV, Lee SR, Nanduri B, Honsinger KL, Stokes JV , Pinchuk LM (2008). Bovine viral diarrhea viruses differentially alter the expression of the protein kinases and related proteins affecting the development of infection and anti-viral mechanisms in bovine monocytes. *Biochim Biophys Acta.* 1784, 1234-1247.

Qu L, McMullan LK , Rice CM (2001). Isolation and characterization of noncytopathic pestivirus mutants reveals a role for nonstructural protein NS4B in viral cytopathogenicity. *J Virol.* 75, 10651-10662.

Reed KE, Gorbalenya AE , Rice CM (1998). The NS5A/NS5 proteins of viruses from three genera of the family flaviviridae are phosphorylated by associated serine/threonine kinases. *J Virol.* 72, 6199-6206.

Ridpath JF, Bolin SR , Dubovi EJ (1994). Segregation of bovine viral diarrhea virus into genotypes. *Virology.* 205, 66-74.

Rinck G, Birghan C, Harada T, Meyers G, Thiel HJ , Tautz N (2001). A cellular J-domain protein modulates polyprotein processing and cytopathogenicity of a pestivirus. *J Virol.* 75, 9470-9482.

Sapay N, Montserret R, Chipot C, Brass V, Moradpour D, Deleage G , Penin F (2006). NMR structure and molecular dynamics of the in-plane membrane anchor of nonstructural protein 5A from bovine viral diarrhea virus. *Biochemistry.* 45, 2221-2233.

Schweizer M , Peterhans E (1999). Oxidative stress in cells infected with bovine viral diarrhoea virus: a crucial step in the induction of apoptosis. *J Gen Virol.* 80 (Pt 5), 1147-1155.

Schweizer M , Peterhans E (2001). Noncytopathic bovine viral diarrhea virus inhibits double-stranded RNA-induced apoptosis and interferon synthesis. *J Virol.* 75, 4692-4698.

Tellinghuisen TL, Paulson MS , Rice CM (2006). The NS5A protein of bovine viral diarrhea virus contains an essential zinc-binding site similar to that of the hepatitis C virus NS5A protein. *J Virol.* 80, 7450-7458.

Weiland E, Stark R, Haas B, Rumenapf T, Meyers G , Thiel HJ (1990). Pestivirus glycoprotein which induces neutralizing antibodies forms part of a disulfide-linked heterodimer. *J Virol.* 64, 3563-3569.

Werling D, Ruryk A, Heaney J, Moeller E , Brownlie J (2005). Ability to differentiate between cp and ncp BVDV by microarrays: towards an application in clinical veterinary medicine? *Vet Immunol Immunopathol.* 108, 157-164.

Yamane D, Zahoor MA, Mohamed YM, Azab W, Kato K, Tohya Y , Akashi H (2009a). Inhibition of sphingosine kinase by bovine viral diarrhea virus NS3 is crucial for efficient viral replication and cytopathogenesis. *J Biol Chem.* 284, 13648-13659.

Yamane D, Zahoor MA, Mohamed YM, Azab W, Kato K, Tohya Y , Akashi H (2009b). Microarray analysis reveals distinct signaling pathways transcriptionally activated by infection with bovine viral diarrhea virus in different cell types. *Virus Res.* 142, 188-199.

Zahoor MA, Yamane D, Mohamed YM, Suda Y, Kobayashi K, Kato K, Tohya Y , Akashi H (2010). Bovine viral diarrhea virus non-structural protein 5A interacts with NIK- and IKKbeta-binding protein. *J Gen Virol.* 91, 1939-1948.

Zhang G, Aldridge S, Clarke MC , McCauley JW (1996). Cell death induced by cytopathic bovine viral diarrhoea virus is mediated by apoptosis. *J Gen Virol.* 77 (Pt 8), 1677-1681.

Part 2

Studying Environmental Complexities

4

Proteomics as a Tool for the Characterization of Microbial Isolates and Complex Communities

Florence Arsène-Ploetze[1], Christine Carapito[2],
Frédéric Plewniak[1] and Philippe N. Bertin[1*]
[1]*Génétique moléculaire, Génomique et Microbiologie,*
Université de Strasbourg, Strasbourg
[2]*Laboratoire de Spectrométrie de Masse Bio-Organique,*
Institut Pluridisciplinaire Hubert Curien, Strasbourg
France

1. Introduction

Proteins may be considered as the main effectors of biological responses of organisms to specific environmental conditions, instead of messenger RNAs. Indeed, a modulation in their activity does not always depend on a modified expression of the corresponding genes but rather on post-translational modifications. Proteome analysis may therefore constitute an appropriate approach to address the question of organism adaptation to environmental stresses or growth under extreme conditions. Recently, the knowledge of the organisms' physiology has led to deep changes in the investigation methods, favouring the use of global analysis methods in complement with conventional strategies. Instead of studying individual proteins or metabolic products, the integral profile of organisms can now be established. This may be of importance when studying adaptive and stress responses in microorganisms because of their multifactorial character. In particular, the differential analysis in various growth conditions of the whole protein content (« proteome »), which allows the simultaneous quantification of gene products in an organism, represents part of the so-called integrative biology (Bertin et al., 2008).

Genomics is a conceptual approach that aims to study the biology of microorganisms by analysing the complete genetic information they contain. This scientific discipline really emerged more than fifteen years ago with the characterization of the first complete genome of autonomous organisms (Bertin et al., 2008). An important reduction of sequencing costs associated with new high-throughput technologies has led to an explosion of genomic programs that now concern organisms in all domains of life (http://www.genomeonline.org). Most of the descriptive and functional genomic efforts initially focused on human pathogens, such as bacteria and parasites, and next, on higher eukaryotes. More recently, there has been a growing interest in microorganisms isolated from various habitats, including extreme ecological niches, to characterize specific properties that allow these organisms to grow in such environments. The field of application of proteomics thus expended in line with genomics. These works should lead not only to a

* Corresponding Author

better understanding of ecosystems themselves, but also to the identification of novel functions that may be exploitable for biotechnological applications, in particular in the bioremediation of contaminated environments. A better understanding of the involved elements, their spatial and temporal distribution, the metabolic pathways they belong to, would allow drawing an integrated picture of biological processes under study. This could lead to an optimal use of microorganism properties, favouring the desired effects. In this review, global proteomics approaches allowing deciphering the physiology of one microorganism or the functioning of a community will be presented, as well as recent advances in targeted proteomics approaches. Finally, the huge amount of data generated by such approaches needs integrative analyses that require specific proteome databases.

2. Global proteomics approaches

In their natural habitat, microorganisms rapidly adapt to environmental changes by modulating their protein content or activity, for instance via post-translational modifications. Therefore, to highlight the physiological state of a microorganism in one particular condition, a large-scale study of its proteome is a widespread approach. In such a workflow, the establishment of global protein profiles (Figure 1) requires protein extraction, separation steps that are often obtained by two-dimensional gel electrophoresis (2DE) followed by mass spectrometry analysis for protein identification.

2.1 Proteomics methodology: From sample preparation to protein identification

Protein extraction and separation in a homogenous population require first to optimize the lysis conditions. Sample preparation is a fundamental step and several protocols are usually tested, such as those described in (Cañas et al., 2007). Physical lysis methods are the most useful methods in the case of microorganisms: vortexing and grinding with glass beads, sonication, freeze/thaw or alternating cycles of high and low pressure. Combining these mechanical methods with enzymatic lysis or use of detergents may improve cell lysis efficiency. 2DE separation has shown to be one of the most common separation techniques used in proteomic studies (Rabilloud et al., 2010). Proteins are separated in a first step according to their charge and in a second step according to their molecular weight. They are then usually visualised by an organic dye (Coomassie blue), by metallic salt reduction (silver nitrate) or fluorescent labelling (Sypro, DeepPurple…). Bidimensional proteome analysis presents however several limitations. Indeed, whatever the detection method used, all proteins of any organism cannot be visualised because some of them are present at very low levels. In addition, some proteins are quite unstable while others are less labile. Moreover, membrane proteins are usually more difficult to detect on 2D gels because of their low solubility. Therefore, other separation techniques may be used such as monodimensional SDS-PAGE, in particular to retain the membrane proteins (Laemmli, 1970), non-gel strategies such as MudPIT approaches (multidimensional protein identification technology) (Fränzel & Wolters, 2011) or any other liquid or affinity chromatography-based separations (Gundry et al., 2009). Many kinds of original chromatography types and combinations have been explored in the field of proteomics to fractionate and separate complex protein mixtures prior to mass spectrometry (MS) analysis either at a protein or at a peptide level. Each of those approaches presents advantages and drawbacks and the choice of the separation method used is a crucial step in the proteomics workflow. Overall, the higher success of one or the other separation method is highly sample-dependent.

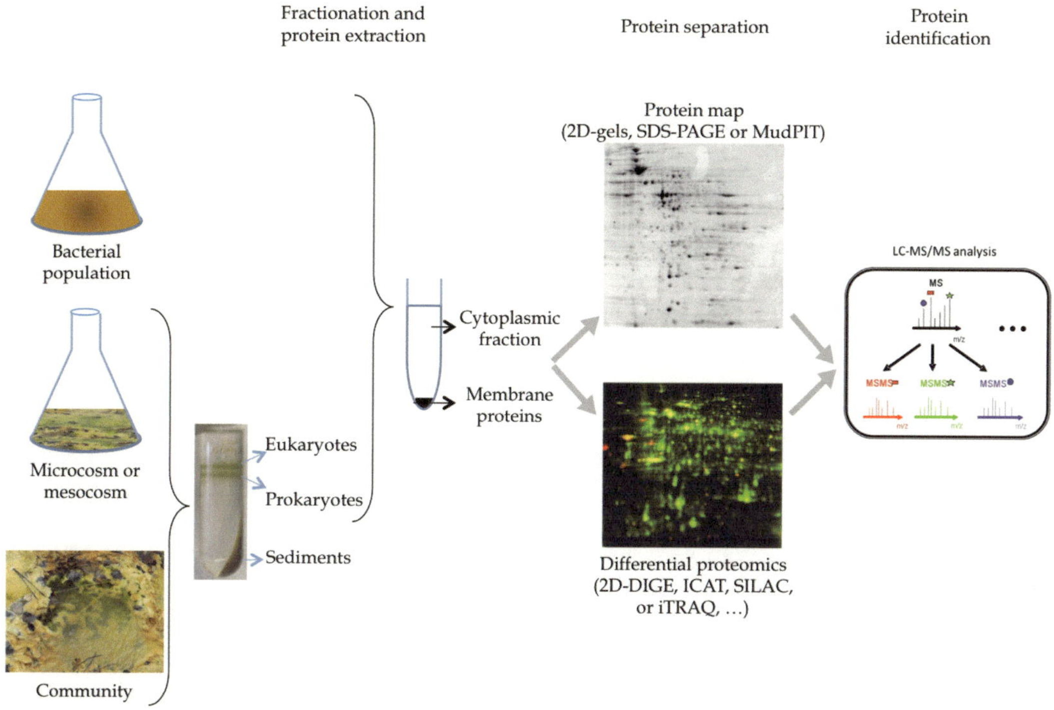

Fig. 1. Classical proteomics workflow to study the physiology of microbial isolates or complex communities.

Once separated, proteins are identified by mass spectrometry. The recent development of functional genomics approaches has led to considerable progress in identification methods (Casado-Vela et al., 2011). Proteins are characterized by mass spectrometers able to ionize and precisely determine the masses of ionized molecules. Proteins of interest are recovered from gels or from any other chromatography separation and enzymatically digested, e.g. by trypsin which specifically cuts the polypeptidic chain at lysine or arginine residues. The whole set of generated peptides are then analysed by MS and most commonly tandem MS (MS/MS) to precisely measure the molecular weight of the peptides and their associated fragments (in MS/MS mode). The experimental MS data are compared to theoretical data calculated from the available protein sequence databases derived from the genome sequence. Historically, the Peptide Mass Fingerprint (PMF) approach, based on the measurement of the peptide masses only, was used to identify proteins (mostly by Matrix-Assisted Laser Desorption Ionization Time of Flight MS, MALDI-TOF-MS) but this approach quickly revealed to be insufficiently specific with the exponentially growing protein sequence databases. MS/MS approaches nowadays constitute the standard method to reliably identify proteins. Over the last 10 years, numerous algorithms, proprietary or open-source, have been developed to compare and score the matching between experimental MS/MS data and theoretical mass lists calculated from expected protein sequences (several of the most commonly used tools are listed in Table 1) (Nesvizhskii, 2010). The MS/MS protein identification workflow is now well established and allows the performance of high throughput and large scale proteomic experiments provided that the genome of the studied organism is sequenced.

Database	Access
Mascot	http://www.matrixscience.com
Phenyx	http://www.genebio.com/products/phenyx
OMSSA	http://pubchem.ncbi.nlm.nih.gov/omssa
X! Tandem	http://www.thegpm.org/TANDEM/

Table 1. Tools useful for MS/MS data analysis.

However, even when genomic information is available, protein identification may be complicated by lacks/errors in the predicted protein sequence databases introduced by automatic genome annotation (translational frameshift, read-through of stop codons) or by post-translational modifications (e.g., glycosylation or phosphorylation) hindering the mass matching procedure. To circumvent those errors widespread in non reference and not thoroughly annotated genomes, original alternative identification strategies have been developed which use the complete unannotated genome sequence to interpret the MS/MS data. These approaches have opened the avenue to proteogenomics, defined as the use of proteomics results to enhance the knowledge of the genome (Delalande et al., 2005; Gallien et al., 2009). Finally, when the genome of the organism under study has not yet been sequenced, *de novo* sequencing is mandatory. This consists in interpreting individually each high quality MS/MS spectrum to derive amino acid sequence tags. These sequence tags are then submitted to MS-BLAST (http://dove.embl-heidelberg.de/Blast2/msblast.html) homology searches in order to identify the proteins by sequence homology with orthologous proteins present in the databases (Carapito et al., 2006).

Altogether, the recent developments in proteomics, in particular in MS instrumentation to gain sensitivity, resolution and mass accuracy, as well as the important increase of genomic data enabled proteomics to become a widespread, useful and robust technique to understand the adaptive capacities of microorganisms, under laboratory conditions but also in their natural habitat within complex communities.

2.2 Proteomics as a tool to understand the physiology of environmental isolates

Proteomics has two major goals. On the one hand, proteomic maps can first make an inventory of functions expressed in an organism under specific conditions. On the other hand, differential proteomic analyses allow studying the response of microorganisms to changes in the environment as well as the underlying regulatory mechanisms. Several examples of these two strategies allowing better understanding of the physiology of environmental isolates are presented in the following sections.

First, by using 2DE or SDS-PAGE separation techniques, the global or partial proteome maps (cytoplasmic, membrane or extracellular fraction) of several microorganisms have been drawn. These often concern model organisms, e.g. *B. subtilis* (Hecker & Völker, 2004) or human pathogens, e.g. *Mycobacterium tuberculosis*, the etiologic agent of tuberculosis (Schmidt et al., 2004). This approach was recently used to list proteins expressed by an arsenic resistant bacterium, *Herminiimonas arsenicoxydans*, which is able to resist and grow in harsh conditions, particularly in the presence of arsenite (Weiss et al., 2009). Another example of bacterium able to adapt to extreme conditions is *Deinococcus geothermalis*, found in geothermal wells. In this bacterium, cytosolic and cell envelope proteome maps revealed that one-fourth of the cell envelope proteome corresponds to *Deinococcus* specific proteins such as V-type ATPases, and that repair enzymes are highly expressed and among the most abundant proteins, even in the absence of stress (Liedert et al., 2010). Finally, these techniques may be used to focus on a particular fraction of the proteome. As an example, using specific staining procedures, a 2DE map of iron-metalloproteins has been drawn in the acidophilic archaeon *Ferroplasma acidiphilum* (Ferrer et al., 2007). Remarkably, the results suggest that the high content of metalloproteins present in this organism represents a relic of early life on Earth, where metals were abundant because of widespread volcanic and hydrothermal activities.

Second, to get further insight into the adaptation capacities of microorganisms, differential proteomic analyses characterize proteins which expression is induced or repressed in response to a particular stimulus, and defines thus a stimulon. Using 2DE, the amount of proteins expressed in different conditions or backgrounds may be compared, which requires robust quantification of each protein in each condition. The use of 2DE coupled to fluorescent labelling (DIGE) makes such quantification easier (Yan et al., 2002). This differential proteomic strategy was extensively used to decipher the adaptive response of pathogens or model bacteria such as *Bacillus subtilis* (Bertin et al., 2008; Hecker & Völker, 2004; Jungblut, 2001). Recently, the physiology of an increasing number of environmental isolates has been studied by proteomics approaches. For example, the adaptation to cold has been studied in psychrophilic microorganisms, such as *Pseudoalteromonas haloplanktis* (Piette et al., 2010) and in the archeon *Methanococcoides burtonii* (Saunders et al., 2005). Similarly, arsenic bacterial metabolism has been investigated in *H. arsenicoxydans* (Carapito et al., 2006; Muller et al., 2007) and *Thiomonas* sp. (Bryan et al., 2009). In *H. arsenicoxydans*, in addition to the proteome map listing the proteins expressed in the presence of arsenite (see above), differential proteomic analyses data were combined with transcriptomics data to study its adaptive response in the presence of arsenite (Cleiss-Arnold et al., 2010; Muller et al., 2007; Weiss et al., 2009). These methodologies revealed that this bacterium is able to grow in the presence of arsenic by inducing the expression of proteins involved in several processes such as oxidative stress, arsenic resistance, energy metabolism or motility/chemotactism. Differential proteomics experiments performed on *Thiomonas* allowed comparing the arsenic response in several strains. Indeed, proteomics has highlighted metabolic differences between *Thiomonas arsenitoxydans* 3As and *Thiomonas arsenivorans* strains. In the presence of As(III), proteins involved in carbon fixation were shown to be preferentially accumulated in *Tm. arsenivorans* but less abundant in *Tm. arsenitoxydans 3As*, supporting the hypothesis that *Tm. arsenivorans* is capable of optimal autotrophic growth in the presence of As(III) when used as an inorganic electron donor. One response shared by these arsenic-oxidizing bacteria is the induction in the presence of arsenite of phosphate transporters, as well as proteins involved in glutathione metabolism, DNA repair and protection against oxidative stress (Bryan et al., 2009; Carapito et al., 2006; Cleiss-Arnold et al., 2010; Weiss et al., 2009).

In differential analyses, the 2DE technology has however one particular limitation, i.e. several proteins may be resolved in the same spot hindering their respective quantification. To avoid such a problem, differential protein patterns can be identified using non-gel protein separations coupled with isotope labelling approaches. To find proteins or peptides with significant differences of concentrations in sampled proteomes, different stable heavy isotope labelling techniques can be applied like ICAT, SILAC, iTRAQ or ICPL. Depending on the sample origin, isotope labelling may be performed on different levels (organism, cell, protein, or peptide) and on different reactive groups (Gevaert et al., 2008). As an example, 2DE and ICAT approaches were combined to study the aromatic catabolic pathways in *Pseudomonas putida* KT 2440. Interestingly, it appears that these two methods are complementary since 110 and 80 proteins were shown to be induced in the presence of benzoate, using ICAT or 2DE, respectively, and only 19 common proteins were identified using both approaches (Kim et al., 2006). Even though those approaches have proven to allow precise quantification of numerous proteins in various applications, each of them presents drawbacks and limitations. For instance, stable isotope labelling with amino acids in cell culture (SILAC) is limited to applications dealing with proteins obtained from cell cultures, free amino-group labelling (like ICPL) induces significant increase of sample complexity leading to an aggravation of undersampling problems, isobaric labelling (like iTRAQ) requires high resolution MS/MS data to be acquired and is often unsuitable for most widespread ion trap MS/MS. The choice of the approach to apply for quantification is therefore very sample-dependent and crucial for the success of the proteomics application.

Once adaptation capacities have been identified, it can be interesting to understand the regulation network allowing microorganisms to respond quickly to changes in their environment. With such an aim, differential proteomics is useful to decipher the role of global regulators and to list genes belonging to the same regulon, i.e. genes that are regulated by the same regulator. Such approaches have helped to highlight unsuspected regulatory networks, revealing that some bacteria have developed sophisticated mechanisms to survive or grow in a large panel of conditions. As an example, the global Crc protein that control the metabolism of carbon sources and catabolite repression of *Pseudomonas aeruginosa* was shown to be involved in the regulation of virulence gene expression (Linares et al., 2010). Finally, proteomics can also be used to highlight post-translational modifications (PTMs) that may be crucial for rapid regulation of protein activity. For instance, a phosphoproteomic study allowed identifying kinases involved in the regulation of several cellular processes in bacteria (Grangeasse et al., 2010).

2.3 Proteomics as a tool to understand the functioning of communities: Metaproteomics or environmental proteomics

Studying microorganisms in laboratory conditions may not reflect their particular adaptation capacities in their environmental niches. For example, in the case of pathogens, symbionts or commensals, it is crucial to identify not only proteins expressed in response to abiotic changes, but also in response to biotic factors, such as those expressed by their host. The major difficulty in such studies is to distinguish microbial and host proteins, a difficulty that is reduced when the genome of both organisms is known. The second problem is to extract sufficient amount of microbial proteins in order to detect them. Recent advances in protein identification have allowed access to such information (see below), as shown by the identification of key proteins involved in virulence in several pathogens such as *Echinococcus granulosus* metacestode (Monteiro et al., 2010), *Clostridium perfringens* (Sengupta & Alam, 2011) or *Anaplasma* when present in the tick vector (Ramabu et al., 2010). Similarly, proteomics has been used to address the complex processes governing the interactions between symbiotic microorganisms and their host and *vice versa*, e.g. the adaptive response of plants interacting with mycorrhizae (Bona et al., 2011). Recently, proteomics approaches have been developed to study the interactions of microorganisms with their host or microbial communities that may contain uncultured microbes, thus extending our knowledge of the diversity of microbial metabolic processes. Microbial communities are complex biological assemblies whose study has been difficult for a long time because of the inability to culture many of their components. However, the taxonomic diversity studies performed by the analysis of 16S rRNA sequences suggest that in any given environment only a small fraction of organisms present can actually be cultivated. These communities can now be explored as a whole by the sequencing of their genomic DNA content, i.e. their metagenome (Bertin et al., 2008). In parallel, a novel proteomic approach called metaproteomics or environmental proteomics has emerged to characterize in a global way the protein content of microbial communities. The metaproteomics approach has some advantages, compared to other functional genomic approaches, such as metatranscriptomics, i.e. the study of mRNA expressed by a community. Indeed, as proteins are more stable than RNA (especially those originating from prokaryotes), the metaproteome content is supposed to be less affected by the extraction procedure, and probably gives a better insight into the biological functions expressed *in situ*. Several examples of recent studies in this environmental proteomics field are presented below.

2.3.1 Environmental proteomics as a tool to characterize uncultured microorganisms: Advantages and limitations

Interestingly, the metaproteomics approach may give taxonomic information complementary to the 16S rRNA gene-based approach, commonly used to analyse the community structure. Previous observations established that it is not always possible to affiliate some bacteria using only 16S rRNA gene sequences (Schleifer, 2009). Indeed, some microorganisms showing very similar 16S rRNA gene sequences turned out to belong to different taxa when other phylogenetic markers were used. For instance, in a recent study, metaproteomics highlighted the expression of proteins involved in conserved biological processes that could be assigned to a specific taxon, at the genus or species level, whereas the RDP (Ribosomal Database Project) analysis allowed the affiliation of only 28% at the genus level (Halter et al., 2011). More generally, the identification of signature peptides in orthologs has enabled the use some proteins as taxonomic markers to describe the active community in the Carnoulès AMD (Bruneel et al., 2011), and to differentiate various ecotypes present in a similar ecosystem (Simmons et al., 2008).

In addition, metaproteomics has enabled the analysis of the role of uncultured microorganisms *in situ*. For example, a study of microbes growing in water plant sludge led to the identification of several proteins belonging to an uncultured organism of the *Rhodocyclus* lineage known to accumulate polyphosphates (Wilmes & Bond, 2004). Similarly, proteins synthesized by microorganisms flourishing in an AMD ecosystem, i.e. inside a biofilm (Ram et al., 2005), have been inventoried. In this study and others, strain-resolved expression patterns indicated that microorganisms belonging to the same species with less than 1% divergence in nucleotide sequences of genes encoding 16S rRNA (ecotypes) coexist in ecosystems. At a functional level, this microdiversity can lead to functional diversity, since these strains may play distinct roles (Denef et al., 2010a, 2010b). In another study, synergistic/antagonistic interactions between fungi and Rhizobacteria were explored (Moretti et al., 2010). Proteomic patterns of a microbial consortium were compared in the presence or the absence of antibiotic, in order to evaluate the bacterial impact on the consortium functioning. Using this strategy, candidate proteins were identified that may provide advantages for the consortium to out-compete pathogen strains such as *Fusarium*.

In some cases, the high level of diversity makes the metaproteomics approach rather difficult to apply. A low number of identified proteins have been observed previously in soil or sediments where a high level of diversity was observed (Benndorf et al., 2007; Halter et al., 2011; Taylor & Williams, 2010). As pointed out by the authors of recent reviews, metaproteomics studies are successful when applied to communities with low levels of diversity. When a high level of diversity is observed, each protein is diluted in a complex mixture and only the most abundant proteins are therefore likely to be identified. Moreover, a large proportion of the bacteria forming such a community have usually never been studied so far *in vitro* and their genome sequences, and hence their protein sequences, which are required for MS identification, are not available in public databases. For example, unlike *Proteobacteria*, only a few *Acidobacteria* or archaeal protein sequences are available in the existing protein sequence databases. To prevent such a limitation, metaproteomics and metagenomics are nowadays often combined.

When the community is too complex or when this community is found in solid phase such as soil or sediments, it may be necessary to fractionate cells, in order to study only a fraction of the community (Figure 1). For instance, metaproteomes of key microbial populations, i.e. *Synechococcus* cells, were drawn after cells separation using microwave fixation and flow cytometric sorting (Mary et al., 2010). In another study, using density gradient, it was possible to separate microorganisms from sediments but also bacteria from the eukaryotic population and to study both populations separately. Indeed, in the Carnoulès arsenic-rich ecosystem, the bacterial community analyses revealed that proteins involved in the biomineralization of iron and arsenic were shown to be expressed by *Acidothiobacillus ferrooxidans* and *Thiomonas,* respectively, which supports a major role of these microorganisms in the natural attenuation of this highly contaminated environment (Bertin et al., 2011). This approach also revealed that most proteins were expressed by uncultured microorganisms belonging to a novel phylum, i.e. "*Candidatus* Fodinabacter communificans". These bacteria may play an indirect but important role in the functioning of the ecosystem by recycling organic matter or providing other members with cofactors such as vitamins (Bertin et al., 2011). An additional study revealed that *Euglena mutabilis,* an abundant protist found in this AMD as well as in other AMDs, produces organic compounds that could serve as nutrients for bacteria (Halter et al., unpublished).

2.3.2 Environmental proteomics as a tool to understand the dynamic and the functioning of ecosystems

To study factors that may influence the community adaptation, metaproteomics approaches are sometimes performed on controlled microcosms (Figure 1). For example, such an approach has been successfully used to study the temporal dynamics of microbial communities subjected to cadmium exposure and to characterize the resulting response in terms of toxicity and resistance (Lacerda et al., 2007). This study illustrates that metaproteomics can be used, not only to describe an ecosystem, but also to study its response to perturbations. For example, the spatial dynamics of bacterioplankton was evaluated along the Chesapeake Bay, the largest estuary in the United States, and the proteins identified were shown to correlate with major microbial lineages, i.e. Bacteroides and Alphaproteobacteria, present in this ecosystem (Kan et al., 2005). Environmental proteomics combined with physiology and geochemical data allowed a description of the ecological distribution of dominant and less abundant organisms, and the changes along environmental gradients in a biofilm within the Richmond mine at Iron Mountain, California, or the effect of the pH on acidophilic AMD microbial communities (Belnap et al., 2011; Mueller et al., 2010). Other studies aimed to elucidate the *Geobacter* physiology during stimulated uranium bioremediation (Callister et al., 2010; Wilkins et al., 2009), or the community responses to different nutrient concentrations on an oceanic scale (Morris et al., 2010). In this study, a shift in nutrient utilization and energy transduction along a natural nutrient concentration gradient was observed, with a dominance of TonB-dependent transporters expressed in these samples. Although it is likely that only the dominant organisms will be visible in metaproteomic studies, results provide evidence that such an approach presents a considerable interest towards a comprehensive analysis of microbial ecosystems. In the future, such approaches will be improved in order to access a large amount of proteins expressed by individual cells within a community.

3. The potential of targeted proteomics approaches

The last 10 years, global proteomic approaches have allowed drawing long lists of hundreds to thousands of identified proteins in all types of proteome fractions. The major weakness of those lists is often the lack of quantitative data for the identified proteins, especially in the case of metaproteomic studies where quantification may be difficult. To alleviate this problem, the proteomics specialists recently initiated a paradigm shift from global approaches towards targeted approaches, trying to find ways to get better quantitative data even if one has to focus on a limited number of proteins of interest. Selected Reaction Monitoring (SRM)-based strategies appear to be the most promising approaches to reach this goal (Picotti et al., 2010) and applications, mostly in the field of protein biomarker research, are starting to become successful (Elschenbroich & Kislinger, 2011).

3.1 Selected Reaction Monitoring-based proteomics workflow

The general SRM-based workflow is described in Figure 2. The first step of a targeted proteomics experiment resides in the definition of a restricted list of proteins of interest. This is the major difference between global approaches (in which the goal is to identify the highest number of peptides/proteins) and targeted approaches in which the targets to focus on have to be defined prior to the experiment itself. Once the targets are defined, developing a SRM-based quantification method relies on the choice of a small series of peptides that will be used as tracers for each protein to be quantified. In the SRM scanning mode, the precursor ion corresponding to the targeted peptide is selected in the first mass filter (Q1) before entering the collision cell (q2) where it undergoes collision-induced dissociation. One fragment ion is then selected in the second mass filter (Q3) and its intensity is monitored. An ion pair of precursor/fragment ions is called a transition and several transitions are recorded for each targeted peptide. The critical steps of the method setup reside in the choice of those so-called proteotypic peptides (unique for the protein and visible in MS) to be used for each protein, in the selection of transitions (number and fragment types) to be followed for each of the selected peptides and, finally, in the

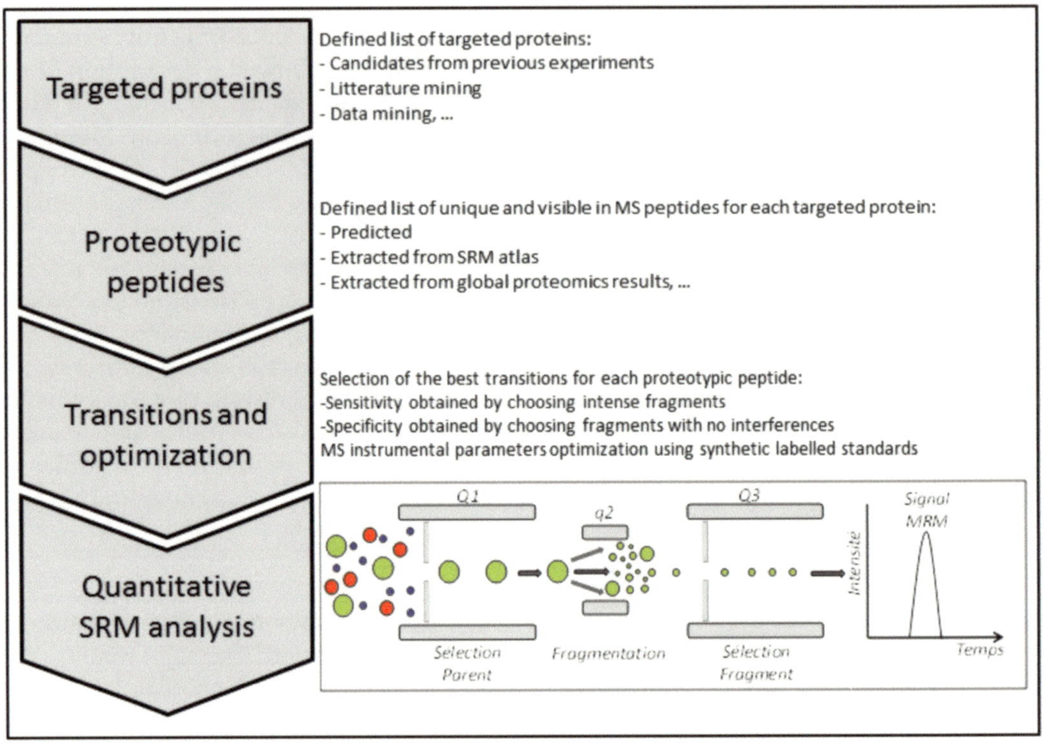

Fig. 2. Targeted SRM-based proteomics workflow

optimisation of the MS instrument parameters. The transitions have to be selected to offer both best sensitivity (intense fragments) and best selectivity (no interferences with other fragments). To accelerate these limiting steps in terms of time and cost, public libraries (atlases) of transitions are being constructed using synthetic peptides for a few proteomes of reference organisms, of which yeast and human (http://www.srmatlas.org). Those atlases will significantly facilitate the choice of the proteotypic peptides as they will contain the lists of 5 peptides for each predicted protein of the reference proteome, along with their optimal transitions and information on instrument parameters. Once the peptides and transitions are established, isotopically heavy labeled standards are required and need to be spiked into the samples in order to be able to quantify the endogenous peptides of interest by calculating heavy/light ratios (Gallien et al., 2009; Lange et al., 2008). The hypothesis-driven nature of such experiments overcomes the bias towards most abundant components and has already allowed previously unreached sensitivity levels using MS techniques.

3.2 SRM quantification in proteomics

The quantitative power of SRM mass spectrometry no longer needs to be proven. This approach has been used for a number of years for the quantification of small molecules such as metabolites of xenobiotics, hormones or pesticides with great precision (CV<5%). However, three main hurdles have hindered its application for the quantification of peptides and proteins. The first major hurdle to be overcome was sensitivity, or more precisely, the capacity to quantify proteins of very low abundance in mixtures in which protein concentrations range over 5 to 10 orders of magnitude, even up to 12 orders in the case of plasma samples. To circumvent this problem, the introduction of fractionation methods permits considerable reduction of the quantification limit provided they are perfectly controlled and reproducible. For instance, SRM methods recently allowed the detection of concentrations typical for candidate protein biomarkers whose abundance can be as low as a few ng/ml or even hundreds of pg/ml in human plasma (Keshishian et al., 2009). The second hurdle was the reproducibility of SRM analyses for proteomics. This hurdle appears to have been overcome today: indeed, as part of a study carried out by the Clinical Proteomic Technology Assessment For Cancer Network Project (CPTAC, (Addona et al., 2009), interlaboratory CVs (including both variations due to sample preparations and MS analysis) between 10 and 23% were obtained across 9 different laboratories. Finally, multiplexing was made possible thanks to significant progress in electronics and acquisition and data processing software developed on triple quadrupole-type instruments. Today the simultaneous quantification in a single analysis of about a hundred peptides can be envisaged using a few hundred transitions.

3.3 SRM-strategies for quantifying proteins in microbial isolates and complex communities

So far, most of the SRM-based applications have dealt with biomarker studies and clinical proteomics (Gallien et al., 2009; Hüttenhain et al., 2009). Nevertheless, it has been proven to be very successfully applicable on *S. cerevisiae* and other whole proteome digests (Picotti et al., 2009). Even though microbial communities are extremely complex protein mixtures, both in terms of number of proteins and dynamic range, protein concentrations ranging over 12 orders of magnitude in the case of plasma samples have revealed the success of the technology. It is therefore reasonable to predict a widespread application of SRM quantification methods in many fields. Actually, a recent study has already demonstrated the possibility to absolutely quantify proteins in complex environmental samples and mixed microbial communities (Werner et al., 2009). An absolute prerequisite for the success of the method is a precise control and reproducibility of the sample preparation and fractionation steps. Additionally, one of the key factors for a reliable quantification will be the use of appropriate quantification standards. Indeed, the use of isotopically labelled standards considerably improves quantification reliability. In any case, absolute quantification of peptides, and thus the proteins producing them, is only possible through the simultaneous LC-SRM measurement of endogenous peptides and isotopically labelled standards added in known quantities. Several choices are possible: synthetic peptides (the AQUA method, Gerber et al. 2003), concatemers of peptides (the QconCAT method, (Beynon et al., 2005)) and protein standards, biochemically identical to the natural proteins to be assayed (the PSAQ method, (Brun et al., 2007)).

4. Data integration and proteome databases

High-throughput proteomics is a rapidly developing field enabling analyses at system level from complexes or cells and organs to environmental communities (Cannon & Webb-Robertson, 2007). With the huge amount of data produced by experiments and subsequent analyses, proteomics repositories will have to take up the challenge of large-scale storage, fast access and easy data retrieval. Furthermore, with the development of Systems Biology, proteomics databases, like other "omics" databases, will require exchange and communication standards which could help to integrate data with related information from other databases or fields (genomics, genetics, metabolomics) into a wider scope. Several proteomics repositories have been established thus far and range from large-scale general databases to more specialized ones (Table 2). However, although some repositories exist that are specialized in microbiology-related proteomics, there is still a lack of proteomics databases dedicated to environmental microbiology.

Database	Description	Access
Swiss 2D-PAGE	1DE and 2DE data	http://world-2dpage.expasy.org/swiss-2dpage/
World 2D-PAGE portal	Federation of 2DE-based databases	http://world-2dpage.expasy.org/portal/
InPACT	Gel-based environmental microbiology database	http://inpact.u-strasbg.fr/
Proteome Database for Microbial Research	Gel and mass spectrometry microbiology database	http://www.mpiib-berlin.mpg.de/2D-PAGE/
GPMDB	Comprehensive mass spectrometry database for validation of identification and protein coverage	http://gpmdb.thegpm.org/
PeptitdeAtlas	Compendium of raw data coming from high-throughput proteomics technologies aiming at the annotation of eukaryotic genomes through a thorough validation of expressed proteins	http://www.peptideatlas.org/
PRIDE	Comprehensive mass spectrometry-derived peptide and protein identifications, MS mass spectra, and associated metadata.	http://www.ebi.ac.uk/pride/
Proteome Commons	Communautary resource for collaborative research and sharing of proteomics data	https://proteomecommons.org/

Table 2. Proteomics repositories useful for the analysis of microbial proteomes.

4.1 Technical challenges

4.1.1 Data storage and the scalability challenge

High sensitivity and high quality mass spectra are delivered by mass spectrometers at an ever-faster rate, yielding an ever-increasing amount of data. For instance, the data available from Proteome Commons (https://proteomecommons.org/) repository has currently (July 2011) a size of 16.7 TB comprising 12,895,832 data files (for the sake of comparison, the 1638 uncompressed flat files of Genbank release 184.0 require approximately 540 GB). Therefore, scalability is expected to become a major issue for those comprehensive proteomics repositories and high capacity storage and efficient data access may require the use of distributed IT technologies similar to those serving large databases on the web (Facebook Cassandra, Google BigTable, Amazon, Dynamo). As a matter of fact, in order to handle data and to facilitate access by users, Proteome Commons based its repository on an implementation of Tranche (https://trancheproject.org/), a free open-source file distributed storage and dissemination software (Falkner et al., 2008).

4.1.2 Data access: Needs for standards

One of the first efforts towards a single access point to proteomics data was the publication in 1996 of guidelines for building federated 2DE databases (Appel et al., 1996). Since then, ExPASy has developed Make2D-DB II, an environment to create, convert, publish, interconnect and keep up-to-date 2DE databases (Mostaguir et al., 2003). More recently, the ProteomeExchange consortium has been established to provide a single point of submission to PRIDE (Vizcaíno et al., 2009), PeptideAtlas (Deutsch et al., 2008) and Tranche (https://proteomecommons.org/tranche/) repositories. This consortium encourages the data exchange and sharing of identifiers between repositories so that the community may easily find datasets. Furthermore, in order to be effective, computational analysis and data-mining require the use of controlled vocabularies or ontologies for data descriptions. The Protein Ontology (Natale et al., 2010) provides a standardized vocabulary for the description of protein evolutionary relatedness (ProEvo), protein forms including isoforms or PTMs (ProForm) and protein-containing complexes (ProComp). On a larger scale, the HUPO proteomics standards initiative (HUPO-PSI) is aiming to define standards to ease data exchange and minimize data loss (Orchard & Hermjakob, 2007) in key areas of proteomics: protein separation, gel electrophoresis, mass spectrometry, molecular interactions, protein modifications and proteomics informatics. The HUPO-PSI is developing the minimum information about proteomics experiments (MIAPE) guidelines defining which information should minimally be reported about a proteomics experiment to allow critical assessment. It also develops data formats for capturing, describing and exchanging MIAPE-compliant data as well as supporting controlled vocabularies. Some of these standards, like MIAPE (Taylor et al., 2007), FuGE (Jones et al., 2007), GelML and mzML have been released or published and, to date, mzData standards for mass spectrometry are widely supported by product manufacturers.

4.1.3 Heterogeneous data integration: Database interoperability

Although standard exchange formats allow easy data exchange between and with repositories, systems level investigations relying on computational analyses require data integration from heterogeneous sources. Instead of trying to duplicate such large amounts of data, integration can be most effectively achieved through connection to repositories. For instance, PeptideAtlas proposes a Distributed Annotation System (DAS) server which allows visualizing data as tracks in the Ensembl Genome Browser (Flicek et al., 2010). Similarly, the PRIDE repository is available through a BioMart service (Smedley et al., 2009). Thus, it can be accessed as a simple REST (Representational State Transfer) web service that involves building an HTTP request including an XML file that encodes the filters and attributes of the request. Web services are indeed being used more and more in bioinformatics, providing remote access to data and tools that can be combined into workflows with workbench environments like Taverna (Hull et al., 2006) or into client applications (Mcwilliam et al., 2009).

4.2 Data repositories

Proteomics data repositories have been made available to the scientific community on the web since the early nineties. Swiss 2D-PAGE (Hoogland et al., 2004) which collects protein identification from 2DE and 1D-PAGE gels was created in 1993. This database is now part of the World 2D-PAGE portal (Hoogland et al., 2008) which federates 9 gel-based proteomics databases for a total of nearly 18,800 identified spots in 141 maps for 22 species, making it the biggest gel-based proteomics dataset accessible from a single interface (June 2011). Other less general two-dimensional electrophoresis databases are also available and give access to gel-based protein identification in different systems. These repositories provide information on identification data (pI, mW, peptides and spot), links between maps and, of course, link to the identified entries in protein databases. Gels are displayed as an interactive image which can be clicked on to visualize spot information. In addition to electrophoresis data, Proteome Database for Microbial Research (Pleißner et al., 2004) also offers access to MS data.

Proteomics repositories which focus on MS data include GPMDB (Craig et al., 2004), PeptitdeAtlas (Deutsch et al., 2008), PRIDE (Vizcaíno et al., 2009), and Proteome Commons (https://proteomecommons.org/) among the most prominent ones. GPMDB is a relational database that was designed to aid in the process of validating peptide-to-mass spectrum assignment and/or protein coverage patterns. Together with data analysis servers it constitutes the open-source system referred to as the Global Proteome Machine. PeptideAtlas is a multi-organism compendium of raw data coming from high-throughput proteomics technologies. Only raw data are accepted and are periodically reprocessed as more advanced interpretation tools for identification and statistical validation are available. The PeptideAtlas project long-term goal is the annotation of eukaryotic genomes through a robust validation of expressed proteins. PRIDE (Proteomics Identifications Database) is a repository of MS derived peptide and protein identifications, MS mass spectra, and associated metadata. Proteome Commons is a public resource for collaborative research and public sharing of proteomics data, tools and news. Permanent storage of data suitable for publication is provided through a distributed repository. Registered users may set up their own or join group projects for easy collaboration with colleagues or partners as project permissions and member responsibilities can be fine-tuned in order to control access to data. As post-translational modifications like phosphorylation play an important role in control of protein activity some proteomics databases focus on phosphorylation and other PTMs (Phospho.ELM (Dinkel et al., 2011), Phospho3D (Zanzoni et al., 2011), Phosida (Gnad et al., 2011) PhosphoSitePlus® (Hornbeck et al., 2004), and PhosphoPep. (Bodenmiller et al., 2008).

4.3 The need for an environmental proteomics database

As metaproteomics allows to study microorganisms' functionality in their natural context, in the coming years it is likely to become the technique of choice for functional characterization of microbial communities. Moreover, this methodology will give invaluable insights into microbial ecology, in particular when associated with metagenomics data. To correlate the observed variations in functions with differences in conditions, comparative studies must also consider environment characteristics such as chemical composition (presence of toxic compounds, organic matter), physical properties (temperature), habitat, sample location and collection dates. High-throughput data-mining would therefore require that databases handling environmental metaproteomics include metadata providing environmental information in a computer-readable form. In the absence of a currently available specific standard for environmental proteomics, the content of these metadata could be inspired by corresponding sections of the minimum information about any sequence standard MIxS

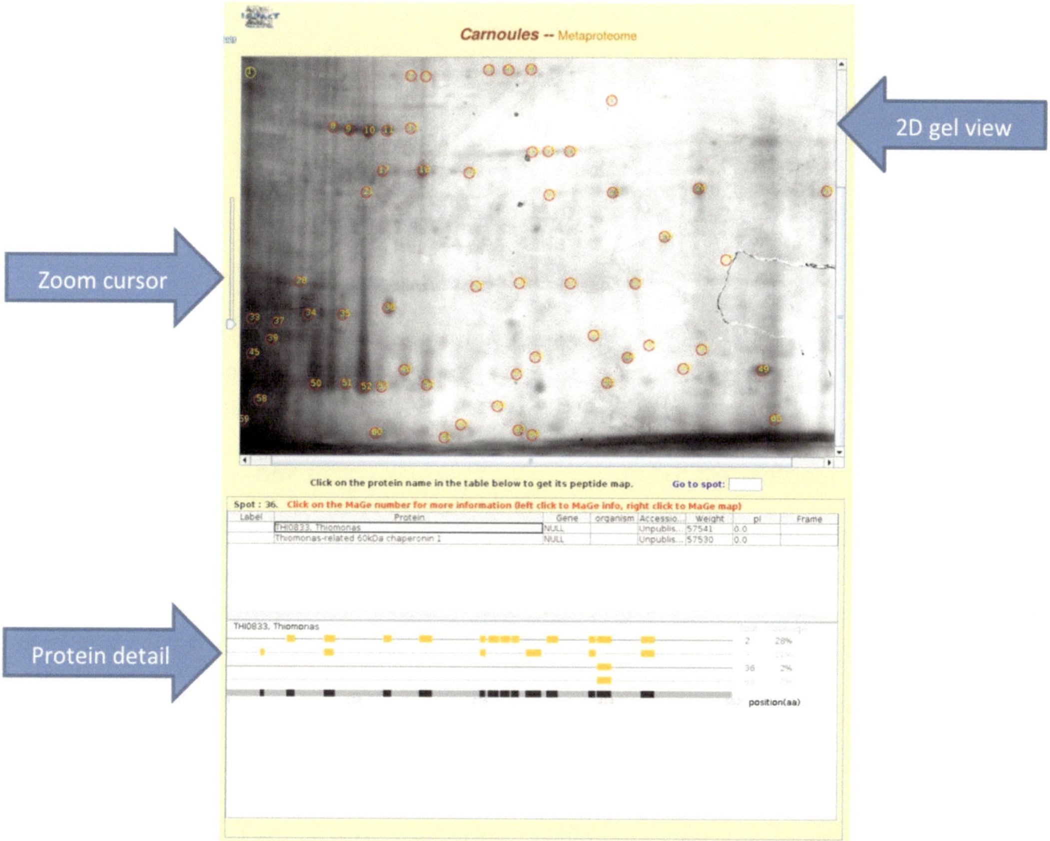

Fig. 3. The InPact proteomic database (http://inpact.u-strasbg.fr). The various functionalities of the interface allow the exploration of specific areas of the 2D gel by using a zoom-in/out function. Spots present in the selected area can be outlined, and the corresponding MS results can be seen for each. In addition, more information can be seen by hovering the mouse over any spot and/or clicking on it (name, Mw, pI, MS peptidic sequence). As an example, one of the numerous GroEL chaperonins identified in the Carnoulès community metaproteome illustrates the data that can be obtained for any protein identified by mass spectrometry, e.g. the label of the corresponding CDS in the genome when available and the spot numbers where the protein has been identified (top), the size and the location within the genome of the MS peptidic fragments obtained as well as their % coverage with respect to the full length CDS.

(Yilmaz et al., 2011) or the IUPAC minimum requirements for reporting analytical data for environmental samples (Egli et al., 2003). For instance, the InPACT environmental microbiology database (http://inpact.u-strasbg.fr/) is a proteomics database dedicated to environmental microbiology providing gel-based data pertaining to microorganisms as well as complex communities (Figure 3). It also provides genomics information thanks to tight links with the MaGe database (Vallenet et al., 2006) and tools for functional profile comparison between gels. InPACT, although still in its infancy, provides tools for gel comparisons at the function level thanks to functional description of proteins using the Gene Ontology (The Gene Ontology Consortium. 2000). Future developments will now focus on the integration of environmental information as metadata and the addition of more comparison tools including multivariate statistical analysis of proteomics data and associated metadata. Integration with external data sources (metabolism, genomic data) will also be reinforced. Finally, in order to increase the accessibility of data, InPACT data access will be offered not only through the web server but also as RESTful web services. In the near future, InPACT and other databases will hopefully prove to be useful proteomics-oriented tools for environmental microbiology.

5. Conclusion

The past few years has seen a huge amount of genomic information published in databases. Associated with functional genomic approaches such as proteomics, those data will greatly improve our knowledge of the structure, the functioning, the diversity and the evolution of microorganisms. Similarly, the study of microbial communities as a whole will be of great interest to investigate complex consortia and to address important questions regarding the role of uncultured microorganisms in microbial ecosystems. Proteomics, when combined not only with other genomic methods such as transcriptomics and metabolomics, but also with more classical methods of genetics, molecular biology and/or biochemistry, will give an integrated view of biological objects present in any environment, their role and their relationships. They will lead to a better understanding of how microorganisms colonize new ecological niches and to the possible use of their specific properties in biotechnology.

6. Acknowledgment

Financial support came from the Université de Strasbourg (UdS), the Centre National de la Recherche Scientifique (EC2CO project and GDR2909 research network, http://gdr2909.alsace.cnrs.fr/) and the Agence Nationale de la Recherche (ANR RARE and MULTIPOLSITE projects).

7. References

Addona, T. A., Abbatiello, S. E., Schilling, B., Skates, S. J., Mani, D. R., Bunk, D. M., Spiegelman, C. H., Zimmerman, L. J., Ham, A.-J. L., Keshishian, H., Hall, S. C., Allen, S., Blackman, R. K., Borchers, C. H., Buck, C., Cardasis, H. L., Cusack, M. P., Dodder, N. G., Gibson, B. W., Held, J. M., Hiltke, T., Jackson, A., Johansen, E. B., Kinsinger, C. R., Li, J., Mesri, M., Neubert, T. A., Niles, R. K., Pulsipher, T. C.,

Ransohoff, D., Rodriguez, H., Rudnick, P. A., Smith, D., Tabb, D. L., Tegeler, T. J., Variyath, A. M., Vega-Montoto, L. J., Wahlander, A., Waldemarson, S., Wang, M., Whiteaker, J. R., Zhao, L., Anderson, N. L., Fisher, S. J., Liebler, D. C., Paulovich, A. G., Regnier, F. E., Tempst, P., & Carr, S. A. (2009). Multi-site assessment of the precision and reproducibility of multiple reaction monitoring-based measurements of proteins in plasma. *Nature Biotechnology*, Vol. 27, No. 7, pp. 633-641.

Appel, R. D., Bairoch, A., Sanchez, J. C., Vargas, J. R., Golaz, O., Pasquali, C., & Hochstrasser, D. F. (1996). Federated two-dimensional electrophoresis database: a simple means of publishing two-dimensional electrophoresis data. *Electrophoresis*, Vol. 17, No. 3, pp. 540-546.

Belnap, C. P., Pan, C., Denef, V. J., Samatova, N. F., Hettich, R. L., & Banfield, J. F. (2011). Quantitative proteomic analyses of the response of acidophilic microbial communities to different pH conditions. *The ISME Journal*, Vol. 5, No. 7, pp. 1152-1161.

Benndorf, D., Balcke, G. U., Harms, H., & von Bergen, M. (2007). Functional metaproteome analysis of protein extracts from contaminated soil and groundwater. *The ISME Journal*, Vol. 1, No. 3, pp. 224-234.

Bertin, P. N., Heinrich-Salmeron, A., Pelletier, E., Goulhen-Chollet, F., Arsène-Ploetze, F., Gallien, S., Lauga, B., Casiot, C., Calteau, A., Vallenet, D., Bonnefoy, V., Bruneel, O., Chane-Woon-Ming, B., Cleiss-Arnold, J., Duran, R., Elbaz-Poulichet, F., Fonknechten, N., Giloteaux, L., Halter, D., Koechler, S., Marchal, M., Mornico, D., Schaeffer, C., Smith, A. A. T., Van Dorsselaer, A., Weissenbach, J., Médigue, C., & Le Paslier, D. (2011). Metabolic diversity among main microorganisms inside an arsenic-rich ecosystem revealed by meta- and proteo-genomics. *International Society for Microbial Ecology*, In press.

Bertin, P. N., Medigue, C., & Normand, P. (2008). Advances in environmental genomics: towards an integrated view of microorganisms and ecosystems. *Microbiology*, Vol. 154, No. 2, pp. 347-359.

Beynon, R. J., Doherty, M. K., Pratt, J. M., & Gaskell, S. J. (2005). Multiplexed absolute quantification in proteomics using artificial QCAT proteins of concatenated signature peptides. *Nature Methods*, Vol. 2, No. 8, pp. 587-589.

Bodenmiller, B., Campbell, D., Gerrits, B., Lam, H., Jovanovic, M., Picotti, P., Schlapbach, R., & Aebersold, R. (2008). PhosphoPep--a database of protein phosphorylation sites in model organisms. *Nature Biotechnology*, Vol. 26, No. 12, pp. 1339-1340.

Bona, E., Marsano, F., Massa, N., Cattaneo, C., Cesaro, P., Argese, E., di Toppi, L. S., Cavaletto, M., & Berta, G. (2011). Proteomic analysis as a tool for investigating arsenic stress in *Pteris vittata* roots colonized or not by arbuscular mycorrhizal symbiosis. *Journal of Proteomics*. In press.

Brun, V., Dupuis, A., Adrait, A., Marcellin, M., Thomas, D., Court, M., Vandenesch, F., & Garin, J. (2007). Isotope-labeled protein standards: toward absolute quantitative proteomics. *Molecular & Cellular Proteomics: MCP*, Vol. 6, No. 12, pp. 2139-2149.

Bruneel, O., Volant, A., Gallien, S., Chaumande, B., Casiot, C., Carapito, C., Bardil, A., Morin, G., Brown, G. E., Personné, C. J., Le Paslier, D., Schaeffer, C., Van Dorsselaer, A., Bertin, P. N., Elbaz-Poulichet, F., & Arsène-Ploetze, F. (2011). Characterization of the Active Bacterial Community Involved in Natural

Attenuation Processes in Arsenic-Rich Creek Sediments. *Microbial Ecology*, Vol. 61, No. Issue 4, pp. 793-810.

Bryan, C. G., Marchal, M., Battaglia-Brunet, F., Kugler, V., Lemaitre-Guillier, C., Lièvremont, D., Bertin, P. N., & Arsène-Ploetze, F. (2009). Carbon and arsenic metabolism in *Thiomonas* strains: differences revealed diverse adaptation processes. *BMC Microbiology*, Vol. 9, pp. 127.

Callister, S. J., Wilkins, M. J., Nicora, C. D., Williams, K. H., Banfield, J. F., VerBerkmoes, N. C., Hettich, R. L., N'Guessan, L., Mouser, P. J., Elifantz, H., Smith, R. D., Lovley, D. R., Lipton, M. S., & Long, P. E. (2010). Analysis of biostimulated microbial communities from two field experiments reveals temporal and spatial differences in proteome profiles. *Environmental Science & Technology*, Vol. 44, No. 23, pp. 8897-8903.

Cannon, W., & Webb-Robertson, B.-J. (2007). Computational proteomics : High-throughput analysis for Systems Biology, *Proceedings of Pacific Symposium on Biocomputing*, Vol. 12, p. 403-408.

Carapito, C., Muller, D., Turlin, E., Koechler, S., Danchin, A., Van Dorsselaer, A., Leize-Wagner, E., Bertin, P. N., & Lett, M.-C. (2006). Identification of genes and proteins involved in the pleiotropic response to arsenic stress in *Caenibacter arsenoxydans*, a metalloresistant beta-proteobacterium with an unsequenced genome. *Biochimie*, Vol. 88, No. 6, pp. 595-606.

Casado-Vela, J., Cebrián, A., del Pulgar, M. T. G., Sánchez-López, E., Vilaseca, M., Menchén, L., Diema, C., Sellés-Marchart, S., Martínez-Esteso, M. J., Yubero, N., Bru-Martínez, R., Lacal, J. C., & Lacal, J. C. (2011). Lights and shadows of proteomic technologies for the study of protein species including isoforms, splicing variants and protein post-translational modifications. *Proteomics*, Vol. 11, No. 4, pp. 590-603.

Cañas, B., Piñeiro, C., Calvo, E., López-Ferrer, D., & Gallardo, J. M. (2007). Trends in sample preparation for classical and second generation proteomics. *Journal of Chromatography. A*, Vol. 1153, No. 1-2, pp. 235-258.

Cleiss-Arnold, J., Koechler, S., Proux, C., Fardeau, M.-L., Dillies, M.-A., Coppee, J.-Y., Arsène-Ploetze, F., & Bertin, P. N. (2010). Temporal transcriptomic response during arsenic stress in *Herminiimonas arsenicoxydans*. *BMC Genomics*, Vol. 11, pp. 709.

Craig, R., Cortens, J. P., & Beavis, R. C. (2004). Open Source System for Analyzing, Validating, and Storing Protein Identification Data. *Journal of Proteome Research*, Vol. 3, No. 6, pp. 1234-1242.

Delalande, F., Carapito, C., Brizard, J.-P., Brugidou, C., & Van Dorsselaer, A. (2005). Multigenic families and proteomics: extended protein characterization as a tool for paralog gene identification. *Proteomics*, Vol. 5, No. 2, pp. 450-460.

Denef, V. J., Kalnejais, L. H., Mueller, R. S., Wilmes, P., Baker, B. J., Thomas, B. C., VerBerkmoes, N. C., Hettich, R. L., & Banfield, J. F. (2010a). Proteogenomic basis for ecological divergence of closely related bacteria in natural acidophilic microbial communities. *Proceedings of the National Academy of Sciences of the United States of America*, Vol. 107, No. 6, pp. 2383-2390.

Denef, V. J., Mueller, R. S., & Banfield, J. F. (2010b). AMD biofilms: using model communities to study microbial evolution and ecological complexity in nature. *The ISME Journal*, Vol. 4, No. 5, pp. 599-610.

Deutsch, E. W., Lam, H., & Aebersold, R. (2008). PeptideAtlas: a resource for target selection for emerging targeted proteomics workflows. *EMBO reports*, Vol. 9, No. 5, pp. 429-434.

Dinkel, H., Chica, C., Via, A., Gould, C. M., Jensen, L. J., Gibson, T. J., & Diella, F. (2011). Phospho.ELM: a database of phosphorylation sites--update 2011. *Nucleic Acids Research*, Vol. 39, No. Database issue, pp. D261-267.

Egli, H., Dassenakis, M., Garelick, H., Van Grieken, R., Peijnenburg, W. J. G. M., Klasinc, L., Kördel, W., Priest, N., & Tavares, T. (2003). Minimum requirements for reporting analytical data for environmental samples (IUPAC Technical Report). *Pure and Applied Chemistry*, Vol. 75, No. 8, pp. 1097-1106.

Elschenbroich, S., & Kislinger, T. (2011). Targeted proteomics by selected reaction monitoring mass spectrometry: applications to systems biology and biomarker discovery. *Molecular bioSystems*, Vol. 7, No. 2, pp. 292-303.

Falkner, J. A., Hill, J. A., & Andrews, P. C. (2008). Proteomics FASTA Archive and Reference Resource. *Proteomics*, Vol. 8, No. 9, pp. 1756-1757.

Ferrer, M., Golyshina, O. V., Beloqui, A., Golyshin, P. N., & Timmis, K. N. (2007). The cellular machinery of *Ferroplasma acidiphilum* is iron-protein-dominated. *Nature*, Vol. 445, No. 7123, pp. 91-94.

Flicek, P., Amode, M. R., Barrell, D., Beal, K., Brent, S., et al. (2010). Ensembl 2011. *Nucleic Acids Research*, Vol. 39, No. Database, pp. D800-D806.

Fränzel, B., & Wolters, D. A. (2011). Advanced MudPIT as a next step towards high proteome coverage. *Proteomics*. In press.

Gallien, S., Perrodou, E., Carapito, C., Deshayes, C., Reyrat, J.-M., Van Dorsselaer, A., Poch, O., Schaeffer, C., & Lecompte, O. (2009). Ortho-proteogenomics: multiple proteomes investigation through orthology and a new MS-based protocol. *Genome Research*, Vol. 19, No. 1, pp. 128-135.

Gevaert, K., Impens, F., Ghesquière, B., Van Damme, P., Lambrechts, A., & Vandekerckhove, J. (2008). Stable isotopic labeling in proteomics. *Proteomics*, Vol. 8, No. 23-24, pp. 4873-4885.

Gnad, F., Gunawardena, J., & Mann, M. (2011). PHOSIDA 2011: the posttranslational modification database. *Nucleic Acids Research*, Vol. 39, No. Database issue, pp. D253-260.

Grangeasse, C., Terreux, R., & Nessler, S. (2010). Bacterial tyrosine-kinases: structure-function analysis and therapeutic potential. *Biochimica Et Biophysica Acta*, Vol. 1804, No. 3, pp. 628-634.

Gundry, R. L., White, M. Y., Murray, C. I., Kane, L. A., Fu, Q., Stanley, B. A., & Van Eyk, J. E. (2009). Preparation of proteins and peptides for mass spectrometry analysis in a bottom-up proteomics workflow. *Current Protocols in Molecular Biology*, Frederick M. Ausubel, Chapter 10, pp. Unit 10.25.

Halter, D., Cordi, A., Gribaldo, S., Gallien, S., Goulhen-Chollet, F., Heinrich-Salmeron, A., Carapito, C., Pagnout, C., Montaut, D., Seby, F., Van Dorsselaer, A., Schaeffer, C., Bertin, P. N., Bauda, P., & Arsène-Ploetze, F. (2011). Taxonomic and functional prokaryote diversity in mildly arsenic-contaminated sediments. *Research in Microbiology*. In press.

Hecker, M., & Völker, U. (2004). Towards a comprehensive understanding of *Bacillus subtilis* cell physiology by physiological proteomics. *Proteomics*, Vol. 4, No. 12, pp. 3727-3750.

Hoogland, C., Mostaguir, K., Appel, R., & Lisacek, F. (2008). The World-2DPAGE Constellation to promote and publish gel-based proteomics data through the ExPASy server. *Journal of Proteomics*, Vol. 71, No. 2, pp. 245-248.

Hoogland, C., Mostaguir, K., Sanchez, J.-C., Hochstrasser, D. F., & Appel, R. D. (2004). SWISS-2DPAGE, ten years later. *Proteomics*, Vol. 4, No. 8, pp. 2352-2356.

Hornbeck, P. V., Chabra, I., Kornhauser, J. M., Skrzypek, E., & Zhang, B. (2004). PhosphoSite: A bioinformatics resource dedicated to physiological protein phosphorylation. *Proteomics*, Vol. 4, No. 6, pp. 1551-1561.

Hull, D., Wolstencroft, K., Stevens, R., Goble, C., Pocock, M. R., Li, P., & Oinn, T. (2006). Taverna: a tool for building and running workflows of services. *Nucleic Acids Research*, Vol. 34, No. Web Server, pp. W729-W732.

Hüttenhain, R., Malmström, J., Picotti, P., & Aebersold, R. (2009). Perspectives of targeted mass spectrometry for protein biomarker verification. *Current Opinion in Chemical Biology*, Vol. 13, No. 5-6, pp. 518-525.

Jones, A. R., Miller, M., Aebersold, R., Apweiler, R., Ball, C. A., Brazma, A., DeGreef, J., Hardy, N., Hermjakob, H., Hubbard, S. J., Hussey, P., Igra, M., Jenkins, H., Julian, R. K., Laursen, K., Oliver, S. G., Paton, N. W., Sansone, S.-A., Sarkans, U., Stoeckert, C. J., Taylor, C. F., Whetzel, P. L., White, J. A., Spellman, P., & Pizarro, A. (2007). The Functional Genomics Experiment model (FuGE): an extensible framework for standards in functional genomics. *Nature Biotechnology*, Vol. 25, No. 10, pp. 1127-1133.

Jungblut, P. R. (2001). Proteome analysis of bacterial pathogens. *Microbes and Infection / Institut Pasteur*, Vol. 3, No. 10, pp. 831-840.

Kan, J., Hanson, T. E., Ginter, J. M., Wang, K., & Chen, F. (2005). Metaproteomic analysis of Chesapeake Bay microbial communities. *Saline Systems*, Vol. 1, pp. 7.

Keshishian, H., Addona, T., Burgess, M., Mani, D. R., Shi, X., Kuhn, E., Sabatine, M. S., Gerszten, R. E., & Carr, S. A. (2009). Quantification of cardiovascular biomarkers in patient plasma by targeted mass spectrometry and stable isotope dilution. *Molecular & Cellular Proteomics: MCP*, Vol. 8, No. 10, pp. 2339-2349.

Kim, Y. H., Cho, K., Yun, S.-H., Kim, J. Y., Kwon, K.-H., Yoo, J. S., & Kim, S. I. (2006). Analysis of aromatic catabolic pathways in *Pseudomonas putida* KT 2440 using a combined proteomic approach: 2-DE/MS and cleavable isotope-coded affinity tag analysis. *Proteomics*, Vol. 6, No. 4, pp. 1301-1318.

Lacerda, C. M. R., Choe, L. H., & Reardon, K. F. (2007). Metaproteomic analysis of a bacterial community response to cadmium exposure. *Journal of Proteome Research*, Vol. 6, No. 3, pp. 1145-1152.

Laemmli, U. K. (1970). Cleavage of structural proteins during the assembly of the head of bacteriophage T4. *Nature*, Vol. 227, No. 5259, pp. 680-685.

Lange, V., Picotti, P., Domon, B., & Aebersold, R. (2008). Selected reaction monitoring for quantitative proteomics: a tutorial. *Molecular Systems Biology*, Vol. 4, pp. 222.

Liedert, C., Bernhardt, J., Albrecht, D., Voigt, B., Hecker, M., Salkinoja-Salonen, M., & Neubauer, P. (2010). Two-dimensional proteome reference map for the radiation-resistant bacterium *Deinococcus geothermalis*. *Proteomics*, Vol. 10, No. 3, pp. 555-563.

Linares, J. F., Moreno, R., Fajardo, A., Martínez-Solano, L., Escalante, R., Rojo, F., & Martínez, J. L. (2010). The global regulator Crc modulates metabolism, susceptibility to antibiotics and virulence in *Pseudomonas aeruginosa*. *Environmental Microbiology*, Vol. 12, No. 12, pp. 3196-3212.

Mary, I., Oliver, A., Skipp, P., Holland, R., Topping, J., Tarran, G., Scanlan, D. J., O'Connor, C. D., Whiteley, A. S., Burkill, P. H., & Zubkov, M. V. (2010). Metaproteomic and metagenomic analyses of defined oceanic microbial populations using microwave cell fixation and flow cytometric sorting. *FEMS Microbiology Ecology*, Vol. 74, No. 1, pp. 10-18.

Mcwilliam, H., Valentin, F., Goujon, M., Li, W., Narayanasamy, M., Martin, J., Miyar, T., & Lopez, R. (2009). Web services at the European Bioinformatics Institute-2009. *Nucleic Acids Research*, Vol. 37, No. Web Server, pp. W6-W10.

Monteiro, K. M., de Carvalho, M. O., Zaha, A., & Ferreira, H. B. (2010). Proteomic analysis of the *Echinococcus granulosus* metacestode during infection of its intermediate host. *Proteomics*, Vol. 10, No. 10, pp. 1985-1999.

Moretti, M., Grunau, A., Minerdi, D., Gehrig, P., Roschitzki, B., Eberl, L., Garibaldi, A., Gullino, M. L., & Riedel, K. (2010). A proteomics approach to study synergistic and antagonistic interactions of the fungal-bacterial consortium *Fusarium oxysporum* wild-type MSA 35. *Proteomics*, Vol. 10, No. 18, pp. 3292-3320.

Morris, R. M., Nunn, B. L., Frazar, C., Goodlett, D. R., Ting, Y. S., & Rocap, G. (2010). Comparative metaproteomics reveals ocean-scale shifts in microbial nutrient utilization and energy transduction. *The ISME Journal*, Vol. 4, No. 5, pp. 673-685.

Mostaguir, K., Hoogland, C., Binz, P.-A., & Appel, R. D. (2003). The Make 2D-DB II package: conversion of federated two-dimensional gel electrophoresis databases into a relational format and interconnection of distributed databases. *Proteomics*, Vol. 3, No. 8, pp. 1441-1444.

Mueller, R. S., Denef, V. J., Kalnejais, L. H., Suttle, K. B., Thomas, B. C., Wilmes, P., Smith, R. L., Nordstrom, D. K., McCleskey, R. B., Shah, M. B., Verberkmoes, N. C., Hettich, R. L., & Banfield, J. F. (2010). Ecological distribution and population physiology defined by proteomics in a natural microbial community. *Molecular Systems Biology*, Vol. 6, pp. 374.

Muller, D., Médigue, C., Koechler, S., Barbe, V., Barakat, M., Talla, E., Bonnefoy, V., Krin, E., Arsène-Ploetze, F., Carapito, C., Chandler, M., Cournoyer, B., Cruveiller, S., Dossat, C., Duval, S., Heymann, M., Leize, E., Lieutaud, A., Lièvremont, D., Makita, Y., Mangenot, S., Nitschke, W., Ortet, P., Perdrial, N., Schoepp, B., Siguier, P., Simeonova, D. D., Rouy, Z., Segurens, B., Turlin, E., Vallenet, D., Van Dorsselaer, A., Weiss, S., Weissenbach, J., Lett, M.-C., Danchin, A., & Bertin, P. N. (2007). A tale of two oxidation states: bacterial colonization of arsenic-rich environments. *PLoS Genetics*, Vol. 3, No. 4, pp. e53.

Natale, D. A., Arighi, C. N., Barker, W. C., Blake, J. A., Bult, C. J., Caudy, M., Drabkin, H. J., D'Eustachio, P., Evsikov, A. V., Huang, H., Nchoutmboube, J., Roberts, N. V., Smith, B., Zhang, J., & Wu, C. H. (2010). The Protein Ontology: a structured representation of protein forms and complexes. *Nucleic Acids Research*, Vol. 39, No. Database, pp. D539-D545.

Nesvizhskii, A. I. (2010). A survey of computational methods and error rate estimation procedures for peptide and protein identification in shotgun proteomics. *Journal of Proteomics*, Vol. 73, No. 11, pp. 2092-2123.

Orchard, S., & Hermjakob, H. (2007). The HUPO proteomics standards initiative--easing communication and minimizing data loss in a changing world. *Briefings in Bioinformatics*, Vol. 9, No. 2, pp. 166-173.

Picotti, P., Bodenmiller, B., Mueller, L. N., Domon, B., & Aebersold, R. (2009). Full dynamic range proteome analysis of *S. cerevisiae* by targeted proteomics. *Cell*, Vol. 138, No. 4, pp. 795-806.

Picotti, P., Rinner, O., Stallmach, R., Dautel, F., Farrah, T., Domon, B., Wenschuh, H., & Aebersold, R. (2010). High-throughput generation of selected reaction-monitoring assays for proteins and proteomes. *Nature Methods*, Vol. 7, No. 1, pp. 43-46.

Piette, F., D'Amico, S., Struvay, C., Mazzucchelli, G., Renaut, J., Tutino, M. L., Danchin, A., Leprince, P., & Feller, G. (2010). Proteomics of life at low temperatures: trigger factor is the primary chaperone in the Antarctic bacterium *Pseudoalteromonas haloplanktis* TAC125. *Molecular Microbiology*, Vol. 76, No. 1, pp. 120-132.

Pleißner, K.-P., Eifert, T., Buettner, S., Schmidt, F., Boehme, M., Meyer, T. F., Kaufmann, S. H. E., & Jungblut, P. R. (2004). Web-accessible proteome databases for microbial research. *Proteomics*, Vol. 4, No. 5, pp. 1305-1313.

Rabilloud, T., Chevallet, M., Luche, S., & Lelong, C. (2010). Two-dimensional gel electrophoresis in proteomics: Past, present and future. *Journal of Proteomics*, Vol. 73, No. 11, pp. 2064-2077.

Ram, R. J., Verberkmoes, N. C., Thelen, M. P., Tyson, G. W., Baker, B. J., Blake, R. C., 2nd, Shah, M., Hettich, R. L., & Banfield, J. F. (2005). Community proteomics of a natural microbial biofilm. *Science (New York, N.Y.)*, Vol. 308, No. 5730, pp. 1915-1920.

Ramabu, S. S., Ueti, M. W., Brayton, K. A., Baszler, T. V., & Palmer, G. H. (2010). Identification of *Anaplasma marginale* proteins specifically upregulated during colonization of the tick vector. *Infection and Immunity*, Vol. 78, No. 7, pp. 3047-3052.

Saunders, N. F. W., Goodchild, A., Raftery, M., Guilhaus, M., Curmi, P. M. G., & Cavicchioli, R. (2005). Predicted roles for hypothetical proteins in the low-temperature expressed proteome of the Antarctic archaeon *Methanococcoides burtonii*. *Journal of Proteome Research*, Vol. 4, No. 2, pp. 464-472.

Schleifer, K. H. (2009). Classification of Bacteria and Archaea: past, present and future. *Systematic and Applied Microbiology*, Vol. 32, No. 8, pp. 533-542.

Schmidt, F., Donahoe, S., Hagens, K., Mattow, J., Schaible, U. E., Kaufmann, S. H. E., Aebersold, R., & Jungblut, P. R. (2004). Complementary Analysis of the *Mycobacterium tuberculosis* Proteome by Two-dimensional Electrophoresis and Isotope-coded Affinity Tag Technology. *Molecular & Cellular Proteomics*, Vol. 3, No. 1, pp. 24 -42.

Sengupta, N., & Alam, S. I. (2011). In vivo studies of *Clostridium perfringens* in mouse gas gangrene model. *Current Microbiology*, Vol. 62, No. 3, pp. 999-1008.

Simmons, S. L., Dibartolo, G., Denef, V. J., Goltsman, D. S. A., Thelen, M. P., & Banfield, J. F. (2008). Population genomic analysis of strain variation in *Leptospirillum* group II bacteria involved in acid mine drainage formation. *PLoS Biology*, Vol. 6, No. 7, pp. e177.

Smedley, D., Haider, S., Ballester, B., Holland, R., London, D., Thorisson, G., & Kasprzyk, A. (2009). BioMart--biological queries made easy. *BMC Genomics*, Vol. 10, pp. 22.

Taylor, C. F., Paton, N. W., Lilley, K. S., Binz, P.-A., Julian, R. K., Jones, A. R., Zhu, W., Apweiler, R., Aebersold, R., Deutsch, E. W., Dunn, M. J., Heck, A. J. R., Leitner, A., Macht, M., Mann, M., Martens, L., Neubert, T. A., Patterson, S. D., Ping, P., Seymour, S. L., Souda, P., Tsugita, A., Vandekerckhove, J., Vondriska, T. M., Whitelegge, J. P., Wilkins, M. R., Xenarios, I., Yates, J. R., & Hermjakob, H. (2007). The minimum information about a proteomics experiment (MIAPE). *Nature Biotechnology*, Vol. 25, No. 8, pp. 887-893.

Taylor, E. B., & Williams, M. A. (2010). Microbial protein in soil: influence of extraction method and C amendment on extraction and recovery. *Microbial Ecology*, Vol. 59, No. 2, pp. 390-399.

Vallenet, D., Labarre, L., Rouy, Z., Barbe, V., Bocs, S., Cruveiller, S., Lajus, A., Pascal, G., Scarpelli, C., & Médigue, C. (2006). MaGe: a microbial genome annotation system supported by synteny results. *Nucleic Acids Research*, Vol. 34, No. 1, pp. 53-65.

Vizcaíno, J. A., Côté, R., Reisinger, F., M. Foster, J., Mueller, M., Rameseder, J., Hermjakob, H., & Martens, L. (2009). A guide to the Proteomics Identifications Database proteomics data repository. *Proteomics*, Vol. 9, No. 18, pp. 4276-4283.

Weiss, S., Carapito, C., Cleiss, J., Koechler, S., Turlin, E., Coppee, J.-Y., Heymann, M., Kugler, V., Stauffert, M., Cruveiller, S., Médigue, C., Van Dorsselaer, A., Bertin, P. N., & Arsène-Ploetze, F. (2009). Enhanced structural and functional genome elucidation of the arsenite-oxidizing strain *Herminiimonas arsenicoxydans* by proteomics data. *Biochimie*, Vol. 91, No. 2, pp. 192-203.

Werner, J. J., Ptak, A. C., Rahm, B. G., Zhang, S., & Richardson, R. E. (2009). Absolute quantification of *Dehalococcoides* proteins: enzyme bioindicators of chlorinated ethene dehalorespiration. *Environmental Microbiology*, Vol. 11, No. 10, pp. 2687-2697.

Wilkins, M. J., Verberkmoes, N. C., Williams, K. H., Callister, S. J., Mouser, P. J., Elifantz, H., N'guessan, A. L., Thomas, B. C., Nicora, C. D., Shah, M. B., Abraham, P., Lipton, M. S., Lovley, D. R., Hettich, R. L., Long, P. E., & Banfield, J. F. (2009). Proteogenomic monitoring of *Geobacter* physiology during stimulated uranium bioremediation. *Applied and Environmental Microbiology*, Vol. 75, No. 20, pp. 6591-6599.

Wilmes, P., & Bond, P. L. (2004). The application of two-dimensional polyacrylamide gel electrophoresis and downstream analyses to a mixed community of prokaryotic microorganisms. *Environmental Microbiology*, Vol. 6, No. 9, pp. 911-920.

Yan, J. X., Devenish, A. T., Wait, R., Stone, T., Lewis, S., & Fowler, S. (2002). Fluorescence two-dimensional difference gel electrophoresis and mass spectrometry based proteomic analysis of *Escherichia coli*. *Proteomics*, Vol. 2, No. 12, pp. 1682-1698.

Yilmaz, P., Kottmann, R., Field, D., Knight, R., Cole, J. R., et al. (2011). Minimum information about a marker gene sequence (MIMARKS) and minimum information about any (x) sequence (MIxS) specifications. *Nature Biotechnology*, Vol. 29, No. 5, pp. 415-420.

Zanzoni, A., Carbajo, D., Diella, F., Gherardini, P. F., Tramontano, A., Helmer-Citterich, M., & Via, A. (2011). Phospho3D 2.0: an enhanced database of three-dimensional structures of phosphorylation sites. *Nucleic Acids Research*, Vol. 39, No. Database issue, pp. D268-271.

5

Life in the Cold: Proteomics of the Antarctic Bacterium *Pseudoalteromonas haloplanktis*

Florence Piette, Caroline Struvay,
Amandine Godin, Alexandre Cipolla and Georges Feller
Laboratory of Biochemistry, Center for Protein Engineering, University of Liège
Belgium

1. Introduction

It is frequently overlooked that the majority (>80%) of the Earth's biosphere is cold and permanently exposed to temperatures below 5 °C (Rodrigues & Tiedje, 2008). Such low mean temperatures mainly arise from the fact that ~70% of the Earth's surface is covered by oceans that have a constant temperature of 4°C below 1000 m depth, irrespective of the latitude. The polar regions account for another 15%, to which the glacier and alpine regions must be added, as well as the permafrost representing more than 20% of terrestrial soils. All these low temperature biotopes have been successfully colonized by cold-adapted microorganisms, termed psychrophiles (Margesin *et al.*, 2008). These organisms do not merely endure such low and extremely inhospitable conditions but are irreversibly adapted to these environments as most psychrophiles are unable to grow at mild (or mesophilic) temperatures. Extreme psychrophiles have been traditionally sampled from Antarctic and Arctic sites, assuming that low temperatures persisting over a geological time-scale have promoted deep and efficient adaptations to freezing conditions. In addition to ice caps and sea ice, polar regions also possess unusual microbiotopes such as porous rocks in Antarctic dry valleys hosting microbial communities surviving at -60 °C (Cary *et al.*, 2010), the liquid brine veins between sea ice crystals harboring metabolically-active microorganisms at -20 °C (Deming, 2002) or permafrost cryopegs, *i.e.* salty water pockets that have remained liquid at -10 °C for about 100 000 years (Gilichinsky *et al.*, 2005). Psychrophiles and their biomolecules also possess an interesting biotechnological potential, which has already found several applications (Margesin & Feller, 2010).
Cold exerts severe physicochemical constraints on living organisms including increased water viscosity, decreased molecular diffusion rates, reduced biochemical reaction rates, perturbation of weak interactions driving molecular recognition and interaction, strengthening of hydrogen bonds that, for instance, stabilize inhibitory nucleic acid structures, increased solubility of gases and stability of toxic metabolites as well as reduced fluidity of cellular membranes (D'Amico *et al.*, 2006; Gerday & Glansdorff, 2007; Margesin *et al.*, 2008; Rodrigues & Tiedje, 2008). Previous biochemical studies have revealed various adaptations at the molecular level such as the synthesis of cold-active enzymes by psychrophiles or the incorporation of membrane lipids promoting homeoviscosity in cold conditions. It was shown that the high level of specific activity at low temperatures of cold-adapted enzymes is a key adaptation to compensate for the exponential decrease in

chemical reaction rates as the temperature is reduced. Such high biocatalytic activity arises from the disappearance of various non-covalent stabilizing interactions, resulting in an improved flexibility of the enzyme conformation (Feller & Gerday, 2003; Siddiqui & Cavicchioli, 2006; Feller, 2010). Whereas membrane structures are rigidified in cold conditions, an adequate fluidity is required to preserve the integrity of their physiological functions. This homeoviscosity is achieved by steric hindrances introduced into the lipid bilayer via incorporation of *cis*-unsaturated and branched-chain lipids, a decrease in average chain length, and an increase both in methyl branching and in the ratio of *anteiso*- to *iso*-branching (Russell, 2007).

More recently, several genomes from psychrophilic bacteria have been sequenced (Danchin, 2007; Casanueva *et al.*, 2010) but only a few of them have been analyzed with respect to cold adaptation (Saunders *et al.*, 2003; Rabus *et al.*, 2004; Medigue *et al.*, 2005; Methe *et al.*, 2005; Riley *et al.*, 2008; Rodrigues *et al.*, 2008; Allen *et al.*, 2009; Ayala-del-Rio *et al.*, 2010). However, the lack of common features shared by all these psychrophilic genomes has suggested that cold adaptation superimposes on pre-existing cellular organization and, accordingly, that the adaptive strategies may differ between the various microorganisms (Bowman, 2008; Piette *et al.*, 2010).

The Gram-negative bacterium *Pseudoalteromonas haloplanktis* is a typical representative of γ-proteobacteria found in cold marine environments and, in fact, strain TAC125 has been isolated from sea water sampled along the Antarctic ice-shell (Terre Adélie). Such strains thrive permanently in sea water at about -2 °C to +4 °C but are also anticipated to endure long term frozen conditions when entrapped in the winter ice pack. The genome of *P. haloplanktis* TAC125 has been fully sequenced and has undergone expert annotation (Medigue *et al.*, 2005). This work has allowed a proteomic study of its cold-acclimation proteins (CAPs)[1], *i.e.* proteins that are continuously overexpressed at a high level during growth at low temperatures (Piette *et al.*, 2010). This has demonstrated that protein synthesis and protein folding are the main up-regulated functions, suggesting that both cellular processes are limiting factors for bacterial development in cold environments. Furthermore, a proteomic survey of cold-repressed proteins at 4 °C has revealed a strong repression of most heat shock proteins (Piette *et al.*, 2011). This chapter describes the various proteomic features analyzed in the context of adaptation to life at low temperature.

2. Temperature dependence of growth

The ability of *P. haloplanktis* to grow at low temperatures is illustrated in Fig. 1. This psychrophilic Antarctic bacterium maintains a doubling time of ~4 h at 4 °C in a marine broth, with an extrapolated generation time of 5 h 15 at 0 °C (Fig. 1a). This can be compared with the behavior of a mesophilic bacterium such as *E. coli*, which displays a doubling time of ~8h at 15 °C and which fails to grow below ~8 °C (Strocchi *et al.*, 2006). When the culture temperature is raised up to 20 °C, the generation time moderately decreases (*e.g.* 1 h 40 at 18 °C) with a concomitant increase in the biomass produced at the stationary phase (Fig. 1b). At temperatures higher than 20 °C, the doubling time of *P. haloplanktis* slightly increases again with, however, a drastic reduction in cell density at the stationary phase (Fig. 1b), indicating a heat-induced stress on the cell. *P. haloplanktis* TAC125 fails to grow above 29 °C, thereby

[1] The abbreviations used are: CAPs, cold acclimation proteins; CRPs, cold repressed proteins; TF, trigger factor; ROS, reactive oxygen species

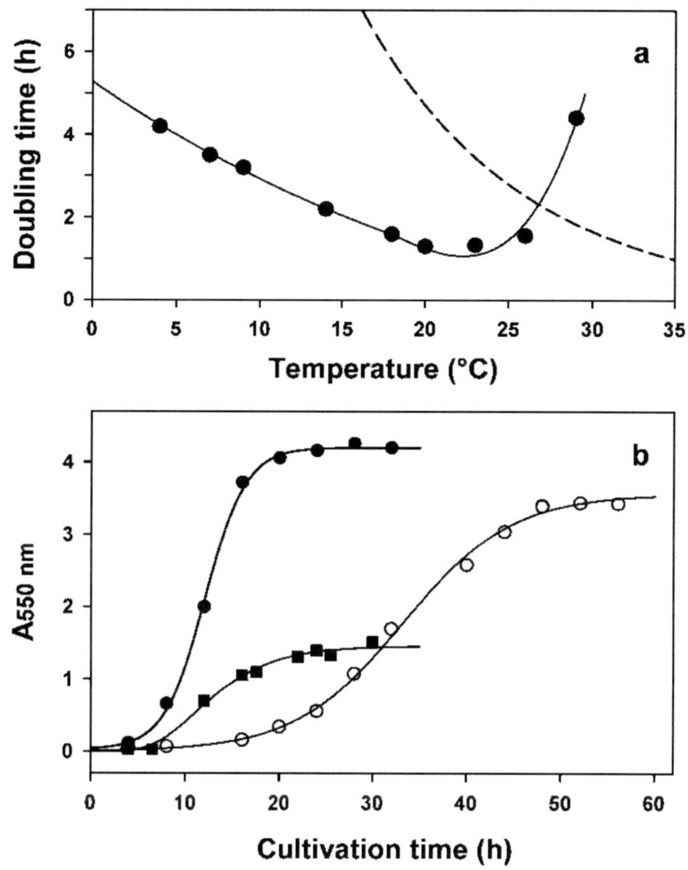

Fig. 1. (a) Temperature dependence of the generation time of *Pseudoalteromonas haloplanktis* TAC125 grown in a marine broth (solid line and circles). A typical curve for *E. coli* RR1 in LB broth is shown for comparison (dashed) (b) Growth curves of *P. haloplanktis* at 4°C (○), 18°C (●) and 26°C (■). Reprinted with permission from Piette *et al.*, 2011. © 2011 American Society for Microbiology.

defining its upper cardinal temperature. According to this growth behavior, the temperatures of 4 °C and 18 °C were selected for the differential comparison of the proteomes, as 18 °C does not induce an excessive stress as far as growth rate and biomass are concerned.

The fast growth rate of the Antarctic bacterium is primarily achieved by a low temperature dependence of the generation times when compared with a mesophilic bacterium, *i.e.* the generation time of *P. haloplanktis* is moderately increased when the culture temperature is decreased (Fig. 1a). It should be stressed that enzymes from cold-adapted organisms are characterized by both a high specific activity at low temperatures and a low temperature dependence of their activity (formally, a weak activation enthalpy), *i.e.* reaction rates of psychrophilic enzymes are less reduced by a decrease in temperature as compared with mesophilic enzymes (D'Amico *et al.*, 2003; Feller & Gerday, 2003). Accordingly, the growth characteristics of the Antarctic bacterium (Fig. 1a) appear to be governed by the properties of its enzymatic machinery: high enzyme-catalyzed reaction rates maintain metabolic fluxes and cellular functions at low temperatures, whereas the weak temperature dependence of enzyme activity counteracts the effect of cold temperatures on biochemical reaction rates.

3. Cold-induced versus cold-repressed proteins

The proteomes expressed by the Antarctic bacterium at 4 °C and 18 °C during the logarithmic phase of growth have been compared by two-dimensional differential in-gel electrophoresis (2D-DIGE), enabling the co-migration in equal amounts of cell extracts obtained from both conditions (labeled by distinct CyDye fluorophores) in triplicate gels (Fig. 2).

Fig. 2. Comparison of intracellular soluble proteins from *P. haloplanktis* grown at 4°C (red-labeled) and 18°C (green-labeled) on 2D-DIGE gels analyzed by fluorescence. From left to right, non-linear gradient from pH 3 to pH 10. From top to bottom, mass scale from ~150 to ~15 kDa. The intense red fluorescence of the trigger factor (TF) spot correlates with its up-regulation at 4°C, whereas the intense green fluorescence of the DnaK spot correlates with its down-regulation Adapted with permission from Piette *et al.*, 2010. © 2010 Wiley.

In a typical single 2D-gel (Fig. 3), 142 protein spots are more abundant at 4 °C. As protein extracts were prepared from cells growing exponentially at this temperature, all up-regulated proteins at 4°C are regarded as CAPs. Furthermore, 309 protein spots are less

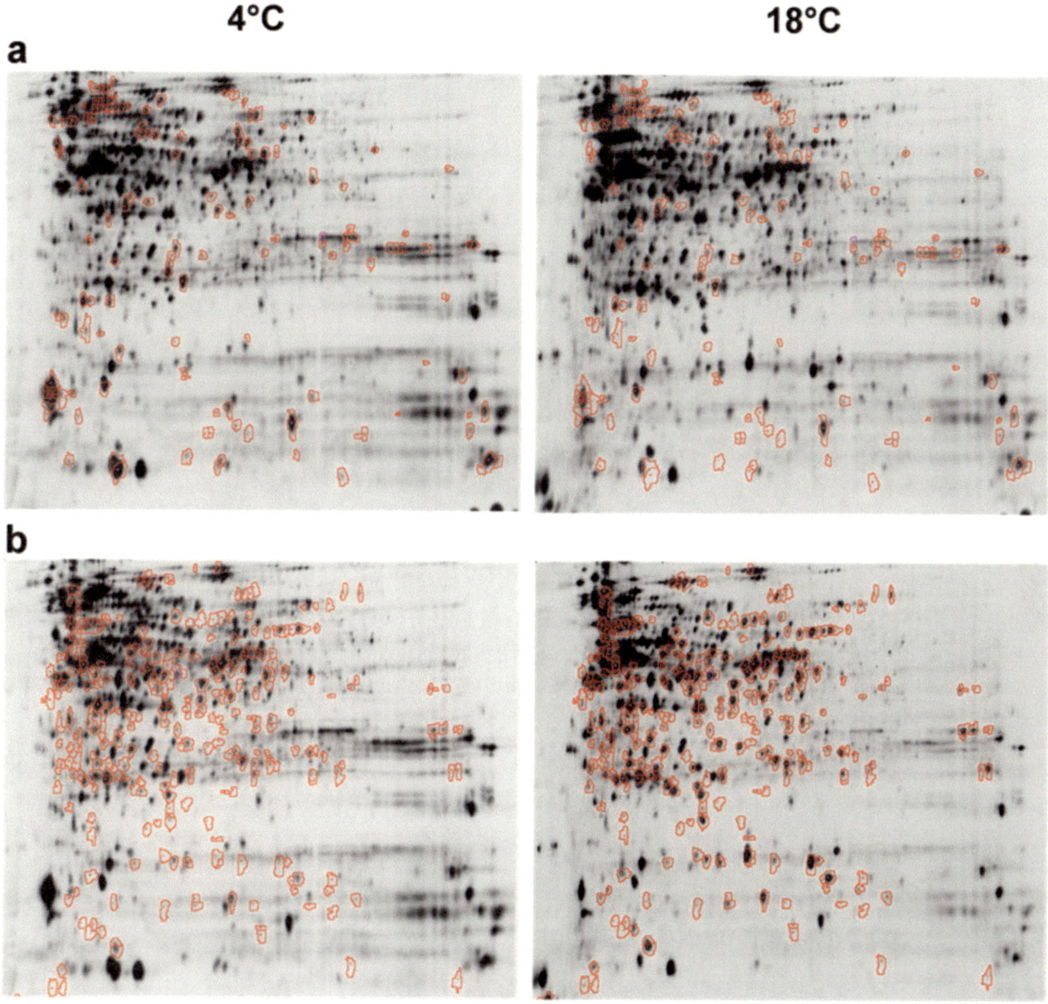

Fig. 3. Differential analyses of soluble cellular proteins from *Pseudoalteromonas haloplanktis* grown at 4°C (left panels) and 18°C (right panels) on 2D-DIGE gels analyzed by fluorescence. **(a)** 142 protein spots that are more intense at 4°C are indicated. **(b)** 309 protein spots that are less intense at 4°C are indicated. Reprinted with permission from Piette *et al.*, 2011. © 2011 American Society for Microbiology.

intense at 4 °C as compared with 18 °C. This unexpected large number of cold-repressed proteins (CRPs) already indicates that numerous cellular functions are down-regulated during growth at low temperature.

The induction factors for CAPs and the repression factors for CRPs, given by the spot volume ratio between 4 °C and 18 °C are illustrated in Fig. 4. This distribution shows that most CAPs and CRPs have a five-time higher or lower relative abundance at 4 °C. However, about 20% of these differentially expressed proteins display up- or down-regulation factors higher than 5, revealing that some key cellular functions are strongly regulated. Amongst all these differentially expressed proteins, 40 CAPs and 83 CRPs were retained, which satisfied both statistical biological variation analysis and mass spectrometry identification scores, as

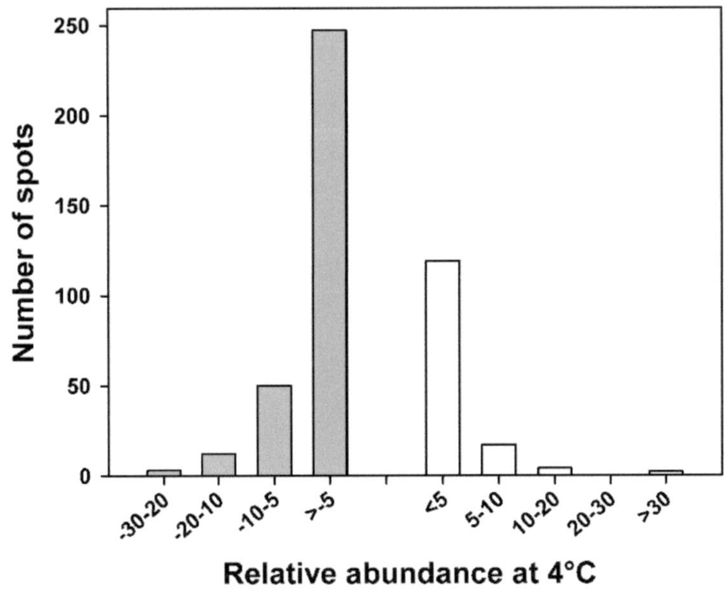

Fig. 4. Distribution of the relative abundance of cold-repressed proteins (dashed, negative values) and of cold acclimation proteins (positive values) in the proteome of *P. haloplanktis* grown at 4°C and 18°C. Reprinted with permission from Piette *et al.*, 2011. © 2011 American Society for Microbiology.

detailed in the original publications (Piette *et al.*, 2010; Piette *et al.*, 2011). Accordingly, the identified proteins should be analyzed as markers of a pathway or of a general function, rather than for their specific function as they represent 27% of the differentially expressed proteins at 4°C.

4. Cold shock and heat shock proteins

One of the most remarkable features of the differentially expressed proteome of *P. haloplanktis* is the strong up-regulation at 4°C of proteins that are regarded as cold shock proteins in mesophilic bacteria, as well as the down-regulation to nearly undetectable levels of proteins classified as heat shock proteins (Fig. 5).

Cold shock proteins that have been identified as CAPs in *P. haloplanktis* include Pnp (+4x), TypA (+5x) and the trigger factor TF (+38x) that are involved in distinct functions (degradosome, membrane integrity and protein folding, respectively). Sustained synthesis of various cold shock protein-homologues has been also reported in other cold-adapted bacteria (Bakermans *et al.*, 2007; Kawamoto *et al.*, 2007; Bergholz *et al.*, 2009). There are therefore striking similarities between the cold shock response in mesophiles and cold adaptation in psychrophiles. From an evolutionary point of view, it can be proposed that one of the adaptive mechanisms to growth in the cold was to regulate the cold shock response, shifting from a transient expression of cold shock proteins to a continuous synthesis of at least some of them. Interestingly, nearly all proteins displaying the highest repression factors at 4°C are heat shock proteins (Rosen & Ron, 2002) including the main chaperones DnaK (-13x) and GroEL (-3.4x), the accessory chaperones such as Hsp90 (-28x), the small heat shock proteins IbpA (-24x) and IbpB (-18x), as well as LysS (-17x).

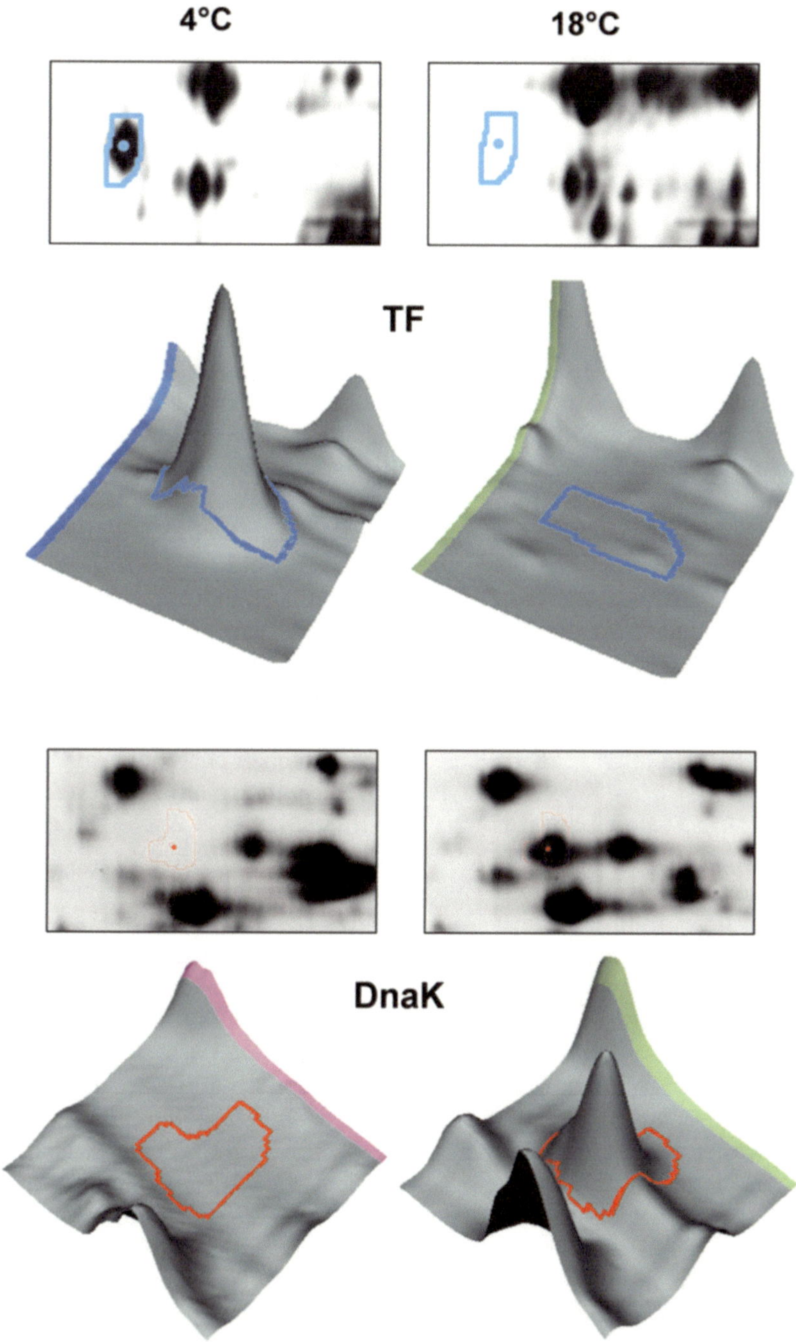

Fig. 5. Comparative analysis of spots containing the trigger factor TF (a cold shock protein) and DnaK (a heat shock protein) from *P. haloplanktis* grown at 4°C (left panels) and 18°C (right panels). Spot views on 2D-gels (circled) and three-dimensional images. Adapted with permission from Piette *et al.*, 2010; Piette *et al.*, 2011. © 2010 Wiley and © 2011 American Society for Microbiology.

In mesophilic bacteria such as *E. coli*, cold shock and heat shock proteins are transiently expressed in response to temperature downshift and upshift, respectively. By contrast, the Antarctic bacterium continuously over-expresses some cold shock proteins (Piette *et al.*, 2010) whereas most heat shock proteins are continuously repressed at 4 °C. It is obvious that regulation of the expression of these proteins involved in thermal stress is a primary adaptation to bacterial growth at low temperatures that remains to be properly explained.

5. Protein folding at low temperature rescued by the trigger factor

In bacteria, the three main chaperones are the trigger factor TF, a cold shock protein that stabilize nascent polypeptides on ribosomes and initiate ATP-independent folding, DnaK that mediates co- or post-transcriptional folding and the GroEL/ES chaperonin that acts downstream in folding assistance (Hartl & Hayer-Hartl, 2009). Both latter chaperones are also well-known heat shock proteins. The trigger factor TF (+38x up-regulated at 4°C) is the first molecular chaperone interacting with virtually all newly synthesized polypeptides on the ribosome. It delays premature chain compaction and maintains the elongating polypeptide in a non-aggregated state until sufficient structural information for productive folding is available and subsequently promotes protein folding (Merz *et al.*, 2008; Hartl & Hayer-Hartl, 2009; Martinez-Hackert & Hendrickson, 2009). Furthermore, TF also contains a domain catalyzing the *cis-trans* isomerization of peptide bonds involving a proline residue (Kramer *et al.*, 2004). This *cis-trans* isomerization is a well-known rate-limiting step in protein folding (Baldwin, 2008). On the other hand the major heat shock proteins were identified as strongly cold-repressed proteins in the proteome of *P. haloplanktis* (or, in other words, they are up-regulated at 18°C). The overexpression of bacterial heat shock proteins at elevated temperatures is well recognized as being indicative of a heat-induced cellular stress (Rosen & Ron, 2002; Goodchild *et al.*, 2005). Although this is obviously relevant for the Antarctic bacterium grown at 18 °C, the implications for the psychrophilic strain appear to be more complex. Indeed, these heat shock proteins are chaperones assisting co- or post-translational protein folding (Hartl & Hayer-Hartl, 2009). Furthermore, it has been demonstrated that GroEL from *P. haloplanktis* is not cold-adapted, it is inefficient at low temperatures as its activity is reduced to the same extent than that of its *E. coli* homologue (Tosco *et al.*, 2003). Accordingly, under this imbalanced synthesis of folding assistants, protein folding at low temperature is apparently compromised in the Antarctic bacterium.
Considering the down-regulation of heat shock chaperones and the inefficiency of GroEL from *P. haloplanktis* at low temperature, as well as the essential function of TF in the initiation of proper protein folding, it can be proposed that TF rescues the chaperone function at low temperatures, therefore explaining its unusual overexpression level. It follows that TF becomes the primary chaperone of the Antarctic bacterium for growth in the cold. Although the psychrophilic bacterium maintains a minimal set of chaperones, this is obviously sufficient to allow bacterial development at low temperature.

6. Possible origins of heat shock protein repression at low temperature

The strong overexpression of TF at low temperature can be understood according to its above mentioned essential function. By contrast, the reasons for the concomitant repression of heat shock chaperones in the Antarctic bacterium remain hypothetical. At least four possible origins, not mutually exclusive, can be mentioned. *i*) Low temperature slows down the folding reaction and is well known to reduce the probability of misfolding and aggregation (King *et al.*, 1996), therefore possibly reducing the need for heat shock chaperones that act downstream from TF. *ii*) In *E. coli*, it has been shown that synthesis of heat shock proteins is repressed during growth at low temperatures, but also that these heat shock proteins are harmful to cells at 4 °C, as their induced expression reduces cell viability at this temperature (Kandror & Goldberg, 1997). Accordingly, the observed cold-repression of heat shock proteins would be beneficial to the psychrophilic bacterium. *iii*) To our knowledge, the reasons for this harmful effect of heat shock proteins have not been investigated. A possible explanation could be found in the second function of these chaperones. Indeed, besides their role in folding assistance, many heat shock proteins bind partly folded polypeptides and promote their fast degradation by proteases Lon and Clp. In addition, TF enhances the binding affinity of some heat shock proteins for these partly folded polypeptides (Kandror *et al.*, 1995; Kandror *et al.*, 1997). In the case of the Antarctic bacterium, heat shock chaperones would then have an increased ability to bind slowly folding polypeptides at low temperatures and to promote their unwanted degradation: this would account for the cold repression of heat shock chaperones by *P. haloplanktis*. *iv*) The observation that GroEL from the Antarctic bacterium is non cold-adapted (Tosco *et al.*, 2003) suggests that this chaperonin is well suited to function during sudden temperature increases of the environment. Indeed, microorganisms subjected to seasonal or local temperature variations (*e.g.* melting sea ice, polar surface soils, etc.) would advantageously maintain a heat shock response involving non cold-adapted chaperones remaining active at transiently high temperatures.

7. Structural properties of the psychrophilic trigger factor

According to its essential function in *P. haloplanktis*, the psychrophilic TF has been analyzed into more details. Its amino acid sequence (47,534 Da) displays 61% identity (85% similarity) on 434 residues with its homologue from *E. coli*. Its sequence is also close to that of some known TF from psychrophilic bacteria. The pronounced sequence similarity and predicted secondary structure conservation with *E. coli* TF suggest that the psychrophilic chaperone should also folds into an extended "crouching dragon" conformation (Ferbitz *et al.*, 2004) comprising three domains (Fig. 6). The N-terminal domain mediates ribosome attachment *via* an exposed loop, the PPIase activity domain located at the opposite end of the molecule (Kramer *et al.*, 2004) and the C-terminal domain forming the body of the protein and bearing the central module of chaperone activity (Merz *et al.*, 2006).
In order to analyze the psychrophilic TF, its gene has been cloned and overexpressed in *E. coli* and the recombinant protein has been purified to homogeneity (Piette *et al.*, 2010). Its thermal stability was investigated by differential scanning calorimetry. Fig. 7 shows that TF from the Antarctic bacterium is a marginally stable protein, exhibiting a melting point T_m at 33°C. It follows that at a typical mesophilic temperature of 37°C, almost all the protein population is already in the unfolded state. In addition, the calorimetric enthalpy is also very weak (ΔH_{cal} = 82.5 kcal mol^{-1}, the sum of all enthalpic contributions to protein stability disrupted during unfolding and calculated from the area under the transition). By comparison, a T_m of 54°C and a calorimetric enthalpy of 178 kcal mol^{-1} have been reported for the *E. coli* trigger factor analyzed by DSC (Fan *et al.*, 2008). Despite its modular structure, *P. haloplanktis* TF unfolds according to a perfect 2-state transition (Fig. 7), *i.e.* without significantly populated intermediates between the native and the unfolded states. This

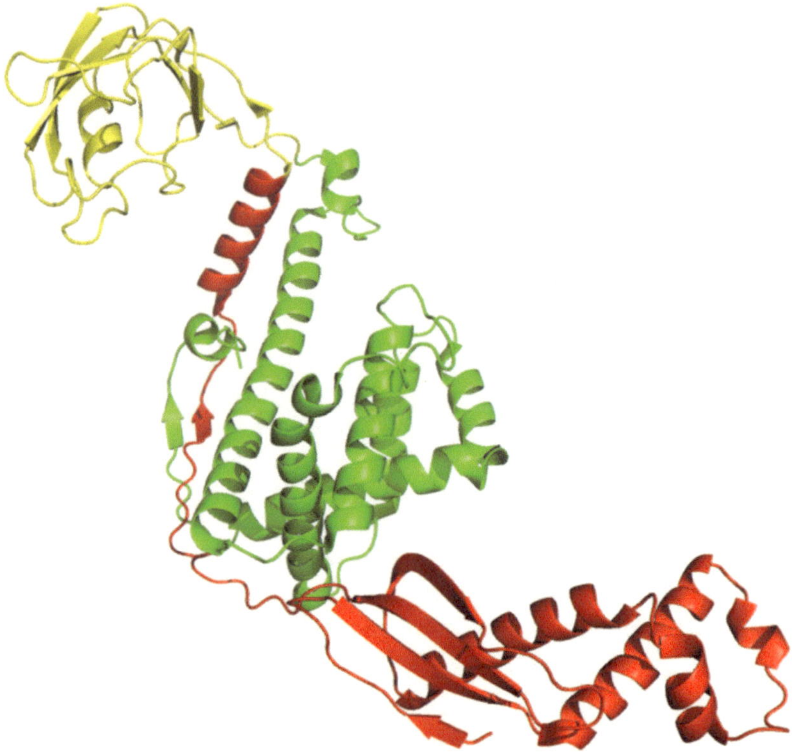

Fig. 6. Domain organization in the trigger factor structure (based on *E. coli* trigger factor: PDB 1W26). The N-terminal domain mediating ribosome attachment (aa 1-144) is in red, the PPiase domain (aa145-247) is in yellow and the C-terminal domain bearing the central module of the chaperone activity (aa 248-432) is in green. As a result of the strong conservation of primary and predicted secondary structures in *P. haloplanktis* TF, a model of its structure built by homology modeling is undistinguishable from the *E. coli* crystal structure. Reprinted with permission from Piette *et al.*, 2010. © 2010 Wiley.

indicates that the psychrophilic TF is uniformly unstable and unfolds cooperatively. To the best of our knowledge, *P. haloplanktis* TF is the least stable protein reported so far. This strongly suggests that the essential chaperone function requires considerable flexibility and dynamics to compensate for the reduction of molecular motions at freezing temperatures.
The *E. coli* trigger factor has been reported to undergo *in vitro* a concentration-dependent dynamic equilibrium between the monomeric and the dimeric forms. Static light scattering experiments performed in batch mode provided a mean particle mass of 51 kDa and 106 kDa for *P. haloplanktis* and *E. coli* TF, respectively. This is in agreement with a monomeric psychrophilic TF and a dimeric *E. coli* TF. However, in dynamic light scattering the particle polydispersity (size distribution) of *P. haloplanktis* TF was twice that of *E. coli* TF, suggesting that the psychrophilic TF may possibly perform transient intermolecular interactions. These observations are in line with a report showing that the trigger factor from the psychrophile *Psychrobacter frigidicola* is a monomeric chaperone (Robin *et al.*, 2009). The interpretation of these differences in oligomerization state remains to be properly explained but suggest noticeable differences between psychrophilic and mesophilic bacteria for the TF function in

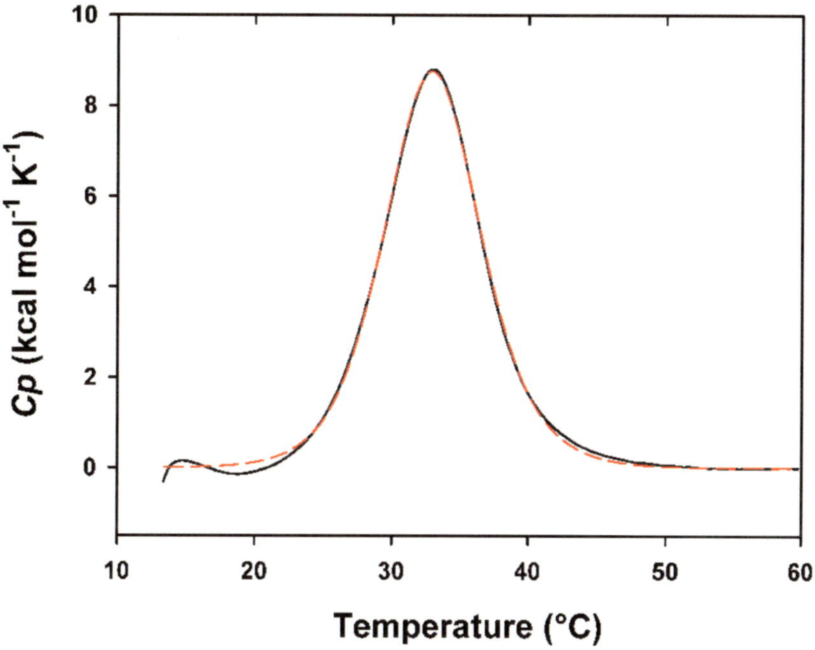

Fig. 7. Microcalorimetric analysis of the trigger factor from *P. haloplanktis*. The melting point T_m corresponds to the top of the transition at 33°C. The calorimetric enthalpy ΔH_{cal} corresponds to the area under the transition. The red dashed line corresponds to the fit of the DSC data to a two-state unfolding transition. Baseline-subtracted data have been normalized for protein concentration (2.6 mg/ml in 30 mM Mops, 250 mM NaCl, pH 7.6.). Adapted with permission from Piette *et al.*, 2010. © 2010 Wiley.

the cytoplasmic fraction, when not bound to the ribosome. Finally, in a typical refolding assay monitoring chaperone activity, it has been found that *P. haloplanktis* TF is inactive at 20°C and recovers partial activity at 15°C. It was also shown that this TF requires near-zero temperatures (Fig. 8) to efficiently bind an unfolded protein (Piette *et al.*, 2010). This illustrates a remarkable cold adaptation of the chaperone function in the psychrophilic TF.

8. Protein synthesis and folding are limiting factors in the cold

Thirty percent of the identified CAPs are directly related to protein synthesis and cover all essential steps, from transcription (including RNA polymerase RpoB) to translation and folding (TF, PpiD). Amongst these CAPs, for instance, genes *pnp* and *rpsA* encode components of the degradosome that regulates transcript lifetimes. The Rho termination factor is a RNA/DNA helicase that can contribute to relieve nucleic acid secondary structures strengthened in cold conditions. Interestingly, mutations in the ribosomal protein L6 (RplF) have been reported to cause loss of *E. coli* cells viability at 0°C (Bosl & Bock, 1981) and it is also a CAP in *P. haloplanktis*. Methionyl-tRNA synthetase MetG displays one of the highest up-regulation ratio (+7.6x): this can be tentatively related to the requirement of an increased pool of initiation tRNA to promote protein synthesis. Two putative proteases were also identified as CAPs and can potentially participate to proteolysis of misfolded proteins.

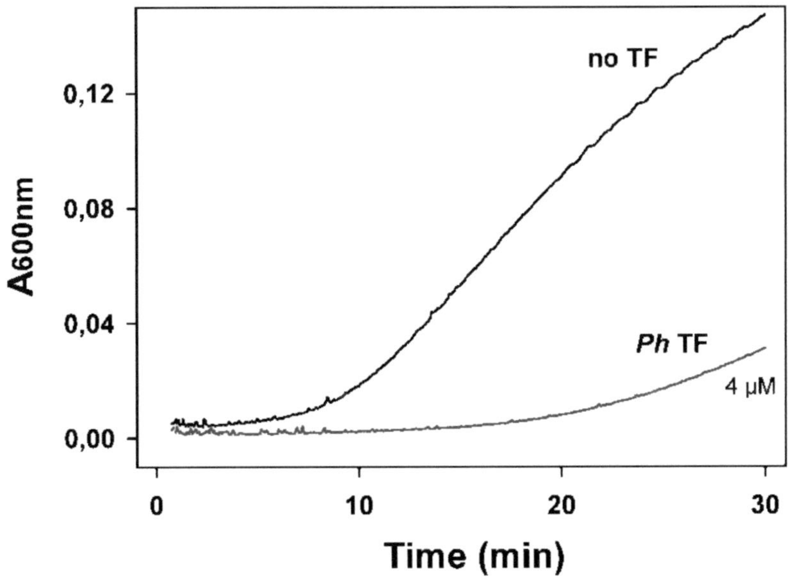

Fig. 8. Detection of the chaperone activity of the psychrophilic TF. Aggregation at 15°C of a chemically unfolded protein (glyceraldehyde-3-phosphate dehydrogenase) is monitored by absorbance at 600 nm after incubation in melting ice. In the absence of TF (black trace), aggregation occurs as soon as the temperature is raised. By contrast, TF suppresses aggregation for 10-15 min (blue trace). Adapted with permission from Piette *et al.*, 2010. © 2010 Wiley

In the last step of protein synthesis, the folding catalyst TF acts on proteins synthesized by the ribosome and also catalyses peptidyl-prolyl *cis-trans* isomerisation (PPiase) while PpiD (another PPiase) is involved in the folding of outer membrane proteins. Peptidyl-prolyl *cis-trans* isomerisation appears therefore as a limiting factor for a wide range of proteins in *P. haloplanktis*. Furthermore, some previous studies on cold-adapted microorganisms have reported either PPiases (Goodchild *et al.*, 2004b; Suzuki *et al.*, 2004) or the trigger factor (Qiu *et al.*, 2006; Kawamoto *et al.*, 2007) as potential CAPs. It seems therefore that the constraints imposed by protein folding in the cold are common traits in several psychrophilic microorganisms.

Altogether, these observations strongly suggest that low temperatures impair protein synthesis and folding, resulting in up-regulation at 4°C of the associated cellular processes.

9. Metabolism depression at low temperatures

Nearly half of down-regulated proteins at 4°C are related to superclasses of function involved in the bacterial general metabolism. This includes the degradation or biosynthesis of compounds and the production of energy. Most of these proteins belong to the oxidative metabolism, in particular to glycolysis, the pentose phosphate pathway, Krebs cycle and electron chain transporters. Accordingly, the Antarctic bacterium depresses its general metabolism when grown at low temperature. This is in agreement with the reduced biomass produced at 4 °C as compared with cultures run at 18 °C (Fig. 1b). As mentioned in the previous section, protein synthesis and folding are limiting factors for the growth of *P.*

haloplanktis at cold temperatures (Piette *et al.*, 2010). However, the proteomic data indicates that when these limitations are alleviated at 18 °C, the bacterium proliferates by activation of its general metabolism and therefore divides actively and produces more biomass. The high number of identified ribosomal proteins and of elongation factors involved in translation indicates that protein synthesis is no longer limiting but is also stimulated at 18 °C.

10. Down-regulation of iron metabolism at low temperatures

Iron uptake and iron-related proteins are clearly down-regulated at 4 °C in *P. haloplanktis*. The uptake of this essential element in an aquatic environment is mediated by several iron transport systems. Two systems were found to be down-regulated at 4 °C: the ABC transporter (FbpA) and a TonB-dependent receptor. The first is involved in the uptake of the weakly soluble ferric ion (Fe^{3+}) directly from the environment and the second is required for the transport of heme complexes and ferric siderophores through the membrane (Clarke *et al.*, 2001). The reduced needs for iron by *P. haloplanktis* at 4 °C can be partly explained by the down-regulation of the Krebs cycle and respiratory chain (and their iron-containing complexes such as SdhB), by the repression of HmgA, which requires Fe^{2+} to degrade cyclic amino-acids or by the strong down-regulation of catalase (which is made up of four heme groups). Hemes are tetrapyrroles that have porphobilinogen as a precursor: this is in agreement with the down-regulation of both GltX (glutamyl-tRNA synthetase) and HemB (5-aminolevulinate dehydratase), which are responsible for porphobilinogen synthesis.

Various metallic ions are essential to the cell metabolism and therefore the fact that proteomic data only points to cold repression of iron-related proteins is puzzling. Iron in a redox-active form (Fe^{2+}) is potentially deleterious, as it is able to induce oxidative cell damage by the Fenton reaction, for instance (Valko *et al.*, 2005). It can be tentatively proposed that, as a result of the improved stability of ROS (reactive oxygen species) at low temperatures, the down-regulation of iron-related proteins could contribute to an avoidance of such detrimental iron-based reactions. In this respect, it should be mentioned that the genome of *P. haloplanktis* entirely lacks the ubiquitous ROS-producing molybdopterin metabolism (Medigue *et al.*, 2005). This suggests that the Antarctic bacterium tends to avoid ROS production involving metallic ions.

11. Oxidative stress-related proteins

The pattern of oxidative stress-related proteins in *P. haloplanktis* is complex because some have been identified as CAPs, while others were found to be CRPs. For instance, glutathione synthetase is the second main up-regulated protein at 4°C (+13.2x) and superoxide dismutase (+1.6x) was also detected as a CAP. This is a clear indication of a cellular response to an oxidative stress arising from increased dioxygen solubility and ROS stability. On the other hand, the second group of proteins that displays the highest repression factors at 4 °C is represented by the oxidative stress-related proteins catalase (-6.5x), glutathione reductase (-8.1x) and peroxiredoxin (-15.7x). At first sight, this may be regarded as a conflicting result because conclusive evidences have indicated that psychrophiles are exposed to a permanent oxidative stress at low temperatures, which originates from improved dioxygen solubility and increased ROS stability (Rabus *et al.*, 2004; Medigue *et al.*, 2005; Methe *et al.*, 2005; Duchaud *et al.*, 2007; Bakermans *et al.*, 2007; Ayub *et al.*, 2009; Piette *et al.*, 2010). In order to reconcile these apparent contradictions, it should be recalled that the general aerobic metabolism of the Antarctic bacterium is stimulated at 18 °C, also resulting in ROS production. Accordingly, the identified oxidative stress-related proteins would be better regarded as being induced at 18 °C, rather than repressed at 4 °C.

The up-regulation of catalase and peroxiredoxin at 18°C shows that the bacterium needs to be protected against ROS like H_2O_2 as both enzymes catalyze its decomposition into O_2 and H_2O. Under oxidative stress, the NADPH supply for reduced glutathione regeneration is also dependent on glucose-6-phosphate dehydrogenase (Zwf) in the first step of the pentose phosphate pathway, and indeed Zwf is positively regulated at 18 °C. Glutathione reductase (Gor) plays a central role in the reoxidation of NADPH from the pentose phosphate pathway, allowing formation of reduced glutathione, an important cellular antioxidant. The up-regulated DNA-binding DPS protein (DpsB) plays a major role in the protection of bacterial DNA from damage by ROS and is induced under stress conditions. Some DPS proteins are also able to bind iron and are involved in its storage and in the protection of the cell (Haikarainen & Papageorgiou, 2010).

There is obviously a finely-tuned balance between the cellular mechanisms protecting against oxidative stresses generated by low temperatures (resulting from ROS stability and oxygen solubility) and by high temperatures (resulting from stimulated metabolic activity). The number of identified proteins does not allow a detailed description of this balance but the strong involvement of glutathione synthetase, glutathione reductase and the

Microorganism	Source	Technique	Ref.
Bacillus psychrosaccharolyticus	Soil, marshes	2D-PAGE	1
Methanococcoides burtonii	Ace Lake, Antarctica	LC/LC-MS/MS	2
Methanococcoides burtonii	Ace Lake, Antarctica	2D-PAGE	3
Methanococcoides burtonii	Ace Lake, Antarctica	ICAT LC MS	4
Methanococcoides burtonii	Ace Lake, Antarctica	LC MS	5
Methanococcoides burtonii	Ace Lake, Antarctica	SDS-PAGE LC-MS/MS	6
Exiguobacterium sibiricum	Siberian permafrost	2D LC MS	7
Shewanella livingstonensis	Antarctic seawater	2D-PAGE	8
Psychrobacter cryohalolentis	Siberian permafrost	2D-PAGE	9
Psychrobacter articus	Siberian permafrost	2-D HPLC & MS	10
Moritella viscosa	Atlantic salmon	2D-PAGE	11
Psychrobacter articus	Siberian permafrost	2D-PAGE	12
Sphingopyxis alaskensis	Alaska seawater	GeLC-MS/MS	13
Methanococcoides burtonii	Ace Lake, Antarctica	LC/LC-MS/MS	14
Lactococcus piscium strain	seafood products	2D-PAGE	15
Pseudoalteromonas haloplanktis	Antarctic seawater	2D-DIGE	16
Methanococcoides burtonii	Ace Lake, Antarctica	8-plex iTRAQ	17
Acidithiobacillus ferrooxidans	Mine drainage, Canada	2D-PAGE	18

Table 1. Proteomic studies performed on psychrophilic microorganisms. References: 1, Seo *et al.*, 2004; 2, Goodchild *et al.*, 2004a; 3, Goodchild *et al.*, 2004b; 4, Goodchild *et al.*, 2005; 5, Saunders *et al.*, 2005; 6, Saunders *et al.*, 2006; 7, Qiu *et al.*, 2006; 8, Kawamoto *et al.*, 2007; 9, Bakermans *et al.*, 2007; 10, Zheng *et al.*, 2007; 11, Tunsjo *et al.*, 2007; 12, Bergholz *et al.*, 2009; 13, Ting *et al.*, 2010; 14, Williams *et al.*, 2010; 15, Garnier *et al.*, 2010; 16, Piette *et al.*, 2010; Piette *et al.*, 2011; 17, Williams *et al.*, 2011; 18, Mykytczuk *et al.*, 2011.

identification of enzymes belonging to the pentose phosphate pathway suggest that regulation of the cytoplasmic redox buffering capacity via glutathione is a key component.

12. Other proteomic studies

A selection of recent proteomic studies on psychrophilic and cold-adapted microorganisms is listed in Table 1. It is worth mentioning that the CAPs and CRPs identified in these studies do not constitute a conserved set of proteins in terms of identification and expression level. Nevertheless, a survey of these data shows that the main upregulated functions for growth at low temperatures are protein synthesis (transcription, translation), RNA and protein folding, membrane integrity and transport, antioxidant activities and regulation of specific metabolic pathways. Such heterogeneous upregulation of CAPs supports the view that cold-adaptation mechanisms are constrained by the species-specific cellular structure and organization, resulting in distinct adaptive strategies. This hypothesis is based on a previous observation made by Bowman (2008). In a review of genome data from psychrophiles, he concluded that the lack of common features shared by these genomes suggests that cold adaptation superimposes on pre-existing cellular organization and, accordingly, that the adaptive strategies may differ between the various microorganisms.

13. Conclusions

The capacity of psychrophilic bacteria to thrive successfully in permanently cold environments obviously requires a vast array of adaptations. At least two prerequisites to this environmental adaptation can be cited: *i*) from a functional standpoint, the synthesis of cold-active enzymes is required to support the bacterial metabolism and its energy production (Feller & Gerday, 2003; Siddiqui & Cavicchioli, 2006; Feller, 2010), and *ii*) from a structural standpoint, the synthesis of cold-adapted lipids is required to maintain the cell membrane integrity, fluidity and functions (Russell, 2007). It should be noted that the first adaptation is genetically encoded in the protein sequence (and results from a long term adaptation), whereas the second adaptation involves regulation of pre-existing biosynthetic pathways. However, neither of these basic adaptations is sufficient because low temperature induces physicochemical constraints that are unavoidable but that can be attenuated by cellular mechanisms. For instance, low temperature reduces molecular diffusion rates and also increases water and cytoplasmic viscosity. It can be proposed that both physicochemical constraints are responsible for the rate limiting steps of *P. haloplanktis* growth in the cold, namely protein synthesis and folding, as deduced from proteomic experiments. Indeed, bacterial protein synthesis is one of the most complex cellular processes and requires diffusion and docking of numerous partners with ribosomes (mRNA, tRNA, initiation factors, elongation factors, GTP…). Assuming a high, cold-active ribosomal efficiency (although this has not been demonstrated to date), its synthetic activity would be nevertheless restricted by diffusion and availability of the required partners. The rate of protein folding is also limited by low temperatures. This is an entropically-driven process governed by the chemical nature of the polypeptide chain and of water molecules. Furthermore, the main protein chaperones are not catalysts *per se* but rather they assist in protein folding and prevent or relieve misfolding. The above mentioned physicochemical constraints exert their effects on all psychrophiles and it can be anticipated that protein

synthesis and folding are also limiting for these microorganisms, unless specific adaptive mechanisms have been developed.
Considering the constraints on protein folding, one would expect the activation of the full set of protein chaperones. By contrast, we found that the Antarctic bacterium strongly over-expresses the trigger factor (a cold-shock protein in *E. coli*) and represses the major chaperones (also HSPs in *E. coli*) at 4 °C. Interestingly, the same trend has been reported for *E. coli* grown at low temperatures (Kandror & Goldberg, 1997). This antagonism between cold-shock and heat-shock chaperones appears to be a common feature in these bacteria, but the origin of this antagonism remains hypothetical as discussed in section 6.
Increased dioxygen solubility and ROS stability is another physicochemical constraint exerted on psychrophilic microorganisms. Indeed, we have noted the activation of oxidative stress protection mechanisms in *P. haloplanktis* grown at 4 °C. Furthermore, the repression of iron-related proteins at 4 °C seems to be related to the avoidance of Fenton-type reactions (Valko *et al.*, 2005). The genome of *P. haloplanktis* also reveals several insights into ROS protection such as deletion of ROS producing pathways, several occurrences of dioxygenases and the repair mechanisms of oxidized compounds (Medigue *et al.*, 2005). Similar observations have been made in the genome and proteome of other psychrophilic microorganisms (Rabus *et al.*, 2004; Medigue *et al.*, 2005; Methe *et al.*, 2005; Duchaud *et al.*, 2007; Bakermans *et al.*, 2007; Ayub *et al.*, 2009; Piette *et al.*, 2010), revealing a general constraint on these bacteria. However, we found that at 18 °C, another type of oxidative stress is induced by stimulation of metabolic activity. This balance between cold-induced and heat-induced oxidative stresses deserves further investigation.
Our proteomic data points to a global reduction of the general metabolism in the Antarctic bacterium at 4 °C. If the behavior of the psychrophilic bacterium were extrapolated at near freezing temperatures, it is anticipated that its metabolism would be further depressed. It is worth mentioning that ancient bacteria survival has been reported in frozen samples of up to half a million years old and such viability has been correlated with the capacity to slowly repair DNA (Johnson *et al.*, 2007). The temperature dependence of the metabolic pattern in psychrophilic bacteria can thus be summarized as follows. During the prevailing cold conditions in polar environments, these bacteria remain metabolically active (Deming, 2002) but their growth can be limited by some temperature-sensitive cellular processes (protein synthesis and folding in the case of *P. haloplanktis*). When the environmental temperature transiently increases, both their metabolism and cell division are stimulated. Besides an elemental thermodynamic effect on the cell unit, this can also be regarded as an adaptive strategy to increase the viable population during short warmer periods. By contrast, at extremely low temperatures or during long-term freezing survival, these bacteria evolve towards a dormancy state with minimal cellular metabolic activity aimed at preserving the cell's genetic program (Johnson *et al.*, 2007). Various exogenic protective mechanisms have also been proposed such as secreted exopolymers (Krembs & Deming, 2008) or particulate matter association (Junge *et al.*, 2004). From an ecological point of view, and in the context of a possible global warming, a rise in the environmental temperature would mainly result in the proliferation of bacteria such as *P. haloplanktis.*

14. Acknowledgments

We thank P. Leprince, G. Mazzucchelli, E. De Pauw, S. D'Amico, A. Danchin, M.L. Tutino and J. Renaut for their contribution to the experimental works and P. Charlier for preparing

Fig. 6. This work was supported by the F.R.S.-FNRS (Fonds National de la Recherche Scientifique, Belgium, FRFC grants to GF). FP was supported by the European Space Agency (Exanam-Prodex Experiment). The Institut Polaire Français Paul Emile Victor is also acknowledged for support at early stages of the work. F.P., A.G. and A.C. were FRIA research fellows and C.S. was a F.R.S.-FNRS research fellow.

15. References

Allen, M. A., Lauro, F. M., Williams, T. J., Burg, D., Siddiqui, K. S., De Francisci, D., Chong, K. W., Pilak, O., Chew, H. H., De Maere, M. Z., Ting, L., Katrib, M., Ng, C., Sowers, K. R., Galperin, M. Y., Anderson, I. J., Ivanova, N., Dalin, E., Martinez, M., Lapidus, A., Hauser, L., Land, M., Thomas, T. & Cavicchioli, R. (2009). The genome sequence of the psychrophilic archaeon, *Methanococcoides burtonii*: the role of genome evolution in cold adaptation. *ISME Journal* 3, 9, 1012-1035.

Ayala-del-Rio, H. L., Chain, P. S., Grzymski, J. J., Ponder, M. A., Ivanova, N., Bergholz, P. W., Di Bartolo, G., Hauser, L., Land, M., Bakermans, C., Rodrigues, D., Klappenbach, J., Zarka, D., Larimer, F., Richardson, P., Murray, A., Thomashow, M. & Tiedje, J. M. (2010). The genome sequence of *Psychrobacter arcticus* 273-4, a psychroactive Siberian permafrost bacterium, reveals mechanisms for adaptation to low-temperature growth. *Applied and Environmental Microbiology* 76, 7, 2304-2312.

Ayub, N. D., Tribelli, P. M. & Lopez, N. I. (2009). Polyhydroxyalkanoates are essential for maintenance of redox state in the Antarctic bacterium *Pseudomonas* sp. 14-3 during low temperature adaptation. *Extremophiles* 13, 1, 59-66.

Bakermans, C., Tollaksen, S. L., Giometti, C. S., Wilkerson, C., Tiedje, J. M. & Thomashow, M. F. (2007). Proteomic analysis of *Psychrobacter cryohalolentis* K5 during growth at subzero temperatures. *Extremophiles* 11, 2, 343-354.

Baldwin, R. L. (2008). The search for folding intermediates and the mechanism of protein folding. *Annual Review of Biophysics* 37, 1-21.

Bergholz, P. W., Bakermans, C. & Tiedje, J. M. (2009). *Psychrobacter arcticus* 273-4 uses resource efficiency and molecular motion adaptations for subzero temperature growth. *Journal of Bacteriology* 191, 7, 2340-2352.

Bosl, A. & Bock, A. (1981). Ribosomal mutation in *Escherichia coli* affecting membrane stability. *Molecular and General Genetics* 182, 2, 358-360.

Bowman, J. B. (2008). Genomic analysis of psychrophilic prokaryotes. In: *Psychrophiles, from Biodiversity to Biotechnology*. Margesin, R., Schinner, F., Marx, J.C., and Gerday,C. (eds), pp. 265-284, Springer-Verlag, ISBN 978-3-540-74334-7,Berlin, Heidelberg.

Cary, S. C., McDonald, I. R., Barrett, J. E. & Cowan, D. A. (2010). On the rocks: the microbiology of Antarctic Dry Valley soils. *Nature Reviews Microbiology* 8, 2, 129-138.

Casanueva, A., Tuffin, M., Cary, C. & Cowan, D. A. (2010). Molecular adaptations to psychrophily: the impact of 'omic' technologies. *Trends in Microbiology* 18, 8, 374-381.

Clarke, T. E., Tari, L. W. & Vogel, H. J. (2001). Structural biology of bacterial iron uptake systems. *Current Topics in Medicinal Chemistry* 1, 1, 7-30.

D'Amico, S., Collins, T., Marx, J. C., Feller, G. & Gerday, C. (2006). Psychrophilic microorganisms: challenges for life. *EMBO Reports* 7, 4, 385-389.

D'Amico, S., Marx, J. C., Gerday, C. & Feller, G. (2003). Activity-stability relationships in extremophilic enzymes. *Journal of Biological Chemistry* 278, 10, 7891-7896.

Danchin, A. (2007). An interplay between metabolic and physicochemical constraints: lessons from the psychrophilic prokaryote genomes. In: *Physiology and Biochemistry of Extremophiles.* Gerday, C. & Glansdorff, N. (eds.), pp. 208-220, ASM Press, ISBN 13 978-1-55581-422-9, Washington, D.C.

Deming, J. W. (2002). Psychrophiles and polar regions. *Current Opinion in Microbiology* 5, 3, 301-309.

Duchaud, E., Boussaha, M., Loux, V., Bernardet, J. F., Michel, C., Kerouault, B., Mondot, S., Nicolas, P., Bossy, R., Caron, C., Bessieres, P., Gibrat, J. F., Claverol, S., Dumetz, F., Le Henaff, M. & Benmansour, A. (2007). Complete genome sequence of the fish pathogen *Flavobacterium psychrophilum. Nature Biotechnology* 25, 7, 763-769.

Fan, D. J., Ding, Y. W., Pan, X. M. & Zhou, J. M. (2008). Thermal unfolding of *Escherichia coli* trigger factor studied by ultra-sensitive differential scanning calorimetry. *Biochimica et Biophysica Acta* 1784, 11, 1728-1734.

Feller, G. (2010). Protein stability and enzyme activity at extreme biological temperatures. *Journal of Physics - Condensed Matter* 22, 323101, doi 10.1088/0953-8984/1022/1032/323101.

Feller, G. & Gerday, C. (2003). Psychrophilic enzymes: hot topics in cold adaptation. *Nature Reviews Microbiology* 1, 3, 200-208.

Ferbitz, L., Maier, T., Patzelt, H., Bukau, B., Deuerling, E. & Ban, N. (2004). Trigger factor in complex with the ribosome forms a molecular cradle for nascent proteins. *Nature* 431, 7008, 590-596.

Garnier, M., Matamoros, S., Chevret, D., Pilet, M. F., Leroi, F. & Tresse, O. (2010). Adaptation to cold and proteomic responses of the psychrotrophic biopreservative *Lactococcus piscium* strain CNCM I-4031. *Applied and Environmental Microbiology* 76, 24, 8011-8018.

Gerday, C. & Glansdorff, N. (Eds.).(2007). *Physiology and Biochemistry of Extremophiles,* ASM Press, ISBN 13 978-1-55581-422-9, Washington, D.C.

Gilichinsky, D., Rivkina, E., Bakermans, C., Shcherbakova, V., Petrovskaya, L., Ozerskaya, S., Ivanushkina, N., Kochkina, G., Laurinavichuis, K., Pecheritsina, S., Fattakhova, R. & Tiedje, J. M. (2005). Biodiversity of cryopegs in permafrost. *FEMS Microbiology Ecology* 53, 1, 117-128.

Goodchild, A., Raftery, M., Saunders, N. F., Guilhaus, M. & Cavicchioli, R. (2004a). Biology of the cold adapted archaeon, *Methanococcoides burtonii* determined by proteomics using liquid chromatography-tandem mass spectrometry. *Journal of Proteome Research* 3, 6, 1164-1176.

Goodchild, A., Saunders, N. F., Ertan, H., Raftery, M., Guilhaus, M., Curmi, P. M. & Cavicchioli, R. (2004b). A proteomic determination of cold adaptation in the Antarctic archaeon, *Methanococcoides burtonii. Molecular Microbiology* 53, 1, 309-321.

Goodchild, A., Raftery, M., Saunders, N. F., Guilhaus, M. & Cavicchioli, R. (2005). Cold adaptation of the Antarctic archaeon, *Methanococcoides burtonii* assessed by proteomics using ICAT. *Journal of Proteome Research* 4, 2, 473-480.

Haikarainen, T. & Papageorgiou, A. C. (2010). Dps-like proteins: structural and functional insights into a versatile protein family. *Cellular and Molecular Life Sciences* 67, 3, 341-351.

Hartl, F. U. & Hayer-Hartl, M. (2009). Converging concepts of protein folding *in vitro* and *in vivo*. *Nature Structural and Molecular Biology* 16, 6, 574-581.

Johnson, S. S., Hebsgaard, M. B., Christensen, T. R., Mastepanov, M., Nielsen, R., Munch, K., Brand, T., Gilbert, M. T., Zuber, M. T., Bunce, M., Ronn, R., Gilichinsky, D., Froese, D. & Willerslev, E. (2007). Ancient bacteria show evidence of DNA repair. *Proceedings of the National Academy of Sciences U S A* 104, 36, 14401-14405.

Junge, K., Eicken, H. & Deming, J. W. (2004). Bacterial Activity at -2 to -20 degrees C in Arctic wintertime sea ice. *Applied and Environmental Microbiology* 70, 1, 550-557.

Kandror, O. & Goldberg, A. L. (1997). Trigger factor is induced upon cold shock and enhances viability of *Escherichia coli* at low temperatures. *Proceedings of the National Academy of Sciences U S A* 94, 10, 4978-4981.

Kandror, O., Sherman, M., Moerschell, R. & Goldberg, A. L. (1997). Trigger factor associates with GroEL *in vivo* and promotes its binding to certain polypeptides. *Journal of Biological Chemistry* 272, 3, 1730-1734.

Kandror, O., Sherman, M., Rhode, M. & Goldberg, A. L. (1995). Trigger factor is involved in GroEL-dependent protein degradation in *Escherichia coli* and promotes binding of GroEL to unfolded proteins. *EMBO Journal* 14, 23, 6021-6027.

Kawamoto, J., Kurihara, T., Kitagawa, M., Kato, I. & Esaki, N. (2007). Proteomic studies of an Antarctic cold-adapted bacterium, *Shewanella livingstonensis* Ac10, for global identification of cold-inducible proteins. *Extremophiles* 11, 6, 819-826.

King, J., Haase-Pettingell, C., Robinson, A. S., Speed, M. & Mitraki, A. (1996). Thermolabile folding intermediates: inclusion body precursors and chaperonin substrates. *FASEB Journal* 10, 1, 57-66.

Kramer, G., Patzelt, H., Rauch, T., Kurz, T. A., Vorderwulbecke, S., Bukau, B. & Deuerling, E. (2004). Trigger factor peptidyl-prolyl cis/trans isomerase activity is not essential for the folding of cytosolic proteins in *Escherichia coli*. *Journal of Biological Chemistry* 279, 14, 14165-14170.

Krembs, C. & Deming, J. W. (2008). The role of exopolymers in microbial adaptation to sea ice. In: *Psychrophiles: from Biodiversity to Biotechnology*. R. Margesin, F. Schinner, J. C. Marx & C. Gerday (eds.), pp. 247-264, Springer-Verlag, ISBN 978-3-540-74334-7, Berlin, Heidelberg

Margesin, R. & Feller, G. (2010). Biotechnological applications of psychrophiles. *Environmental Technology* 31, 8-9, 835-844.

Margesin, R., Schinner, F., Marx, J. C. & Gerday, C. (2008). *Psychrophiles, from Biodiversity to Biotechnology*, Springer-Verlag, ISBN 978-3-540-74334-7, Berlin, Heidelberg.

Martinez-Hackert, E. & Hendrickson, W. A. (2009). Promiscuous substrate recognition in folding and assembly activities of the trigger factor chaperone. *Cell* 138, 5, 923-934.

Medigue, C., Krin, E., Pascal, G., Barbe, V., Bernsel, A., Bertin, P. N., Cheung, F., Cruveiller, S., D'Amico, S., Duilio, A., Fang, G., Feller, G., Ho, C., Mangenot, S., Marino, G., Nilsson, J., Parrilli, E., Rocha, E. P., Rouy, Z., Sekowska, A., Tutino, M. L., Vallenet, D., von Heijne, G. & Danchin, A. (2005). Coping with cold: The genome of the versatile marine Antarctica bacterium *Pseudoalteromonas haloplanktis* TAC125. *Genome Research* 15, 10, 1325-1335.

Merz, F., Boehringer, D., Schaffitzel, C., Preissler, S., Hoffmann, A., Maier, T., Rutkowska, A., Lozza, J., Ban, N., Bukau, B. & Deuerling, E. (2008). Molecular mechanism and structure of trigger factor bound to the translating ribosome. *EMBO Journal* 27, 11, 1622-1632.

Merz, F., Hoffmann, A., Rutkowska, A., Zachmann-Brand, B., Bukau, B. & Deuerling, E. (2006). The C-terminal domain of *Escherichia coli* trigger factor represents the central module of its chaperone activity. *Journal of Biological Chemistry* 281, 42, 31963-31971.

Methe, B. A., Nelson, K. E., Deming, J. W., Momen, B., Melamud, E., Zhang, X., Moult, J., Madupu, R., Nelson, W. C., Dodson, R. J., Brinkac, L. M., Daugherty, S. C., Durkin, A. S., DeBoy, R. T., Kolonay, J. F., Sullivan, S. A., Zhou, L., Davidsen, T. M., Wu, M., Huston, A. L., Lewis, M., Weaver, B., Weidman, J. F., Khouri, H., Utterback, T. R., Feldblyum, T. V. & Fraser, C. M. (2005). The psychrophilic lifestyle as revealed by the genome sequence of *Colwellia psychrerythraea* 34H through genomic and proteomic analyses. *Proceedings of the National Academy of Sciences U S A* 102, 31, 10913-10918.

Mykytczuk, N. C., Trevors, J. T., Foote, S. J., Leduc, L. G., Ferroni, G. D. & Twine, S. M. (2011). Proteomic insights into cold adaptation of psychrotrophic and mesophilic *Acidithiobacillus ferrooxidans* strains. *Antonie Van Leeuwenhoek* 100, 2, 259-277.

Piette, F., D'Amico, S., Mazzucchelli, G., Danchin, A., Leprince, P. & Feller, G. (2011). Life in the cold: a proteomic study of cold-repressed proteins in the Antarctic bacterium *Pseudoalteromonas haloplanktis* TAC125. *Applied and Environmental Microbiology* 77, 11, 3881-3883.

Piette, F., D'Amico, S., Struvay, C., Mazzucchelli, G., Renaut, J., Tutino, M. L., Danchin, A., Leprince, P. & Feller, G. (2010). Proteomics of life at low temperatures: trigger factor is the primary chaperone in the Antarctic bacterium *Pseudoalteromonas haloplanktis* TAC125. *Molecular Microbiology* 76, 1, 120-132.

Qiu, Y., Kathariou, S. & Lubman, D. M. (2006). Proteomic analysis of cold adaptation in a Siberian permafrost bacterium *Exiguobacterium sibiricum* 255-15 by two-dimensional liquid separation coupled with mass spectrometry. *Proteomics* 6, 19, 5221-5233.

Rabus, R., Ruepp, A., Frickey, T., Rattei, T., Fartmann, B., Stark, M., Bauer, M., Zibat, A., Lombardot, T., Becker, I., Amann, J., Gellner, K., Teeling, H., Leuschner, W. D., Glockner, F. O., Lupas, A. N., Amann, R. & Klenk, H. P. (2004). The genome of *Desulfotalea psychrophila*, a sulfate-reducing bacterium from permanently cold Arctic sediments. *Environmental Microbiology* 6, 9, 887-902.

Riley, M., Staley, J. T., Danchin, A., Wang, T. Z., Brettin, T. S., Hauser, L. J., Land, M. L. & Thompson, L. S. (2008). Genomics of an extreme psychrophile, *Psychromonas ingrahamii*. *BMC Genomics* 9, 210. doi210.1186/1471-2164-1189-1210.

Robin, S., Togashi, D. M., Ryder, A. G. & Wall, J. G. (2009). Trigger factor from the psychrophilic bacterium *Psychrobacter frigidicola* is a monomeric chaperone. *Journal of Bacteriology* 191, 4, 1162-1168.

Rodrigues, D. F., Ivanova, N., He, Z., Huebner, M., Zhou, J. & Tiedje, J. M. (2008). Architecture of thermal adaptation in an *Exiguobacterium sibiricum* strain isolated from 3 million year old permafrost: a genome and transcriptome approach. *BMC Genomics* 9, 547.doi510.1186/1471-2164-1189-1547.

Rodrigues, D. F. & Tiedje, J. M. (2008). Coping with our cold planet. *Applied and Environmental Microbiology* 74, 6, 1677-1686.

Rosen, R. & Ron, E. Z. (2002). Proteome analysis in the study of the bacterial heat-shock response. *Mass Spectrometry Reviews* 21, 4, 244-265.

Russell, N. J. (2007). Psychrophiles: membrane adaptations. In: *Physiology and Biochemistry of Extremophiles*. C. Gerday & N. Glansdorff (eds), pp. 155-164, ASM Press, ISBN 13 978-1-55581-422-9, Washington, D.C.

Saunders, N. F., Goodchild, A., Raftery, M., Guilhaus, M., Curmi, P. M. & Cavicchioli, R., (2005). Predicted roles for hypothetical proteins in the low-temperature expressed proteome of the Antarctic archaeon *Methanococcoides burtonii*. *Journal of Proteome Research* 4, 2, 464-472.

Saunders, N. F., Ng, C., Raftery, M., Guilhaus, M., Goodchild, A. & Cavicchioli, R. (2006). Proteomic and computational analysis of secreted proteins with type I signal peptides from the Antarctic archaeon *Methanococcoides burtonii*. *Journal of Proteome Research* 5, 9, 2457-2464.

Saunders, N. F., Thomas, T., Curmi, P. M., Mattick, J. S., Kuczek, E., Slade, R., Davis, J., Franzmann, P. D., Boone, D., Rusterholtz, K., Feldman, R., Gates, C., Bench, S., Sowers, K., Kadner, K., Aerts, A., Dehal, P., Detter, C., Glavina, T., Lucas, S., Richardson, P., Larimer, F., Hauser, L., Land, M. & Cavicchioli, R. (2003). Mechanisms of thermal adaptation revealed from the genomes of the Antarctic Archaea *Methanogenium frigidum* and *Methanococcoides burtonii*. *Genome Research* 13, 7, 1580-1588.

Seo, J. B., Kim, H. S., Jung, G. Y., Nam, M. H., Chung, J. H., Kim, J. Y., Yoo, J. S., Kim, C. W. & Kwon, O. (2004). Psychrophilicity of *Bacillus psychrosaccharolyticus*: a proteomic study. *Proteomics* 4, 11, 3654-3659.

Siddiqui, K. S. & Cavicchioli, R. (2006). Cold-adapted enzymes. *Annual Review of Biochemistry* 75, 403-433.

Strocchi, M., Ferrer, M., Timmis, K. N. & Golyshin, P. N. (2006). Low temperature-induced systems failure in *Escherichia coli*: insights from rescue by cold-adapted chaperones. *Proteomics* 6, 1, 193-206.

Suzuki, Y., Haruki, M., Takano, K., Morikawa, M. & Kanaya, S. (2004). Possible involvement of an FKBP family member protein from a psychrotrophic bacterium *Shewanella* sp. SIB1 in cold-adaptation. *European Journal of Biochemistry* 271, 7, 1372-1381.

Ting, L., Williams, T. J., Cowley, M. J., Lauro, F. M., Guilhaus, M., Raftery, M. J. & Cavicchioli, R. (2010). Cold adaptation in the marine bacterium, *Sphingopyxis alaskensis*, assessed using quantitative proteomics. *Environmental Microbiology* 12, 10, 2658-2676.

Tosco, A., Birolo, L., Madonna, S., Lolli, G., Sannia, G. & Marino, G. (2003). GroEL from the psychrophilic bacterium *Pseudoalteromonas haloplanktis* TAC 125: molecular characterization and gene cloning. *Extremophiles* 7, 1, 17-28.

Tunsjo, H. S., Paulsen, S. M., Mikkelsen, H., L'Abee-Lund, T. M., Skjerve, E. & Sorum, H. (2007). Adaptive response to environmental changes in the fish pathogen *Moritella viscosa*. *Research in Microbiology* 158, 3, 244-250.

Valko, M., Morris, H. & Cronin, M. T. (2005). Metals, toxicity and oxidative stress. *Current Medicinal Chemistry* 12, 10, 1161-1208.

Williams, T. J., Burg, D. W., Raftery, M. J., Poljak, A., Guilhaus, M., Pilak, O. & Cavicchioli, R. (2010). Global proteomic analysis of the insoluble, soluble, and supernatant fractions of the psychrophilic archaeon *Methanococcoides burtonii*. Part I: the effect of growth temperature. *Journal of Proteome Research* 9, 2, 640-652.

Williams, T. J., Lauro, F. M., Ertan, H., Burg, D. W., Poljak, A., Raftery, M. J. & Cavicchioli, R. (2011). Defining the response of a microorganism to temperatures that span its complete growth temperature range (-2 degrees C to 28 degrees C) using multiplex quantitative proteomics. *Environmental Microbiology* 13, 8, 2186-2203.

Zheng, S., Ponder, M. A., Shih, J. Y., Tiedje, J. M., Thomashow, M. F. & Lubman, D. M. (2007). A proteomic analysis of *Psychrobacter articus* 273-4 adaptation to low temperature and salinity using a 2-D liquid mapping approach. *Electrophoresis* 28, 3, 467-488.

Part 3

Diverse Impacts in Plant Proteomics

6

Plant Protein Analysis

Alessio Malcevschi and Nelson Marmiroli
Division of Environmental Biotechnology,
Department of Environmental Sciences,
University of Parma, Parma
Italy

1. Introduction

The influx of data from the past ten years of large-scale plant genomes sequencing projects have yielded the sequence, complete or in its final assembly level, of several plant genomes, including *Arabidopsis thaliana, Oryza sativa, Zea mays , Brachypodium distachyon, Cucumis sativus, Populus trichocarpa, Medicago truncatula, Glycine max, Malus domestica, Physcomitella patens, Selaginella moellendorfii, Sorghum bicolour, Theobroma cacao, Vitis vinifera, Prunus pumice, Rricinus communis* and *Vigna radicata.* This knowledge, combined with the implementation of classical and innovative parallel high-throughput proteomic technologies associated to new protein search algorithms, has triggered a growing interest in plant proteomics to address a comprehensive analysis of cellular functions from the level of the plant to the whole organisms in different physiological and environmental conditions. A number of reviews have been recently written providing detailed insights into the basic lines of plant proteomics studies (Baginsky, 2009; Rose et al. 2004). In addition a number of initiatives such as the International Plant Proteomics Organization (INPPO) and The Plant Proteomics Database (PPDB) have been launched recently to organize the massive amount of information that emerged within the field of plant proteomics (Agrawal et al. 2011, Sun et al. 2009). Figure 1 highlights the rapid increase of scientific interest in plant proteomics that has occurred in the last ten years with model species including *Arabidopsis* (Van Norman & Benfey, 2009) and rice (Agrawal & Rakwal, 2011) which opened the way also for studying non-model plants species.

The majority of plant proteomics studies to date can be divided into two basic categories: the first involves protein annotation and profiling with the aim of separating and cataloguing as many proteins extracted from whole cells and organelles as possible to provide a snapshot of the major constituents of the proteome. The most notable examples of descriptive plant proteomics are studies carried out in different organs of *Arabidopsis* (Giavalisco et al. 2005, Baerenfaller et al. 2008, Joshi et al. 2011) and in rice (Agrawal et al. 2009, Koller et al.2002, Ferrari et al. 2011) where, respectively, 13,029 and 2,528 unique proteins have been identified from several tissues. However it should be noted that entire proteomes of single cell types cannot yet be fully mapped, as will be explained later, and to date the number of protein entries in the UniprotKB database for plant organisms is still limited to just above 500,000 which corresponds to less than 1/10th of the total number of entries (Schneider et al. 2009). The second category of proteome analysis aims at revealing

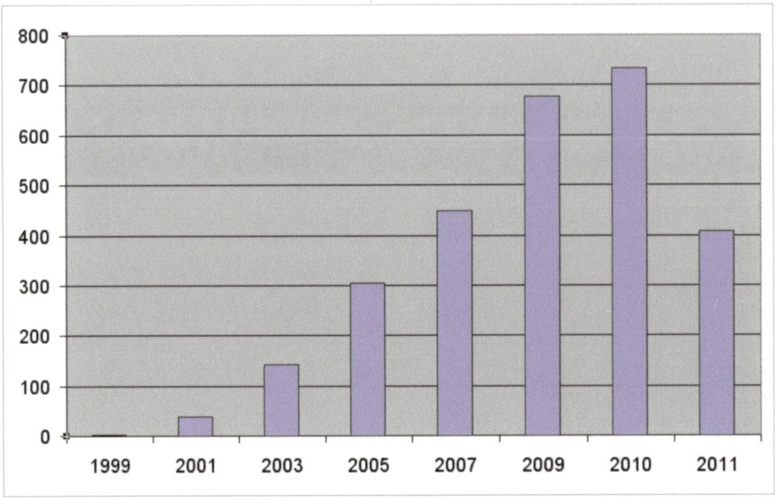

Fig. 1. Growth index of scientific papers using the word plant proteome within the search engine Pubmed (updated to July 2011).

changes in protein expression in response to physiological and environmental stimuli and is commonly termed comparative proteomics. It has been employed in a variety of studies including wood formation (Gion et al. 2005), response to cold stress (Neilson et al.. 2011) drought (Riccardi et al. 2004), heavy metal toxicity (Villiers et al.. 2011, Visioli et al. 2010a), flower development (Theissen et al. 2001) and seed development (Hajduch et al. 2005). Only a few plants have been intensively studied among the many plant species sequenced, including *Arabidopsis thaliana,* the first plant to be sequenced (Kaul et al. 2000), which has a short life cycle and it is easy to handle; rice (*Oryza sativa*), which is used as a model for cereal monocots (Matsumoto et al. 2005); maize (Schnable et al. 2009) and poplar which is being used as a model plant for woody species and for its economic and eco-physiological relevance (Tuskan et al. 2006). Although the analysis of the "green" proteome has grown rapidly we are still far away from an integrated understanding of plant proteome and identification of the role of the many proteins involved in cross-talk between cross-linked metabolic pathways. A challenge in comparative proteomics is the difficulty in delivering large-scale protein quantification (Schulze & Usadel 2010) to assay global protein changes elicited by biotic /abiotic events. A second problem is the inadequacy of current technologies for analysing a representative proportion of the expressed proteins present in a plant sample (Patterson 2004). This is mainly due to the dynamic range of protein concentrations within plant cells which is estimated to be as wide as 10^5-10^6 (Pattersons & Aebersold 2003). Abundant proteins such as RuBisCO (1,5-biphosphate carboxylase/oxygenase), the world's most abundant protein, can comprise up to 40% of total protein content in green tissue. The same is true for seed storage proteins or other housekeeping proteins which can be present at levels of 10^5 -10^7 molecules per cell. These highly abundant proteins hinder the detection of the low abundance proteins such as kinases, phosphatases, regulatory protein, transcription factors and rare membrane proteins whose concentrations are below 10-100 molecules per cell. To deplete the more abundant proteins from plant samples, many protocols require selective precipitations such as sucrose density gradient centrifugation or FPLC anion-exchange chromatography (XI et al. 2006). Unfortunately many of these approaches can be laborious, time consuming or require expensive equipment. Furthermore, proteomes are much more dynamic that genomes resulting in a considerable increase in complexity when gene expression is analysed at the protein level. While the human genome consists of approximately 30,000 genes, the corresponding proteome is expected to include between 200,000 and 2 millions proteins due to splicing and post-translational modifications (Gygi et al.1999). A similar situation is expected in plants. For instance thousands of phosphorylation sites have been characterized in plant proteins (Heazlewood et al. 2008) and it is likely that different post-translational modifications of storage proteins could explain the discrepancy between these proteins and the corresponding mRNAs abundances found in many plants living in temperate climates (Dai et al. 2007, Holdsworth et al. 2008, Rose et al. 2004). Recently protein arrays, which allow fast and parallel data analysis with miniaturization and automation, are emerging as a tool to supplement classical proteomics concepts. Protein arrays are able to profile and functionally characterize recombinant proteins encoded by globally or differentially expressed cDNA clones (Bussow et al. 2001,) or by the high-throughput sub-cloning of ORFs (Jahn et al. 2001).

2. Protein extraction

Isolation of intact total protein is the first and the most critical step toward any proteomics study, in fact analysis of plant proteomes present very specific problems when compared to other organisms. Proteins in plant cells are present at relatively low concentrations and constitute highly heterogeneous populations as a consequence of their functional diversity. Polypeptide molecular size, complexes (e.g. "clusters" or "modules" of interacting molecules that carry out cellular functions), spatial and time-dependent concentrations (e.g. proteins in the nucleus for transcription or in the mitochondrion for energy regeneration), charge (pI ranges from 3 to 12) proteins present in compartments like the cytosol or distinct organelles like the mitochondrion or plastid, to highly hydrophobic proteins embedded within the different cell membranes are some aspects of this complexity. As a consequence a multi-step procedure is often necessary to extract subsets of specific proteins. The key to protein isolation is the efficient solubilization of different protein types, including membrane proteins, with a minimum of handling time. The technique also needs to be suitable for downstream proteomics analysis procedures with minimal post-extraction artefacts and non-proteinaceous contaminants. The presence in plant cells of multiple interfering substances such as proteases, polyphenols, tannins, pigments, waxes, high carbohydrate/protein ratio further complicates the eventual extraction, solubilisation and separation procedures, that even under optimal conditions, results in the reduction of approximately 25% of the expected proteome (Patterson 2004). No single protein extraction protocol can capture an entire proteome, consequently a range of different extraction protocols, involving many permutations of physical and chemical treatments, solvent and buffers have been reported in literature (Rose et al. 2004, Baginsky 2009). A schematic outline of protein extraction methods is shown in Figure 2.

In some cases specialized protocols have been developed to extract a specific subset of proteins such as membrane or cell wall-associated proteins (Everberg et al. 2004). Specific mass spectrometry compatible protein extraction protocols have been developed (Sheoran et al. 2009). In addition, sequential extraction of tissues with a series of solvents can be effective in decreasing protein complexity and in enhancing the detection of low abundant proteins (Maltman et al. 2002). Extraction of plant proteins generally involves physical disruption by

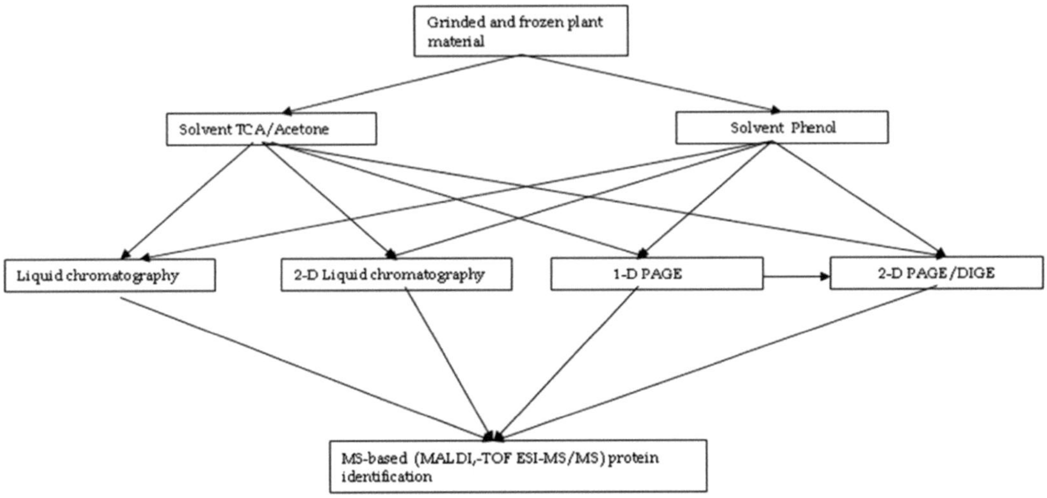

Fig. 2. Flow chart of proteins extraction methods which are highly dependent on cell type, tissue and organs to be analyzed.

mechanical means, grinding, sonication, chemical or enzymatic lysis of the cell and release of the contents into an extraction medium. Proteins are dissolved into a buffer solution as close as possible in composition to the original intracellular medium with respect to pH and ionic strength. To help protein solubilization, to protect them from hydrolysis or oxidation and to remove non-proteinaceous constituents from the aqueous extract, additional components are also added. Often subsequent separation and analytical steps may be intolerant of these additives: for instance inorganic salts may interfere in electrospray mass spectrometry, detergents in chromatographic and electrophoretic separations and in MALDI mass spectrometry, while protease inhibitors cocktails may interfere in the digestion of the proteins by trypsin. Thus it is essential to design extraction strategies with full knowledge of the nature and sensitivities of further processing and analytical steps. Two excellent and complementary methods currently in use to prepare a total plant protein extract are: i) trichloroacetic acid (TCA)/acetone precipitation and ii) phenol extraction in combination with different extraction buffers. Homogenization of the sample in 10% TCA dissolved in acetone almost immediately inactivates proteases and precipitates proteins, in addition it provides a means for delipidating membranes and releasing membrane associated proteins. This procedure also allows interfering substances to be washed out from the precipitated proteins and provides a clean sample for isoelectric focusing. While the TCA/acetone procedure is extremely effective for many plant tissues, particularly for young growing vegetative tissues, the method can sometimes result in the co-extraction of polymeric contaminants such polysaccharides and phenolic compounds. In this case the second protocol involving protein solubilization in phenol, with or without SDS, and subsequent precipitation with methanol and ammonium acetate is preferred (Hurkman & Tanaka 1986). A way to identify rare or hydrophobic proteins and increase the overall detectable proportion of the proteome is to reduce the protein complexity. Protein profiling of isolated organelles provides information about their enzymatic inventory and allows conclusions to be made about the compartmentalization of metabolic pathways. A number of studies have analyzed the proteomes of plant sub-cellular organelles including plastids (chloroplast, amyloplast, etioplast) (Baginsky et al. 2007, Ferro 2010), mitochondria (Heazlewood et al. 2004), vacuoles (Schmidt et al. 2007), peroxisomes (Reumann et al. 2007). A significant contribution of organelle proteomics to cell biology comes from the sub-cellular localization of protein and enzymes that can not be inferred from genome sequences. Chloroplast proteome analysis, for example, revealed that many proteins in the organelle were imported into the chloroplast via the secretory pathway, without a predictable N-terminal transit peptide (Friso et al. 2004). An exceptionally surprising finding given that it may explains why some chloroplast proteins are glycosylated (Villarejo et al. 2005). Similarly a recent proteome survey of *Arabidopsis* peroxisomes revealed the presence of unexpected proteins in the peroxisomal matrix. Additional validation with GFP-tagged proteins allowed the characterization of a novel peroxisomal targeting sequence (Reumann et al. 2007). A list of references of the most common extraction methods for different plant tissues is shown in table 1.

Tissue/organ	Extraction methods	reference
Suspension culture	TCA/Acetone	Laukens et al. 2007
cereal seeds	TCA/Acetone	Brandlard and Bancel 2007
Xylem and Phloem sap	TCA/Acetone	Kher and Rep 2007
Wood and other recalcitrant plant tissues	Phenol	Faurobert et al. 2007
chloroplasts	Sorbitol/Percoll	van Wijk et al. 2007
mitochondria	Mannitol/Percoll	Eubel et al. 2007
nucleus	Glycerol/Ficoll	Gonzales-Camacho and Medina 2007
cell wall	LiCl	Watson and Summer 2007
pollen	TCA/Acetone	Chen et al.2007
plasma membrane	Glycerol/Dextran/PEG	Santoni 2007

Table 1. Most common extraction methods of proteins from different plant tissues/organs

3. Protein separation

Two approaches have been generally used for analyses of plant proteins. Gel-based analysis methods involve the separation of proteins from a complex mixture and are typically accomplished by 2D-PAGE. With gel-free approaches, protein fractionation is carried out using liquid chromatography devices. Both techniques involve the subsequent identification and characterization of proteins by mass spectrometry. Initial analyses were carried out by separating protein samples in the first dimension using self constructed isoelectro focusing (IEF), followed by second dimension PAGE. In the last few years' reproducibility, sample loading and resolution of 2D gel electrophoresis have significantly improved with the introduction of immobilised pH gradient strips in the first dimension. After separation proteins are visualized by different staining techniques such as Silver staining and Coomassie Brilliant Blue (CBB) and quantified by densitometry. An example of plant sample arrayed by 2D- PAGE is outlined in figure 3.

By employing 2D-PAGE analyses it has been possible to analyse the rice and *Arabidopsis* proteomes (Kamo et al. 1995, Tsugita et al. 1994) and undertake comparative quantification

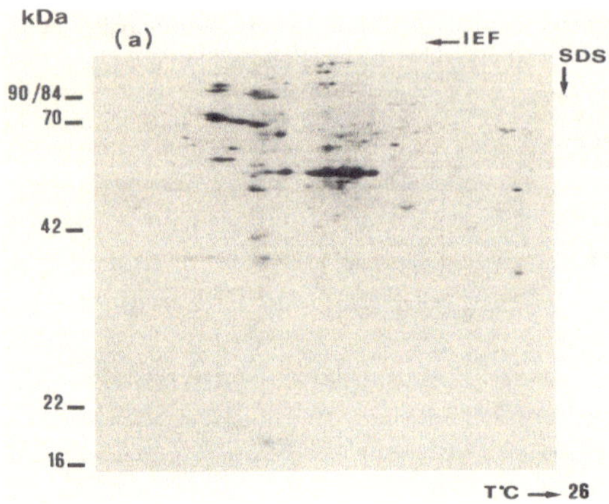

Fig. 3. Two-dimensional gel electrophoretogram of a total protein extract from barley tissue (Marmiroli et al. 1993).

of cold or salt-stressed plants and identify the responsive protein mediators of stress signal (Hajheidari et al. 2005). Recently 2D-PAGE has been used for establishing a protein reference map for soybean root hair cells (Brechenmacher et al 2009). Although 2D-PAGE is a robust and relatively straightforward technique and allows for the separation of up to 10,000 discrete proteins it has been criticized for being cumbersome and labor-intensive due to the time consuming process of image analysis and gel-to-gel variations that can complicate reproducibility (Taylor et al. 2011). Even with advanced 2D-PAGE analysis software, a high number of computationally generated 2D-PAGE spots have to be compared in a manual validation to get reliable accuracy (Hajduch et al. 2006). Moreover 2D-PAGE provides only a rough estimate of a proteins quantity due to variations in staining efficiency of individual gels and of its dependency on samples processing. 2D gels of plant proteins are also problematic due to post-translational modifications, such phosphorylation, glycosylation and myristoylation which cause proteins encoded by the same gene to migrate at different locations on the gel. The same holds true for multiple protein isoforms arrayed by 2D-PAGE. Low copy number proteins such as transcription factors, which are of considerable interest in plant biology, are liable to lie beyond detection limits of 2D-PAGE. Furthermore the number of spots resolved varies depending on the chosen tissue and plant species and often a single spot can contain multiple proteins species complicating protein identifications. Larger integral membrane proteins tend to be poorly soluble under common experimental conditions and are thus under-represented in the 2D-gels. Sometimes reactions of carbamylation, deamidation and isoaspartate formation occur during denaturing IEF resulting in changes in a proteins isoelectric point and causing horizontal strings of spots seen on 2D gels. 2D-PAGE is also notoriously difficult to automate which limits throughput and results in greater experimental variability. In addition, the 2D-PAGE approach is generally more suitable for analysis of soluble and peripheral membrane proteins. Recently proteome analyses have been performed using "gel less" procedures based entirely on liquid chromatography (LC). The main advantage of LC is that crude protein extracts can be analysed after few purification steps thus achieving a higher level of reproducibility than most of the chemical procedures, allowing a better comparison of protein patterns (Lambert 2005). The use of LC or two-dimensional liquid chromatography (2D-LC) separations is a robust methods for characterizing large numbers of total plant protein samples and proteins from plant organelles or sub-cellular compartments, followed by selective intact-protein analysis by MS (Pirondini et al. 2006) Among the different LC approaches a 2D-LC separation technique called PF-2D, based on chromatofocusing (CF) in the first dimension and high performance reversed phase (HPRP) liquid chromatography in the second dimension, has been recently developed allowing a fine separation of high amount of heterogeneous proteins. A dedicated software package then converts complex chromatograms of a large number of fractions into easily visualized 2-D maps, "virtual gels", in which pH is plotted against the retention time (Figure 4).

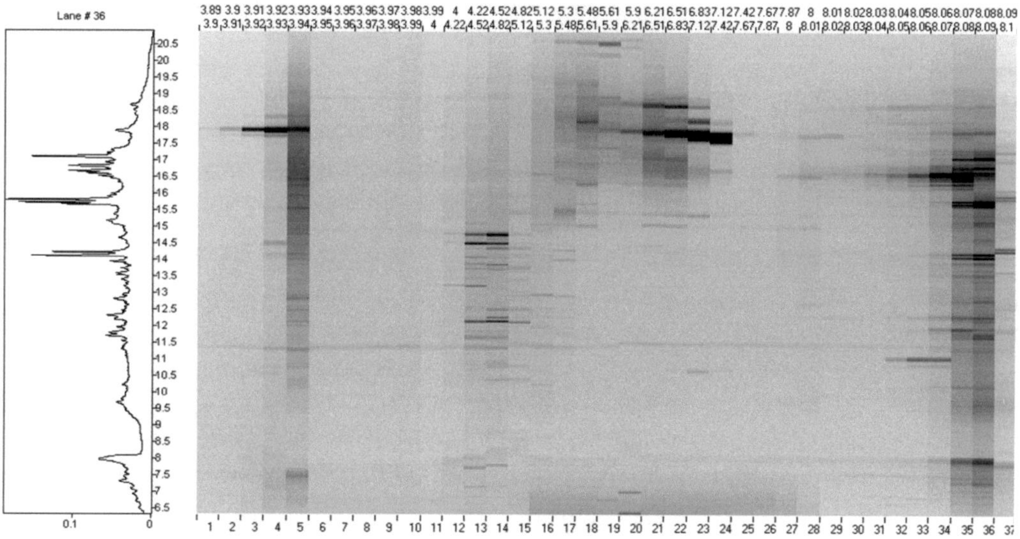

Fig. 4. PF-2D virtual separation gel of total protein extract from *Arabidopsis thaliana* (Pirondini et al. 2006)

In silico analysis of different "virtual gels" can be used to generate a complete catalogue of the qualitative and quantitative differences existing between different proteomes. Such an approach has been successfully applied to the identification of proteins involved in plant proteomic response to heavy metals and viruses (Larson et al. 2008, Visioli et al. 2010 b). Affinity chromatography has also demonstrated its potential in plant proteomics to overcome challenges associated with the enrichment of low-abundance proteins or to deplete high-abundance proteins. Many tags are currently used in plant protein purification including green fluorescent protein (Peckham et al. 2006), gluthatione S-transferase (Sridhar et al.2006), hexahistidine (Koroleva et al. 2009), maltose binding protein (Koroleva et al. 2009 To improve the purification of plant protein complexes new protein tags (TAP tags) based on Biotin carboxyl carrier domain have been developed (Qi & Katagiri 2009). One important application of these techniques has been the investigation of post-translational modifications (PTMs) in plant proteins, for example protein phosphorylation is one of the most extensively studied PTMs in plants where -immobilized metal affinity chromatography (Fe-IMAC) is widely used to enrich phosphopeptides from complex peptide mixtures (Kersten et al. 2009). The same approaches can be employed to study other PMTs such as glycosylation and ubiquitination (Morelle 2008). Affinity chromatography has also been applied to map protein-protein interactions by isolating protein complexes (Morris 2008).

4. Protein identification

Over the past decade the increasing availability of ESTs and genomic sequence data along with the rapid advances in MS have paved the way for a new era of protein identification and quantification. Generally two forms of mass spectrometry are used for protein identifications, both of which employ "soft" ionization techniques (Fenn 2002, Tanaka et al. 1988). The first is matrix-assisted laser desorption ionization (MALDI)-time of flight (TOF) mass spectrometry, used to perform peptide mass fingerprinting (PMF). The second is electrospray (ESI), which is usually coupled to high performance liquid chromatography (HPLC) sample separation, and is often used in tandem mass spectrometry to undertake peptide fragmentation. With the rapid increase in MS popularity, an assortment of instruments developed for different budgets and needs have become available (e.g. Waters, AB Sciex, Bruker Daltonics, Shimadzu, Agilent Technologies and Thermo Scientific). The improved mass accuracy, mass resolution and sensitivity allow for the rapid identification of picomoles or even femtomoles of proteins and peptides if matching genomic sequence data is available. The principle of mass spectrometry is outlined in figure 5.

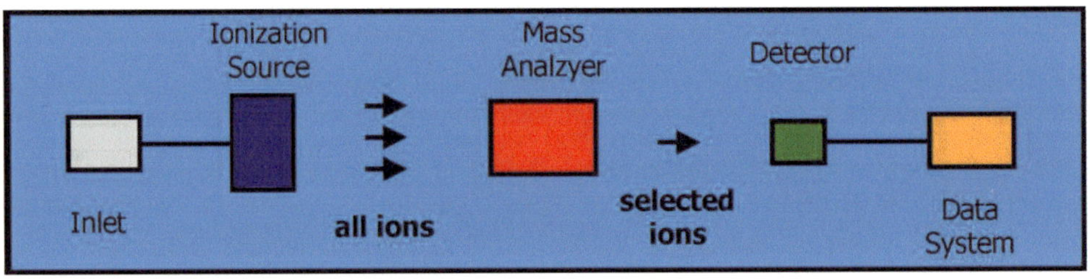

Fig. 5. Main functions of a mass spectrometer

4.1 Matrix assisted laser desorption/ionization (MALDI)

Among the different MALDI-based MS techniques, MALDI-TOF has been for many years the most widespread MS analysis approach. Though it is not the most rigorous approach to protein identification, it still represents an economically convenient alternative to more complex MS systems especially when proteomic analyses are carried out on plants species whose complete genome/protein databases are complete or well annotated. In typical MALDI- TOF analysis the first step is excision of 2-D gel plugs containing the selected protein spot of interest or a low-complexity fraction resulting from sample purification. The second step involves protein digestion (Shevchenko et al. 2007), with a site specific protease (e.g., trypsin or CNBr). The resulting mixture of ionized peptides is then mixed with a matrix solution of α-cyano-4- hydroxycinnamic acid (CHCA) whose function is to absorb most of the energy coming from a UV laser fired at the sample. Lighter ions travel faster in the TOF analyzer than heavier ions and thus the time taken to travel down the analyzer and reach the detector varies according their mass-to-charge ratio to produce a mass spectrum. Finally, the list of masses produced from the mass spectrum, is interrogated against a protein database (e.g. SwissProt, NCBInr) using a software package (e.g. MASCOT) with experimental mass accuracy of ca. 10 ppm. The peptide masses derived from the spectrum are compared to proteins in the database that have been "*in silico*" digested to produce a list of possible matches. This approach is referred to as peptide mass fingerprinting (PMF), it is relatively straightforward to perform and the spectra are usually simple to interpret. A scoring mechanism is employed to assess the likelihood of a correct identification. Robust protein identification requires the correct assignment of the molecular weights of at least four or five peptides. In absence of exhaustive protein or genomic databases information, large expressed sequence tag (EST) databases have been used for protein identification. An example of a peptide mass fingerprint experiment is shown in figure 6.

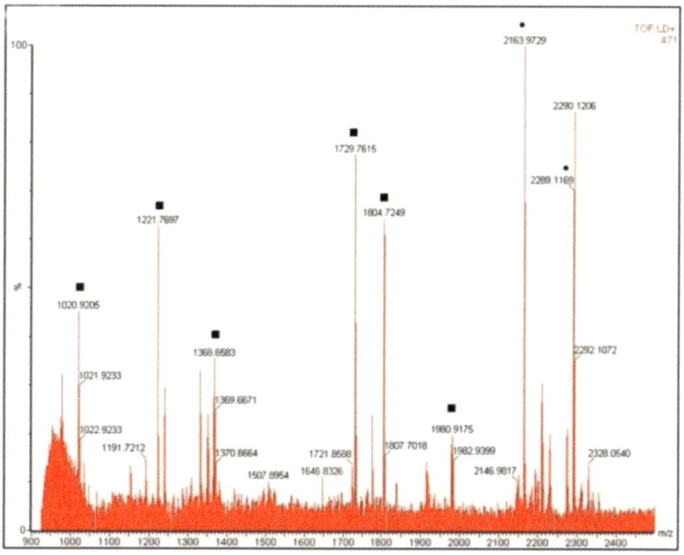

Fig. 6. MALDI-TOF MS spectrum of a Thlaspi protein (Visioli et al. 2010 a.).

The peaks marked with • represent trypsin autolysis peaks that were used to internally calibrate the mass spectrum. The peaks marked with ■ represent peaks identified as peptides produced by the trypsin digestion of a protein of interest and finally MS analysis of the eluted proteins. PMF has been used for proteome analyses from model as well as crop plants (Colas et al. 2010, Glinski & Weckwerth 2006, Hajduch et al. 2005, Mooney et al. 2004, Oeljeklaus et al. 2008). For instance MALDI-TOF mass spectrometry has been used in the characterization of the *Arabidopsis thaliana* proteome (Giavalisco et al. 2005). The authors report the identification of 2,943 spots from 2-DE from 663 different gene products. This is a small number considering there are more than 35,000 proteins coded by the *Arabidopsis* genome. A survey of the proteomes of six tissues from the model legume *Medicago truncatula* produced 2D-PAGE reference maps from which 551 proteins were identified (Watson et al. 2003). In this case, the overall successful identification rate was 55%, a figure that is considered good in absence of a fully sequenced genome, although the figure depended on the tissue in question. For example, identification was achieved for 43% of the proteins extracted from root tissue, while the figure for leaves was 76%. The difference presumably reflects the differences in the quality of the separations and the information in the databases and availability of ESTs. An investigation of soybean seed filling successfully provided 679 2-DE protein spots at five sequential developmental stages (Hajduch et al. 2005). Analysis of each of these protein spots by MALDI-TOF yielded the identity of 422 of these proteins, representing 216 non-redundant proteins. In nuclei isolated from rice suspension cell culture cell, from a total of 549 proteins resolved on 2-DE, 190 proteins were identified by MALDI-TOF MS from 257 major protein spots (Khan & Komatsu 2004). In *Populus nigra* cultivated under cadmium stress a subset of 20 out of 126 spots were identified by MALDI-TOF MS. Proteins that were more abundant in the metal exposed plants were located in the chloroplast and in the mitochondrion, suggesting the importance of these organelles in response and adaptation to metal stress (Visioli et al. 2010b). MALDI-TOF MS has been used also for the identification of differentially expressed proteins of rice leaves in presence of arsenic (Ahsan et al. 2010) and alteration of barley root proteome in response towards salt stress conditions (Witzel et al. 2009). To reduce the influence of ion-suppression effects in MALDI-TOF/MS measurements and obtain more peptide peaks, separation of the tryptic peptides can be obtained using an off-line combination of capillary reverse-phase HPLC column with MALDI-TOF. LC-MALDI techniques do not suffer from the time constrains imposed by the transient presence of peptides eluting from a column and each sample can be analyzed more than once. LC-MALDI has been used for instance for identification of proteins involved in different plant signaling processes (Karlova et al. 2006). The beneficial features of MALDI have led this ionization technique to be incorporated into tandem instruments such as those with quadrupole ion trap/TOF, quadrupole TOF and TOF/TOF geometries. The advantage of hyphenated MS over single MALDI-TOF fingerprinting is that the precise sequence of amino acids in each peptide can be determined, allowing a more reliable identification. Examples of application in plant proteome analysis of tandem MALDI vary from characterization of *Medicago truncatula* cell wall proteome (Gokulakannan & Niehaus 2010), to the analysis of the glycoproteome of tomato and barley (Català et al. 2011). Identification of proteins involved in metabolic pathways affected by different cropping regimes (Nawrocki et al. 2011), in Cadmium response in poplars (Kieffer et al. 2008) and in salt stress effect on sorghum leaves (Swami et al. 2011) were also carried out taking advantage of hyphenated MALDI MS analysis. An emerging technique in plant biology based on MALDI, and made possible because of advances in instrumentation, is MALDI-imaging MS (MSI). This technique can be applied at both the tissue and single-cell level providing information on spatial distribution of specific molecules (Kaspar et al.2011). Whereas many plant metabolite profiles have been described so far, no comparable plant protein analyses are available; the only application of this technique is the identification of a precursor of a secreted peptide hormone identified in *Arabidopsis* (Kondo et al. 2006). To summarize MALDI-TOF analysis is extremely fast with regard to data acquisition, requires little expertise, is tolerant to contaminants such as salts and detergents, is easy to automate and allows the analysis of large number of samples in a short period of time, the protein identification relies purely upon the matching of the peptide masses accurately, and it can be relatively inexpensive. Unfortunately the data can be ambiguous and rely heavily on availability of a proteomic or genomic sequence or at least a substantial EST collection for the species being studied. Cross-species PMF studies from four plant species (Mathesius et al. 2002) for instance concluded that PMF data are not particularly useful for inter-species protein identification except for the highly conserved proteins.

4.2 Electrospray ionization (ESI)

A different method for protein identification by mass spectrometry is peptide fragmentation by means of electrospray ionization tandem mass spectrometry (ESI-MS/MS). The technique provides structural information about the peptide which can be used for more reliably protein identifications when analysed against protein databases (Grossmann et al. 2005). The first step of tandem MS involves sample digestion (e.g. with trypsin), the resulting peptides are loaded onto an HPLC coupled to an ESI mass spectrometer which allows the analysis of ionized molecules in solution. During electrospray ionization peptides enter the ion source as a fine mist of droplets via a needle which is surrounded by an accompanying flow of nitrogen gas. A high voltage is applied to the needle through which the solution arrives in the source causing the droplets produced to be charged on the surface. This whole process results in the ions being released from the liquid droplet to produce gas phase ions that are drawn into the first mass analyzer and separated according to their mass-to-charge ratio. ESI is commonly used as ionization technique in tandem mass spectrometry (MS/MS) which adds a second dimension to mass spectrometric selection improving the specificity of the technique and allowing structural analysis of peptides. Multiple stages of mass analysis separation can be accomplished by individual mass analyzer elements separated in space by a fragmentation cell. Examples include TOF, Fourier-transform ion Cyclotron Resonance (FTCIR), ion trap, quadrupole, orbitrap and linear quadrupole ion trap. (Cotter et al. 2007, Douglas et al. 2005, Hardman& Makarov 2003, Marshall et al. 1998). The first mass analyzer detects the whole spectrum of peptide ions present in the sample (MS scan) then precursor ions of interest are fragmented by collision, inside a collision cell, with inert gas molecules (e.g. argon or nitrogen) in a process called collision-induced dissociation (CID) to produce a fragmentation spectrum of the selected peptide. This process produces a series of fragments ions that can differ by single amino acids, allowing a portion of the peptide sequence subsequently used in a bottom-up approach for protein identification by database interrogation. Tandem mass spectrometry has been used to analyze proteomes of *Arabidopsis thaliana, Oryza sativa* and *Medicago truncatula* by taking advantage of their extensively sequenced and/or annotated genomes and proteomes. For instance the most extensive plant proteomic analysis reported to date was conducted with *Arabidopsis thaliana* and led to the identification of 13,029 proteins on the basis of 86,456 unique peptides which represent approximately the 50% of the predicted expressed genes (Baerenfaller et al. 2008). Due to the limited applicability of MALDI-TOF to study the proteomes of organisms with un-sequenced genomes *de novo* sequence data derived from peptide fragmentation has been particularly useful for proteome analysis of non-model plants. *De novo* sequences can be searched against protein databases of relatives of the organism under investigation using MS-BLAST on the basis of close protein identities. *De novo* sequencing in plant proteomics has been employed for the analysis of barley thylakoid membrane proteins (Granvogl et al. 2006), proteome analysis of opium poppy cell cultures (Zulak et al. 2009), oak (*Quercus ilex)* (Jorge et al.2006*)* and banana (Liska & Scevchenko 2003). The combination of LC-MS/MS analysis with new single and two-step affinity purification methods of plant protein has triggered the interest for the isolation and characterization of plant protein complexes (Pflieger et al.2011). Even if the results are far from exhaustive and the structure-function relation of these protein assemblies are still poorly understood, the identification and characterization of these plant complexes are necessary to fully understand the cellular dynamics and homeostasis. Over the past decade MS techniques have advanced and alternative non-gel approaches have developed to address technical limitations inherent in 2D-PAGE/MS/MS. This "shotgun" approach, referred to as multidimensional protein identification technology (MudPIT) (Link et al. 1999), consists of a two-dimensional chromatography separation, prior to electrospray mass spectrometry followed by database searching. Shot gun proteomics refers to direct and rapid analysis of the entire protein complement of whole organelles, cells and tissues starting from chemical or enzymatic digestion of proteins to generate a highly complex set of peptides that is well beyond the separation capacity of 2D-PAGE. The theoretical peak capacity of MudPIT system has been calculated to be ca. 23,000 proteins (Wolters et al. 2001) making this system a powerful tool for proteomics. The rationale behind this method is that since the properties of peptides are more approachable than proteins, standardized protocols can be developed to face with proteome wide measurements by means of peptide analysis only. Sample preparation is relatively straightforward, the proteins are denatured, the cysteines reduced and alkylated and then the proteins are digested producing complex mixture of peptides. Peptides are then separated prior to analysis by tandem MS. The first dimension is normally a strong cation exchange (SCX) column with high loading capacity and high- resolution separation capacity. Peptides are stepped from the cation exchanger in a series of salt steps that increase in concentration onto the second dimension a reverse phase chromatography (RP) column. A subsequent RP gradient separates the eluting peptides relative to their hydrophobicity and delivers them, after each salt step, into a tandem mass spectrometer for selection and fragmentation. In contrast to the traditional 2-DE/MS/MS approach the shotgun method is largely unbiased providing a strategy for the efficient detection of low-abundant and hydrophobic proteins. A typical qualitative shotgun plant protein analysis in the range of 200 to 1,000 proteins for plants such as rice and *Arabidopsis thaliana* is theoretically achievable (Froehlich et al. 2003). Application of this "shotgun" approach has allowed the identification of more than 1,000 distinct proteins from rice leaf and root samples (Breci & Haynes 2007) and 294 ubiquitines in *Arabidopsis thaliana* (Maor et al. 2007). Shotgun protein analysis has led also to the identification of 44 differentially expressed proteins, out of a set of 3,004 non-redundant proteins previously identified, in the rice *reduced culm number1* mutant when compared to wild-type rice (Lee et al. 2011).

5. Protein quantitation

Determination of relative abundances of proteins in organisms or tissues subjected to a variety of environmental or physiological conditions is the final goal of any plant proteomic study. Techniques such as difference gel electrophoresis (2D-DIGE) (Timms & Cramer 2008,) which permit changes in protein abundance to be more readily assessed, has partially overcome limitations caused by inter-gel variations. Another advantage of 2D-DIGE is that it requires low amounts of protein (0.025 mg) as compared to the requirements of standard 2D-PAGE (ca. 0.2 to 1 mg) This technique involves covalent labelling of two different protein samples with fluorescent cyanine dyes (for example, Cy2, Cy3 and/or Cy5 which fluoresce at different wavelengths) prior to two-dimensional electrophoresis and produce sub-nanogram sensitivity. The intensity of fluorescence at each of the wavelengths for Cy3 and Cy5 is measured and after employing gel matching software, intensity ratios are used to evaluate relative abundance of proteins in the two different samples. A variety of plant proteomic studies have used DIGE (Granlund et al. 2009, Schenkluhn et al. 2010) to investigate abiotic stresses such as freezing, effect of UV on maize, aluminium stress in tomato, the effects of abscissic acid (ABA) and beta-aminobutyric acid (BABA) on *Malus pumila*. Generally, MS analysis of proteins by MALDI or LC-MS/MS is not quantitative because of the different physical and chemical properties of the tryptic peptides: difference in charge state, ionization competition, peptide length, non-homogeneous sample introduction, amino acid composition or post-translational modification and limitations in sample handling all result in variations in ion intensity for the peptides even when they belong to the same protein. As a consequence MS signals are notoriously variable, unpredictable, and therefore a potential source of significant error in quantitative proteomic studies. Despite these hurdles a number of comparative strategies have been adopted and have been categorized as either stable-isotope-labelling or label free approaches. Protein quantification by means of stable-isotope-labelling is based on the fact that when a peptide is labelled with different isotopic mass tags (^{2}H, ^{13}C, ^{15}N, ^{18}O) it differs from the unlabeled peptide only in terms of its mass but exhibits the same chemical properties during chromatography. In MS spectra obtained from peptide samples after their chromatographic separation, the ratio of MS signal intensities or peak areas of differentially labelled species extracted from the relative mass spectra between the labelled and unlabelled peptide permits an accurate relative quantification of differences. Labelling can be introduced at different steps during sample preparation. In metabolic labelling whole cell or organisms are labelled *in vivo* through the growth medium. In chemical post extraction labelling the isotopic modification is added to proteins or tryptic peptides through a chemical reaction. The most common strategies for chemical labelling include isotope-coded affinity tag (ICAT) and isobaric tag for relative and absolute quantitation (iTRAQ). In the ICAT method two protein mixtures representing two cell states are treated with different reagents consisting of a biotin affinity tag, heavy and light isotopologues and a cysteine-reactive group. To minimize error both samples are then pooled, digested with a protease and subjected to avidin affinity chromatography to isolate labelled biotinylated peptides in order to reduce the sample complexity by about 10-fold. Subsequently LC-MS/MS analysis is performed to determine the abundance ratio for each identified peptide. So far there have been few reports using ICAT in plant proteomics (Dunkley et al. 2004, Majeran et al. 2005) due to the fact that ICAT can only distinguish between protein samples containing cysteine, a rare amino acid present only in a fraction of proteins or peptides. In plant proteomics studies ICAT was employed to study organellar proteomes using fractionation of cellular organelles in *Arabidopsis thaliana* and maize (Dunkley et al. 2004, Majeran et al. 2005). ICAT has also been used in a recent study with proteins from solubilized mitochondria of *Arabidopsis thaliana* in order to investigate protein complexes (Hartman et al. 2007). A similar method, called Isotope Coded Protein Label (ICPL), based on labelling of more frequent amino acid groups has been developed (Kellermann 2008) and it has been used for comparative quantification of cell-wall proteins of *Medicago truncatula* plants interacting with nitrogen-fixing bacteria *Rhizobia* (Hahner et.al 2007). The iTRAQ method (Applied Biosystems) is based on isobaric tags, i.e. tags that have the same mass and are primarily designed for the chemically labelling the N-terminus of peptides generated from protein digests that have been isolated from cells in, for example, two different physiological conditions. The labelled samples are combined, fractionated by nanoLC and analyzed by tandem mass spectrometry. Database searching of peptide fragmentation data results in the identification of the labelled peptides and hence of the corresponding proteins. Fragmentation of the tag attached to the peptides generates a low molecular mass reporter ion that is unique to the tag used to label each of the digests. The reporter ion intensities enable relative quantification of the peptides in each digest and hence the proteins from which they originate. In quantitative plant proteomics iTRAQ has been used to quantify 45 proteins from *Arabidopsis thaliana* cells treated with bacterial pathogen *Pseudomonas syringae* (Kaffarnik et al. 2009). Labelling with iTRAQ has been also used for investigating the proteome of guard cells in *Arabidopsis thaliana* mutants impaired in the Gα subunit of GPA1 in order to understand the signalling role played by trimeric G-proteins in plants. (Zhao et al. 2010). This study has allowed the identification of 18 proteins which are differentially expressed in the mutant. These proteins included ATP synthase, enzyme of the Calvin Cycle and proteins involved in the stress response. In a study of grape proteomes at different stages of ripeness identification of between 1,000 and 1,400 proteins, 91 in the exocarp and 58 in the mesocarp were up-regulated during fruit maturation (Lucker et al. 2009). iTRAQ has been also widely used to study phosphoproteomics responses of elicitor treatment by comparing several time points post-treatment (Nuhse et al. 2007), protein degradation in chloroplasts and developmentally induced changes in chloroplasts proteomes in maize and *Brassica*. Unlike chemical labelling which is typically applied after protein extraction, fractionation and digestion, metabolic labelling takes place at the very first stage, i.e. at the level of protein biosynthesis. In this quantification procedure, called stable isotope labelling by amino acid in culture (SILAC), labelled essential amino acids (usually deuterated leucine) are added to amino acid deficient culture media, and thus become incorporated into all proteins. In general, SILAC has the advantage of a simpler analysis compared with metabolic labeling with ^{15}N. Usually, a single amino acid is used for SILAC. If the supplied amino acid is Lys or Arg, analysis of peptides from a trypsin digest that cleaves after these two amino acids will result in peptides containing only a single difference from the labeled amino acid. Therefore, the mass difference between peptides in the MS scan will be known and consistent. Experimental cell populations are treated in a specific way, such as cytokine stimulation, with different isotopologues then protein populations are harvested and compared. Because the label is embedded directly into the amino acid sequence of every protein, the extracts can be pooled directly. Purified proteins or peptides will preserve the exact ratio of labelled to unlabelled proteins as no more synthesis is taking place. Relative quantitation takes place at the level of the peptide mass spectrum or peptide fragment mass spectrum exactly as in any other stable isotope method by calculating the MS peak intensity, or area, ratio of the light and heavy peptides. In some plants SILAC gives label incorporation of approximately 70% which is not satisfying for many global proteomics applications, this is because plants are very versatile autotrophs and are able to generate all the 20 amino acids necessary for protein synthesis. The other disadvantage of SILAC is that the labelled amino acids are expensive when used in amounts needed for efficient labelling, so this method is likely to be limited to plant cell cultures. The only organisms of the plant kingdom that have been efficiently SILAC labelled are auxotrophic mutants of *Chlamydomonas* (Naumann et al. 2007) and cultured cells of *Arabidopsis* (Grhuler et al. 2005). Nevertheless the ability of plants to synthesize amino acids from inorganic salts provides an opportunity for a simpler labelling strategy. The use of ^{15}N-KNO_3 was first used successfully in potato plants where 98% of the total protein was labelled with ^{15}N. *Arabidopsis thaliana* plants can be also be successfully labelled because it does not affect plant development (Ippel et al. 2004). In another study hydroponic isotope labelling of entire plants was used for relative protein quantification of seven-week-old *Arabidopsis thaliana* plants treated with oxidative stress (Bindschedler et al. 2008). Label-free quantification strategies are becoming increasingly popular to compare samples (Schulze and Usadel 2010). The rationale behind these methods relies on the comparison of peptide abundance as a measure for the corresponding protein between multiple LC-MS/MS analyses (Proll et al. 2007). Ideally samples for label-free comparisons are run consecutively on the same LC-MS/MS setup to avoid variations due to the system setup (column properties, temperatures) and thereby allow precise reproduction of retention times. Label free approaches are inexpensive with high proteome coverage of quantified proteins since every protein that is identified by one or more peptide spectra can be quantified. There are currently two different label-free strategies which use either MS1 precursor ion (i.e. MS survey scan) data or MS2 tandem mass spectrometry data (i.e. MS/MS) to estimate changes in relative abundance or proteins between samples. The MS1 based methods associate changes in relative protein abundance from direct measurement and comparison of the mass spectrometric signal intensity of peptide precursor ion belonging to a particular protein (Wiener et al. 2004). The rationale in this approach is that the height or peak area with a given m/z is a measure of the number of ions of that particular mass detected within a given time interval. This process of determining the peak area is referred to as ion extraction and results in a so-called extracted ion chromatogram of a given ion species. Such extracted ion chromatograms can be produced for each m/z across all the LC-MS/MS runs within an experiment, and the resulting peak areas can then be compared quantitatively provided that only the same ion species can be compared between samples due to the differences in ionization efficiency among different peptide species. On the other hand the MS2 based methods estimate differences in relative protein expression by either accounting for the extent of protein sequence coverage or the number of tandem mass spectra generated, a technique also known as spectral counting (Zybailov et al 2009). This quantitation method does not require any protein labelling and uses a simple additive procedure for quantitative evaluation and does not rely on chromatographic peak integration or retention time alignment. The relative quantification through spectral counting is achieved by comparing the number of MS/MS spectra for the same protein between two or more MS/MS analyses. The absolute concentration of each protein within the sample is derived from an exponentially modified abundance index (emPAI) which is calculated from the number of observed spectra for each protein divided by the number of possibly observable peptides, a fraction that has been described as a protein abundance index (PAI) (Rappsilber et al. 2002). The emPAI index along with another similar index for protein expression profiling (APEX) have been used to analyze differential protein expression in root nodules of *Medicago truncatula* (Larrainzar et al. 2007) in response to drought and to determine the abundance of stromal proteins in chloroplasts from *Arabidopsis thaliana* (Zybailov et al. 2008). Analysis of sucrose-induced changes in the phosphorylation levels of *Arabidopsis* plasma membrane proteins has been also carried out by exploiting spectral counting (Niittyla et al. 2007).

6. Concluding remarks and future perspectives

Proteome analysis along with profiling tools such as transcriptomics and metabolomics are becoming essential components of the emerging "systems biology" approach. It is clear from most of the current literature (Ning et al. 2011) that all proteomics including plant proteomics are changing in scale and focus, from their initial objective of identifying as many individual proteins as possible in a given biological sample to the development of high-throughput parallel and quantitative technologies for analyzing proteomes in a dynamic context. Methods such as metabolic labelling using, for instance, CO_2 via photosynthesis or inexpensive nitrogen salts in protein synthesis offer new ways to quantify plant proteomes and can even be exploited for labelling organisms that feed on plant materials. Several proof-of-principle studies have demonstrated the linearity and/or reproducibility of label-free quantification for the analysis of complex mixtures (Wang et al. 2003). Comparative studies have also shown that results obtained with both methods are generally in good accordance (Wienkoop et al. 2006), with spectral counting covering a slightly higher dynamic range and measurements of ion abundance being more accurate for the determination of protein ratios. With the development of modern high-precision mass spectrometers, the label free quantification is becoming an appealing alternative as mass accuracies increase and the reliability of mapping peptides across samples due to more narrow mass-to-charge windows. However reproducibility of the retention times over different LC-MS/MS runs remains crucial for precision in label-free quantification using peptide ion intensities. In addition evaluation of proteomics data is facilitated if experimental variations are minimized between experiments. In this context plants are also well-suited experimental organisms for achieving lower statistical variability through their clonal reproduction and their ability to grow in highly standardized and controlled environments. Not surprisingly most quantitative plant proteomics studies performed so far have utilized *Arabidopsis thaliana* as model organism. This plant has excellent features for proteomics studies, including: its genome is fully sequenced, genetic mutants for comparative experiments are available, it has a relatively short life cycle and can conveniently be cultivated under laboratory conditions, making it readily amenable to metabolic stable isotope labelling. With the completion of further plant sequencing projects and the advent of high-throughput global proteome analysis via non-gel-based shotgun, proteomics studies will become more and more appealing for an increasing number of plant species. Moreover, the combination of new intriguing methods in quantitative MS with biochemical, biological and genetic approaches are adding new dimensions to the characterization of cellular processes resulting in improved knowledge of (plant) biological systems. This is exemplified by the combinatorial use of advanced protein quantification strategies and elaborate phosphopeptides enrichment techniques (e.g. LC-FTCIR-MS), which have promoted phosphoproteomics as a tool with extraordinary potential for spatio-temporal analysis of entire signalling pathways in plants. The main current bottleneck in plant proteomic studies is still the wide dynamic range of proteins. Global abundance measurements in *Saccharomyces cerevisiae* have revealed a bell-shaped distribution of proteins spanning approximately six orders of magnitude in abundance (Ghaemmaghami et al. 2003), while only approximately three or four orders of magnitude can be covered by modern LC-MS/MS methods for complex samples. Proteins identified represent only a small fraction of the complete proteome or sub-proteome of plants and organelles. For this reason proteome fractionation and intelligent strategies of enrichment of protein targets have to be developed. For example the estimated total number of genes in the rice genome lies in the range of 32,000 to 50,000 for *Oryza japonica*, whereas the comprehensive display analysis of rice leaf, root and seed tissue using 2-DE followed by tandem MS and MudPIT have led to the identification of 5.1% to 7.9% of the expected number of protein. These data clearly demonstrate that further developments are needed to increase the resolving power of this method to allow the detection of the low abundance proteins present in the "extractome". On the technical side improvements in pre-electrophoretic fractionation and in mass spectrometry scan speed will likely contribute to deeper proteome coverage in the future. For example an atmospheric MALDI (APMALDI) has been developed (Doroshenko et al. 2002) which is relatively simple to interface to mass analyzers. Surface enhanced desorption ionization (SELDI) has been shown to be very powerful for selective ionization of peptides and protein fractions, although it has not applied to plant systems so far (Poon 2007). Another challenge in the large-scale, quantitative plant proteomics experiments lies in the application of new data-mining strategies. Irrespective of the applied methods for protein identification, advanced bioinformatics and statistical tools for data evaluation are essential to extract biologically meaningful data from the plethora of qualitative and quantitative information obtained in global-scale experiments. Recently a single, centralized, authoritative resource for protein sequences and functional informatics, UniProt has been created by joining the information contained in the SwissProt, Translation of the EMBL nucleotide sequence (TrEMBL) and the protein Information Resource-Protein Sequence Database (PIR-PSD) (Schneider et al. 2004). To conclude, qualitative and quantitative plant proteomics, especially MS-based proteomics, will be applied to more and more non-model plant species for comprehensive and in-depth characterization of plant-environment interactions and plant growth and differentiation to provide more reliable basis to the emerging phenomena of phenotypic plasticity and epigenetic variation.

7. Acknowledgment

The authors gratefully acknowledge Caterina Agrimonti, Mariolina Gulli, Elena Maestri, Marta Marmiroli, Giovanna Visioli for their support in the preparation of the manuscript and University of Parma for providing financial support in buying the Lab PF-2D and for access to MALDI-TOF/MS and Orbitrap facilities at CIM (Centre for Interfaculty Measures) necessary for the plant proteomic studies carried out in our laboratory. Thanks also to Foundation, AGER and RISINNOVA for the financial support of the project "Integrated genetic and genomic systems for updating rice varieties along the Italian rice food chain".

8. Abbreviations

2D-PAGE, two dimensional gel electrophoresis; 2D-LC, two dimensional liquid chromatography; APEX, absolute protein expression; CF, chromatofocusing; CNBr, cyanogen bromide; DIGE, difference gel electrophoresis; emPAI, exponentially modified abundance index; ESI, electrospray; ESTs, expressed sequence tags; FTCIR, Fourier transform ion cyclotron resonance; HPLC, high performance liquid chromatography; HPRP, high performance reversed phase; ICAT, isotope-coded affinity tag; ICPL, isotope coded protein label; IEC, ion exchange liquid chromatography; IEF, isoelectro focusing; IMAC, immobilized metal affinity chromatography; iTRAQ, isobaric tag for relative and absolute quantitation; LC, liquid chromatography; MALDI, matrix assisted laser desorption ionization; MS, mass spectrometry; MudPIT, multidimensional protein identification technology; PF-2D, two dimensional protein fractionation; PMF, peptide mass fingerprinting; PTM, post-translational modifications; SELDI, surface enhanced desorption ionization; SILAC, stable isotope labelling by amino acid in culture; TOF, time of flight; TPP, three phase partitioning; TrEMBL , translated European molecular biology laboratory.

9. References

Agrawal, G.; Jwa, N.S.; Rakwal, R. (2009). Rice proteomics: ending phase I and the beginning of phase II. *Proteomics*, 9, 935-963

Agrawal, G.K.; Rakwal, R. (2011). Rice proteomics: a move towards expanded proteome coverage to comparative and functional proteomics uncovers the mysteries of rice and plant biology. *Proteomics*, 11, 1630-1649

Agrawal, G.K.; Job, D.; Zivy, M.; Agrawal, V.P.; Bradshaw, R.A.; Dunn, M.J.; Haynes, P.A.; Van Wijk, K.J.; Kikuchi, S.; Renaut, J.; Weckwerth, W.; Rakwal, R. (2011). Time to articulate a vision for the future of plant proteomics-A global perspective for establishing the International Plant Proteomics Organization (INPPO). *Proteomics*, 11, 1559-1568

Ahsan, N.; Lee, D.G.; Kim, K.H.; Alam, I.; Lee, S.H.; Lee, B.H. (2010). Analysis of arsenic stress-induced differentially expressed proteins in rice leaves by two-dimensionally gel electrophoresis coupled with mass spectrometry. *Chemosphere*, 78, 224-231

Baerenfaller, K.; Grossmann, J.; Grobei, M.A.; Hull, R.; Hirsch-Hoffmann, M.; Yalovsky, S.; Zimmermann, P.; Grossniklaus, U.; Gruissem, W.; Baginsky, S. (2008). Genome-scale proteomics reveals *Arabidopsis thaliana* models and proteome dynamics. *Science*, 320, 938-941

Baginsky, S.; Grossmann, J.; Gruissem,W. (2007).Proteome analysis of chloroplast mRNA processing and degradation. *J.Proteome Res.*, 6, 809-820

Baginsky, S. (2009). Plant proteomics: concepts, applications and novel strategies for data interpretation. *Mass Spectrometry Reviews*, 28, 93-120

Bindschedler, L.V.; Palmblad, M.; Cramer, H. (2008). Hydroponic isotope labelling of entire plants (HILEP) for quantitative plant proteomics: an oxidative stress case study. *Phytochemistry*,69, 1962-1972

Brandlard , G.; Bancel, E. (2007). Plant Proteomics Methods and Protocols. In *Methods in Molecular Biology* Thiellement, H.; Zivy, M.; Damerval, C.; Mechin, V. (eds). 355, Humana Press, Totowa, New Jersey

Brechenmacher, L:; Lee, J.; Sachdev, S.; Song, Z.; Nguyen, T.H.N.; Joshi, T.; Oehrle, N.; Libault, M.; Mooney, B.; Xu, D.; Cooper, B.; Stacey, G. (2009). Establishment of a protein reference map for soybean root hair cells. *Plant Physiol.*, 149,670-682

Breci, L.; Haynes, P.A., (2007). Two-dimensional nanoflow liquid chromatography-tandem mass spectrometry of proteins extracted from rice leaves and roots. *Methods Mol.Biol.*, 355, 249-266

Bussow, K.; Konthur, Z.; Lueking, A.; Lehrach, H.; Walter, G. (2001). Protein array technology. *Am.J.Pharmacogenomics*,1, 37-43

Català, C.; Howe, K.J.; Hucko, S.; Rose, J.K.; Thannhauser, T.W. (2011). Towards characterization of the glycoproteome of tomato (*Solanum lycopersicum*) fruit using Concavalin A lectin affinity chromatography and LC-MALDI-MS/MS analysis. *Protemics*, 11, 1530-1544 Chen, T.;

Chen, T.; Wu, X.; Chen, Y.; Bohm, N.; Lin, J.; Samaj, J. (2007). Pollen and pollen tube proteomics, in *Plant Proteomics*, Samaj, J.& Thelen, J.J. eds , 270-282,Springer-Verlag Berlin Heidelberg

Colas, I.; Koroleva, O.; Shaw, P.J. (2010). Mass spectrometry in plant proteomic analysis. *Plant Biosystems*, 144, 703-714

Cotter, R.J.; Griffith, J.; Nardhoff, E. (2005). High-throughput proteomics using matrix-assisted laser desorption/ionization reflectron. *J.Chromatog.B Analyst Technol.Biomed.Life Sci.*, 855, 2-13

Dai, S.; Chen, T.; Chong, K.; Xue, Y. (2007). Proteomics identification of differentially expressed proteins associated with pollen germination andtube growth reveals characteristics of germinates *Oryza sativa* pollen. *Mol. Cell.Proteomics, 6, 207-230*

Doroshenko, V.M.; Laiko, V.V.; Taranenko, N.I.; Berkout, V.D.; Lee, H.S. (2002). Recent developments in atmospheric pressure MALDI mass spectrometry. *International J.Mass.Spec.*, 221, 39-58

Douglas, D.J.; Frank, A.J.; Mao, D. (2005). Linear ion traps in mass spectrometry. *Mass Spectrom.Rev.*, 24, 1-29

Dunkley, T.P.J.; Dupree, P.; Watson, R.B.; Lilley, K.S. (2004). The use of isotope-coded affinity tags (ICAT) to study organelle proteomes in *Arabidopsis thaliana. Biochem.Soc.Trans.*, 32, 520-523

Eubel, H.; Heazlewood, J.L.; Millar, A.H..(2007). Plant Proteomics Methods and Protocols. In *Methods in Molecular Biology* Thiellement, H.; Zivy, M.; Damerval, C.; Mechin, V. (eds). 355, Humana Press, Totowa, New Jersey

Everberg, H.; Sivars, U.; Emanuelsson, C.; Persson, C.; Englund, A.K.; Haneskog, L.; Lipniunas, P.; Jornten-Karlsson, M.; Tjerneld, F. (2004). Protein pre-fractionation in detergent-polymer aqueous two-phase systems for facilitated proteomic studies of membrane proteins. (2004). *J.Chromatogr.A.*, 1029, 113-124

Fenn, J.B. (2002). Electrospray ionization mass spectrometry: how it all began. *J.Biomol.Tech.*, 13, 101-118

Ferrari, F.; Fumagalli, M.; Profumo, A.; Viglio, S.; (2009). Deciphering the proteomic profiling of rice (*Oryza sativa*) bran: a pilot study. *Electrophoresis*, 30, 4083-4094

Ferro, M.; Brugiere, S.; Salvi, D.; Seigneurin-Berny, D.; Court, M.; Moyet, L.; Ramus, C.; Miras, S.; Mellal, M.; LeGall, S.; Kieffer-Jaquinod, S.;Bruley, C.;Garin, J.; Joynard, J.;Masselon, C.; Rolland, N. (2010). AT_CHLORO, comprehensive chloroplast proteome database with subplastidial localization and curated information on envelope proteins. *Mol.Cell.Proteomics*, 9,1063-1084

Friso, G.; Giacomelli, L.; Ytterberg, A.J.; Peltire,J.B.; Rudella, A.; Sun, Q.; vanWijk, K.J. (2004). In-depth analysis of the thylakoid membrane proteome of *Arabidopsis thaliana* chloroplasts: new proteins, new functions and a plastid proteome database. *Plant Cell*, 16 , 478-499

Froehlich, J.E.; Wiolkerson, C.G.; Ray, W.K.; McAndrew, R.S.; Osteryoung, K.W.; Gage, D.A.; Phinney, B.S. (2003). Proteomic study of the *Arabidopsis thaliana* chloroplastic envelope membrane utilizing alternatives to traditional two-dimensional electrophoresis. *J.Proteome Res.*, 2, 413-425

Ghaemmaghami, S.; Huh, W.K.; Bower, K.; Howson, R.W.; Belle, A. (2003). Global analysis of protein expression in yeast. *Nature*, 425, 737-741

Giavalisco, P.; Nordhoff, E.; Kreitler, T.; Kloppel, K.D.; Lehrach,H.; Klose, J.; Gobom,J. (2005). Proteome analysis of *Arabidospis thaliana* by two-dimensional gel electrophoresis and matrix-assisted laser desorption-time of flight mass spectrometry. *Proteomics*, 5, 1902-1913

Gion, J.M; Lalanne C.; Le Provost, G.; Ferry-Dumazet, H.; Paiva, J.; Chaumeil, P.; Frigerio, J.M.; Brach J., Barré A.; de Daruvar A.; Claverol, S.; Bonneu, M.; Sommerer, N.;

Negroni, L. ; .Plomion, C. (2005). The proteome of maritime pine wood forming tissue. *Proteomics*, 5, 3731–3751

Glinski, M.; Weckwerth, W. (2006). The role of mass spectrometry in plant systems biology. *Mass Spec.Rev.*, 25, 173-214

Gonzales-Camacho , F.; Medina F.J. (2007). Plant Proteomics Methods and Protocols. In *Methods in Molecular Biology* Thiellement, H.; Zivy, M.; Damerval, C.; Mechin, V. (eds). 355, Humana Press, Totowa, New Jersey

Gokulakannan, G.G.; Niehaus, K. (2010). Characterization of the *Medicago truncatula* cell wall proteome in cell suspension culture upon elicitation and suppression of plant defencse. *Journal of Plant Physiol.*, 167, 1533-1541

Granlund, I.; Hall, M.N.; Kieselbach, T.; Schroder, W.P. (2009). Light induced changes in protein expression and uniform regulation of transcription in the thylakoyd lumen of *Arabidopsis thaliana*. *PLoS One*, 4(5):e5649.

Granvogl, B.; Reisenger, V.; Eichacker, L.A. (2006). *Proteomics*, 3681-3695

Grossmann, J.; Ross, F.F.; Cielibak, M.; Liptak, Z.; Mathis, L.K.; Muller, M.; Gruissem, W.; Baginsky, S. (2005). AUDENS: a tool for automated peptide de novo sequencing. *J.Proteome Res.*, 4, 1768-1774

Grhuler, A.; Schulze, W. X.; Matthiesen, R.; Mann, M.; Jensen, O.N. (2005). Stable isotope labelling of *Arabidopsis thaliana* cells and quantitative proteomics by mass spectrometry. *Mol.Cell.Proteomics*, 4, 1697-1709

Gygi, S.P.; Rochon,Y.; Franza, B.R.; Aebershold, R. (1999). Correlation between protein and mRNA abundance in yeast. *Moll.Cell.Biol.*, 19, 1720-1730

Hajduch, M.; Ganapathy, A.; Stein, J.W.; Thelen, J.J. (2005). A Systematic Proteomic Study of Seed Filling in Soybean. Establishment of High-Resolution Two-Dimensional Reference Maps, Expression Profiles, and an Interactive Proteome Database. *Plant Physiol*, 137, 1397-1419

Hajduch, M.; Casteel, J.E.; Hurrelmeyer, K.E.; Song, Z.; Agrawal, G.K.; Thelen, J.J. (2006). Proteomic analysis of seed filling in *Brassica napus*: Developmental characerization of metabolic isozymes using high-resolution two-dimensional gel electrophoresis. *Plant Physiol.*, 141, 32-46

Hajheidari, M.; Abdollahian-Noghabi, M.; Askari, H.; Heidari, M.; Sadeghian, S.Y., Ober, S.E.; Salekdh, G.H. (2005). Proteome analysis of sugar beet leaves under drought stress. *Proteomics*, 5, 950-960

Hahner, S.; Jabs, W.; Brand, S.; Dickler, S.; Gounder, G.; Niehaus, K.; Suckau, D. (2007). P201-T ICPL duplex and triplex technology for quantitative proteomics. *J.Biomol.Tech*, 18,70

Hardman, M.; Makarov, A.A. (2003).Interfacing the orbitrap mass analyzer to an electrospray ion source. *Anal.Chem.*, 75, 1699-1705

Hartman, N.T.; Sicilia, F.; Lilley, K.S.; Dupree, P.(2007). Proteomic complex detection using sedimentation. *Anal. Chem.*, 79, 2078-2083

Heazlewood, J.L.; Tonti-Filippini, J.S.; Gout, A.M.; Day, D.A.; Wehelan, J.; Millar, A.H. (2004). Experimental analysis of the Arabidopsis mitochondrial proteome highlights signalling and regulatory components, provides assessment of targeting prediction programs and indicates plant-specific mitochondrial protein. *Plant Cell*, 16, 241-256

Heazlewood, J.L.; Durek, P.; Hummel, J.; Selbig, J.; Weckwerth, W.; Walther, D.; Schulze, W.X. (2008). PhosPhAt: a database of phosphorylation sites in *Arabidopsis thaliana* and a plant-specific phosphorylation site predictor. *Nucleic Acids. Res.*, 36, D1015-D1021

Holdsworth, M.J.; Finch-Savage, W.E.; Grappun, P.; Job, D. (2008). Post-genomics dissection of seed dormancy and germination. Trends Plant. Sci., 13, 7-13

Hurkman, W.J.; Tanaka, C.K.(1986). Solubilization of plant membrane proteins for analysis by two-dimensional electrophoresis. *Plant Physiol.*, 81, 802-805

Ippel, J.H.; Pouvreau, L.; Kroef, T. ; Gruppen, H. ;Versteeg, G.; van den Putten, P. (2004). In vivo uniform (15)N-isotope labelling of plants: using the greenhouse for structural proteomics. *Proteomics*, 4, 226-234

Jahn, T.; Dietrich, J.; Anderse, B.; Leidvik, B.; Otter, C.; Briving, C.; Kuhlbrandt, W.; Palmgren, M.G. (2001). Large scale expression, purification and 2D-crystallization of recombinant plant plasma membrane H^+-AtPase. *J.Mol.Biol.*, 309, 465-476

Jorge, I.; Navarro, R.M.; Lenz, C.; Ariza, D.; Jorrin, J. (2006). Variation in the holm oak leaf proteome at different plant developmental stages, between provenances and in response at drought stress. *Proteomic.* 6, 207-214

Joshi, H.; Hirsch-Hoffmann, M.; Baerenfaller, K.; Gruissem, W.; Baginsky, S.; Schmidt, R.; Schulze, W.X.; Sun, Q.; Van Wijk, K.; Egelhofer, V.; Wienkoop, S.; Weckwerth, W.; Bruley, C.; Rolland, R.; Toyoda, T.; Nakagam, H.; Jones, A.; Briggs, S.P.; Castleden, I.; Tanz, S.; Millar, A.H.; Heazlewood, J.L. (2011). MASCP gator: an aggregation portal for the visualization of Arabidopsis proteomics data. *Plant Physiol.*, 155, 259-270

Kaffarnik, F.; Jones, A.M.; Rathjen, J.P.; Peck, S.C. (2009). Effector proteins of the bacterial pathogen *Pseudomonas syringae* alter the extracellular proteome of the host plant *Arabidopsis thaliana*. Mol.Cell. *Proteomics*, 8, 145-156

Kamo, M.; Kawakami, T.; Miytake, N; Tsugita, A. (1995). Separation and characterization of *Arabidopsis thaliana* proteins by two-dimensional gel electrophoresis. *Electrophoresis*, 16, 423-300

Karlova,R.; Boeren, S.; Russinova, E.; Aker, J.; Vervoort, J.; de Vries, S. (2006). The *Arabidopsis* somatic embryogenesis receptor-like kinase1 protein complex includes brassinosteroid-insensitive1.*The Plant Cell*, 18, 626-638

Kaspar, S.; Peukert, M.; Svatos, A.;Matros, A.; Mock, H.P. (2011). MALDI_imaging mass spectrometry- An emerging technique in plant biology. *Proteomics*, 11, 1840-1850

Kaul, S.; Koo, H.L.; Jenkins, J.; Rizzo, M. (2000). Analysis of the genome sequence of the flowering plant *Arabidopsis thaliana*. *Nature*, 408, 796-815

Kellermann, J. (2008). ICPL – Isotope-coded protein label. *Methods Mol.Biol.*, 424, 113-123

Kersten, B.; Agrawal, G.K.; Durek, P.; Neigenfind, J. (2009). Plant phosphoproteomics: an update. *Proteomics*, 9, 964-988

Khan, M.M.; Komatsu, S. (2004). Rice proteomics: recent developments and analysis of nuclear proteins. *Phytochemistry*, 65, 1671-1681

Kher, J.; Rep, M. (2007). Plant Proteomics Methods and Protocols. In *Methods in Molecular Biology* Thiellement, H.; Zivy, M.; Damerval, C.; Mechin, V. (eds). 355, Humana Press, Totowa, New Jersey

Kieffer, P.; Dommes, J.; Hoffmann, L.; Hausman, J.F.; Renaut, J. (2008). Quantitative changes in protein expression of cadmium-exposed poplar plants. *Proteomics*, 8, 2514-2540

Koller, A.; Washburn, M.P.; Lange, B.M. (2002). Proteomic survey of metabolic pathways in rice. *Proc:Natl.Acad.Sci.USA*, 99, 11969-11974

Kondo, T:; Sawa, S.; Kinoshita, A.; Mizuno, S. (2006). A plant peptide encoded by CLV3 identified in situ MALDI-TOF MS analysis. *Science,* 313, 845-848

Koroleva, O.A.; Calder, G.; Pendle, A.F.; Kim, S.H.; Lewandowska, D.; Simpson, C.G. (2009). Dynamic behaviour of *Arabidopsis* eIF4A-III, putative core protein of exon junction complex: fast relocation to nucleolus and splicing speckles under hypoxia. *Plant Cell,* 21, 1592-1606

Lambert,J.P.; Ethier, M.; Smith, J.C..; Figeys, D. (2005). Proteomics: from gel based to gel free. *Anal. Chem.*, 77, 3771-3787

Larrainzar, E.; Wienkoop, S.; Weckwerth, W.; Ladrera, R.; Arrese-Igor, C.; Gonzales, E.M. (2007). *Medicago truncatula* root nodule proteome analysis reveals differential plant and bacteroid responses to drought stress. *Plant Physiol.*, 144, 1495-1507

Larson, R.; Wintermantel, W.M.; Hill, A.; Fortis, L.; Nunez, A. (2008). Proteome changes in sugar beet in response to beet necrotic yellow vein virus. *Physiol. and Mol. Plant Pathology,* 72, 62

Laukens, K.; Remmerie, N.; De Vijlder, T.; Hendrickx, K.; Witters, E. (2007). Proteome analysis of *Nicotiana tabacum* suspension cultures, in *Plant Proteomics,* Samaj, J.& Thelen, J.J. eds , 155-168,Springer-Verlag Berlin Heidelberg

Lee, J.; Jiang, W.; Qiao, Y.; Cho, Y.; Woo, M.O.; Chin, J.H.; Kwon, S.W.; Hong, S.S.; Choi, Y.; Koh, H.J. (2011). Shotgun proteomic analysis for detecting differentially expressed proteins in the reduced culm number rice. *Proteomics*, 11, 455-468

Link, A.J.; Eng, J.; Schieltz, D.M.; Carmak, E.; Mize, G.I.; Morris, D.R.; Garvik, B.M.; Yates J.R. (1999). Direct analysis of protein complexes using mass spectrometry. *Nat. Biotechnol.*, 17, 676-682

Liska, J.; Shevchenko, A. (2003). Expanding the organismal scope of proteomics: Cross-species protein identification by mass spectrometry and its implications. *Proteomics*, 3, 19–28

Lucker, J.; Laszczak, M.; Smith, D.; Lund, S.T. (2009). Generation of a predicted protein database from EST data and application to iTRAQ analyses in grape (*Vitis vinifera* cv Cabernet Sauvignon) berries at ripening initiation. *BMC Genomics*, 10, 50

Majeran, W.; Cai, Y.; Sun, Q.; van Wijk, K.J. (2005). Functional differentiation of bundle sheath and mesophyll maize chloroplasts determined by comparative proteomics. *Plant Cell*, 17, 3111-3140

Maltman, D.J.; Simon, W.J.; Wheeler, C.H.; Dunn, M.J.; Wait, R.; Slabas, A.R. (2002). Proteomic analysis of the endoplasmic reticulum from developing and germinating seed of castor (*Ricinus communis*). *Electrophoresis*, 23, 626-639

Maor, R.; Jones, A.; Nuhse, T.; Studholme, D.J.; Peck, S.C.; Shirasu, K. (2007). Multidimensional protein identification rechnology (MudPIT) analysis of ubiquinated proteins in plants. *Mol. Cell.Proteomics*, 6, 601-610

Marmiroli, N.; Pavesi, A.; Di Cola, G.; Harting, H.; Raho, G.; Conte, M.; Perrotta, C. (1993). Identification, characterization and analysis of cDNA and genomic sequences encoding two different small heat shock proteins in *Hordeum vulgare.. Genome*, 36, 1111-1118

Marshall, A.G.; Hendricksom, C.L.; Jackson, G.S. (1998). Fourier transform ion cyclotron resonance mass spectrometry: a primer. *Mass.Spectrom.Rev.*, 17, 1-35

Mathesius, U.; Imin, N.; Chen, H. (2002). Evaluation of proteome reference maps for cross-species identification of proteins by mass fingerprinting. *Proteomics*, 2, 1288-1303

Matsumoto, T.; Wu, J.; Kanamori, H.; Katayose, Y. (2005). The map based sequence of the rice genome. *Nature*, 436 ,793-800

Mooney, B.P.; Krishan, H.B.; Thelen, J.J. (2004). High-throughput peptide mass fingerprinting of soy-bean seed proteins: automated workflow and utiliy of Unigene expressed sequence tag data-bases for protein identification. *Phytochemistry*, 65, 1733-1744

Morelle, W. (2008). Analysis of the N-glycosylation of proteins in plants, In *Plant Proteomics: technologies, strategies and applications*. Agrawal, G.K.; Rakwal, R. (Eds), 455-467, Wiley, NJ Morris, P.C. (2008). 14-3-3 proteins: regulators of key cellular functions, In *Plant Proteomics: technologies, strategies and applications*. Agrawal, G.K.; Rakwal, R. (Eds), 515-524, Wiley, NJ

Naumann, B.; Bausch, A.; Allmer, J.; Osterdorf, E.; Zeller, M. (2007). Comparative quantitative proteomics to investigate the remodelling of bioenergetic pathways under iron deficiency in *Chlamydomonas reinhardtii*. *Proteomics*, 7, 3964-3979

Nawrocki, A.; Thorup-Kristensen, K.; Jensen, O.N. (2011). Quantitative proteomics by 2DE and MALDI MS/MS uncover the effects of organic and conventional cropping methods on vegetable products. *J.Proteomics*, epub ahead of prints

Neilson, K.A.; Mariani, M.; Haynes, P.A. (2011). Quantitative proteomic analysis of cold-responsive proteins in rice.*Proteomics*, 11, 1696-1706

Niittyla, T.; Fuglsang, A.T.; Palmgren, M.G.; Frommer, W.B.; Schulze W.X. (2007). Temporal analysis of sucrose-induced phosphorylazion changes in plasma membrane proteins of *Arabidospis*. *Moll.Cell.Proteomics*, 6, 1711-1726

Ning, Z.; Zhou,H.; Wang, F.; Abu-Farha, M.; Figeys, D. (2011). Analytical aspects of proteomics: 2009-2010. *Anal.Chem*, 83, 4407-4026

Nuhse, T.S.; Bottrill, A.R.; Jones, A.M.; Peck, S.C. (2007). Quantitative phosphoproteomic analysis of plasma membrane proteins reveals regulatory mechanisms of plant innate immune system responses. *Plant J.*, 18, 931-940

Oeljeklaus; S.,Meyer, H.E.,Warscheid, B. (2009). New dimensions in the study of protein complexes using quantitative mass spectrometry. *FEBS Lett*; 583, 1674.

Patterson, S.D.; Aebersold, R.(2003). Proteomics: the first decade and beyond. *Nat.Genet.*, 33, 311-32

Patterson, S.D.(2004). How much of the proteome do we see with discovery-based proteomics methods and how much do we need to see? *Curr.Proteomics*, 1, 3-12

Peckham, G.D.; Bugos, R.C.; Su, W.W. (2006). Purification of GFP fusion proteins from transgenic plant cell cultures. *Protein Expr. Purif.*, 49, 183-189

Pirondini, A.; Visioli, G.; Malcevschi, A.; Marmiroli, N. (2006). A 2-D liquid-phase chromatography for proteomic analysis in plant tissues. *Journal of chromatography B*, 833, 91-100

Pflieger, D.; Bigeard, J.; Hirt, H. (2011). Isolation and characterization of plant protein complexes by mass spectrometry. *Proteomics*, 11, 1824-1833

Faurobert, M.; Pelpoir, E.; Chaib, J.. (2007). Phenol extraction of proteins for proteomic studies of recalcitrant plant tissues. In *Methods in Molecular Biology* Thiellement, H.; Zivy, M.; Damerval, C.; Mechin, V. (eds). 355, Humana Press, Totowa, New Jersey

Poon, T.C. (2007). Opportunities and limitations of SELDI-TOF-MS in biomedical research: practical advices. *Expert Rev.Proteomics*, 4, 51-65

Proll, G.; Steinle, L.;Proll,F.; Kumpf, M.;Moehrle, B.; Mehlmann, M.; Gauglitz, G. (2007). Potential of label-free detection in high-content-screening applications. *J.Chromatog. A*, 1161, 2-8

Qi,Y.; Katagiri, F. (2009). Purification of low-abundance *Arabidopsis* plasma-membrane protein complexes and identification of candidate component. Plant J., 57, 932-944

Rappsilber, J.; Ryder, U.; Lamon, A.I.; Mann, M. (2002). Large-scale proteomic analysis of the human spliceo-some. Genome Res., 12, 1231-1245

Reumann, S.; Babujee, L.; Ma, C.; Wienkoop, S.; Siemsen, T.; Antonicelli, G.E.; Rasche, N.; Luder, F.; Weckwerth, W.; Jahn, O. (2007). Proteome analysis of Arabidopsis leaf peroxisomes reveals targeting peptides, metabolic pathways and defence mechanisms. *Plant Cell*, 19, 3170.3193

Riccardi, F.; Gazeau, P.; Jacquemot, M.P.; Vincent , D.; Zivy, M. (2004). Deciphering genetic variations of proteome responses to water deficit in maize leaves. *Plant Physiology and Biochemistry*, 42, 1003-1011

Rose, J.K.C.; Bashir, S.; Giovannoni, J.J.; Jahn, M.J.; Saravnan, R.S. (2004). Tackling the plant proteome: practical approaches, hurdles and experimental tools. *The plant journal*, 39, 715-733

Santoni, V. (2007). Plant Proteomics Methods and Protocols. In *Methods in Molecular Biology* Thiellement, H.; Zivy, M.; Damerval, C.; Mechin, V. (eds). 355, Humana Press, Totowa, New Jersey

Schenkluhn, L.; Hohnjec, N.; Niehaus, K.; Schmitz, U.; Colditz, F. (2010). Differential gel electrophoresis (DIGE) to quantitatively monitor early symbiosis and pathogenesis-induced changes of the *Medicago truncatula* root proteome. *J.Proteomics*, 73, 753-768

Shevchenko, A.; Matthias, W.; Vorm, O.; Mann, M. (1996). Mass spectrometric sequencing of proteins from silver-stained polyacrilamide gels. *Analytical Chemistry*, 76, 144-154

Shevchenko, A.; Tomas, H.; Havli, H., J.; Olsen, J.V.; Mann, M. (2007). In-gel digestion for mass spectrometric characterization of proteins and proteomes *Nature Protocols*, 1, 2856 - 2860

Schnable, P.S.; Ware, D.; Fulton, R.S.; Stein, J.C. (2009). The B73 maize genome: complexity, diversity and dynamics. *Science*, 326, 1112-1115

Schneider, M.; Tognolli, M.; Bairoch, A. (2004). The Swiss-Prot protein knowledgebase and ExPASy: providing the plant community with high quality proteomic data and tools. *Plant Physiol. and Biochem.*, 42, 1013-1021

Schneider, M.; Lane, L.; Boutet, E.; Lieberherr, D. (2009). The UniProtKB/Swiss-Prot knowledgebase and its plant proteome annotation program. *J.Proteomics*, 72, 567-573

Schulze, W.X.; Usadel, B. (2010). Quantitation in mass-spectrometry-based proteomics. *Annu.Rev.Plant.Biol.*, 61, 491-516

Schmidt, U.G.; Endler, A.; Schelbert, S.; Brunner, A.; Schnell, M.; Neuhaus, H.E.; Marty-Mazars, D.; Marty, F.; Baginsky, S.; Martinoia, E. (2007). Novel tonoplast transporters identified using a proteomic approach with vacuoles isolated from cauliflower buds. *Plant Physiol.*, 145, 216-229

Schulze, W.X.; Usadel, B. (2010). Quantitation in mass-spectrometry-based proteomics. *Annu.Rev.Plant Biol.*, 61, 491-516

Sheoran, I.S.; Ross, A.R.S.; Olson, D.J.H.; Sawheney, V.K. (2009). Compatibility of plant protein extraction methods with mass spectrometry for proteome analysis. *Plant Sci.*, 176, 99-104

Sridhar, V.V.; Surendrarao, A.; Liu, Z. (2006). APETALA1 and SEPALLATA3 interact with SEUSS to mediate transcription repression during flower development. *Development*, 133, 3159-3166

Sun, Q.; Zybaliov, B.; Wojciech, M.; Friso, G.; Olinares, P.D.B.; van Wijk, K. (2009). PPDB, the Plant Proteomics Database at Cornell. *Nucleic Acids Res.*, 37, D969-D974

Swami, A.K.; Alam, S.I.; Sengupta, N.; Sarin, R. (2011). Differential proteomic analysis of salt stress response in Sorghum bicolour leaves. *Env. And Exp. Botany*, 71, 321-328

Tanaka, K.W.; Waki, H.; Ido, Y.;Akita, S.; Yoshida, Y.; Yoshida, T. (1988).Protein and polymer analysis up to m/z 100.000by laser ionization time-of-flight mass spectrometry. *Rapid Commun.Mass.Spectrom.*, 2, 151-153

Taylor, N.L.; Haezlewood, J.L.; Millar, A.H. (2011). The *Arabidopsis thaliana* 2D gel mitochondrial proteome refining the value of reference maps for assessing protein abundance, contaminants and post-translational modifications. *Proteomic*, doi: 10.1002/pmic201000620. epub ahead of print

Theissen, G. (2001). Development of floral organ identity: stories from the MADS house. *Current Opinion in Plant Biology*, 4, 75-85

Timms, J.F. ; Cramer, R. (2008). Difference gel electrophoresis. Proteomics, 8, 4886-4897

Tsugita, A.; Kawakami, T.; Uchiyama, Y.; Kamo, M.; Miyatake, N., Nozu, Y. (1994). Separation and characterization of rice proteins. *Electrophoresis*, 15, 708-720

Tuskan, G.A.; DiFazio, S.; Jansson, S.; Bohlmann, J.; (2006). The genome of black cottonwood *Populus trichocarpa .Science*, 313, 1596-1604

Van Norman, J.M.; Benfey, P.N. (2009). Arabidopsis thaliana as a model organism in system biology. *Interdiscip.rev.Syst.Biol.Med.*, 1, 372-379

van Wijk, K.; Peltier J-B.; Giacomelli, L. (2007). Plant Proteomics Methods and Protocols. In *Methods in Molecular Biology* Thiellement, H.; Zivy, M.; Damerval, C.; Mechin, V. (eds). 355, Humana Press, Totowa, New Jersey

Villarejo, A.; Buren, S.; Larsson, S.; Dejarden, A.; Monne, M.; Radhe, C.; Karlsson, J.; Janssom, S.; Lerouge, P.; Rolland,N.; von Heijne, G.; Grebe, M.;Bako, L.; Samuelsson,G. (2005). Evidence for a protein transported through the secretory pathway en route to the higher plant chloroplast. *Nat.Cell.Biol.*, 7, 1124-1131

Villiers, F.; Ducruix, C.; Hugouvieux, V. Nolwenn, J. ; Ezan, E. ; Garin, J. ; Junot, C. ; Bourguignon, J. (2011). Investigatine the plant response to cadmium exposure by proteomic and metabolomic approaches. *Proteomics*, 11, 1650-1663

Visioli, G.; Pirondini, A.; Malcevschi A.;, N. (2010a). Comparison of Protein Variations in *Thlaspi Caerulescens* populations from metalliferous and non-metalliferous soils. *International Journal of Phytoremediation*, 12, 805-819

Visioli, G.; Marmiroli, M.; Marmiroli, N. (2010b). Two-dimensional liquid chromatography technique coupled with mass spectrometry analysis to compare the proteomic response to cadmium stress in plants. *Journal of Biomedicine and Biotechnology*, doi: 10.155/2010/567510, 1-11

Wang, G.; Zhou, H.; Lin, H.; Roy, S.; Shaler, T.A., Hill, L.R. (2003). Quantification of proteins and metabolites by mass spectrometry without isotopic labelling or spiked standards. *Nat. Chem.*, 75, 4818-4826

Watson, B.S.; Asrivatham, V.S.; Wang, L.; Summer, L.W. (2003). Mapping the proteome of Barrel medic. *Plant Physiol.*, 131, 1104-1123

Watson, B.S.; Summer L.W. (2007). Plant Proteomics Methods and Protocols. In *Methods in Molecular Biology* Thiellement, H.; Zivy, M.; Damerval, C.; Mechin, V. (eds). 355, Humana Press, Totowa, New Jersey

Wiener, M.C.; Sachs, J.R.; Deyanova, E.G.; Yates, N.A. (2004). Differential mass-spectrometry: a label-free LC-MS method for finding significant differences in complex peptide and protein mixtures. *Anal. Chem.*, 76, 6085-6096

Wienkoop, S.; Larrainzar, E.; Niemann,M.; Gonzales, E.M.; Lehmann U.; Weckwerth, W. (2006). Stable isotope-free quantitative shotgun proteomics combined with sample pattern recognition for rapid diagnostic. *J.Sep.Sci.*, 29, 2793-2801

Witzel, K.; Weidner, A..; Surabhi, G.K.; Borner, A.; Mock, H.P. (2009). Salt-stress-induced alterations in the root proteome of barley genotypes with contrasting response towards salinity. *J.Exp.Bot.*, 60, 3545-3557

Wolters, D.A.; Washburn, M.P.; Yates, J.R. (2001). An automated multidimensional protein identification technology for shotgun proteomics. *Anal.Chem.*, 73, 5683-5690

Xi, J.; Wang, X.; Shanyu, L.; Zhou, X.; Yue, L.; Fan, Y.; Hao, D. (2006). Polyethilene glycol improved detection of low-abundant proteins by two-dimensional electrophoresis analysis of plant proteome. *Phytochemistry*, 67, 2341-234

Zhao, Z.X.; Stanley, B.A.; Zhang, W.; Assmann, S.M. (2010). ABA-regulated G protein signalling in arabidopsis guard cell: a proteomic perspective. *J.Proteome Res.*, 9, 1637-1647

Zulak, K.G.; Khan,M.F.; Alcantara, J.; Schriemer, D.C.; Facchini, P.J. (2009). Plant defence responses in opium poppy cell cultures revealed by liquid chromatography-tandem mass spectrometry proteomics. *Mol.Cell.Proteomics, 8, 86-98*

Zybailov, B.; Rutschow, H.; Friso, G.; Rudella, A.; Emanuelsson., O. Sun, Q. Van Wijk, K.J. (2008). Sorting signals, N-terminal modifications and abundance of the chloroplast proteome. *PLoS One, 3(4):e1994.*

Zybailov, B.; Friso, G.; Kim, J.; Rudella, A.; Ramirez Rodriguez, V.; Asakura, Y.; Sun, Q.; van Wijk, K.J. (2009). Large scale comparative of a chloroplast Clp protease mutant reveals folding stress, altered protein homeostasis, and feed back regulation of metabolism. *Mol. Cell. Proteomics,* 8, 1789-1810

7

Assessment of Proteomics Strategies for Plant Cell Wall Glycosyltransferases in Wheat, a Non-Model Species: Glucurono(Arabino)Xylan as a Case Study

Faik Ahmed
Ohio University
USA

1. Introduction

The proteome of a tissue is a snapshot of the entire set of proteins expressed by that tissue at a given developmental stage. This set of proteins can be considered as an integrated and complex response of the tissue to a set of environmental and growth conditions. Thus, the analysis of a proteome would provide valuable clues of the physiological status of a tissue (or a cell) at a given time. Proteomics, an expanding scientific field, represents a promising tool to investigate such proteomes in a high-throughput manner. Although proteomics proved to be helpful in elucidating difficult cellular processes, its use in plant cell wall polysaccharides biosynthesis in non-model plants remains challenging. In this chapter, we evaluated the capabilities of two proteomics strategies in identifying specifically three Golgi-resident glycosyltransferases (GTs), TaGT43-4, TaGT47-13, and TaGT75-4, involved in glucurono(arabino)xylan (GAX) biosynthesis in wheat, an economically important non-model crop plant (Zeng et al., 2008, 2010). GAX polymer is the second most abundant polymer in the biomass from grass plants and there is an urgent need to elucidate its biosynthetic pathways to allow engineering of plant biomass for biofuel and other human needs (Faik, 2010).

1.1 Proteomics in plant cell wall polysaccharides biosynthesis

In the model plant Arabidopsis (*Arabidopsis thaliana*) about 10% of the genome (~2,500 genes) is dedicated to cell wall metabolism and function (Carpita 2011). Among these genes 17% are putative GTs identified through a homology search, and only a dozen have characterized biochemical functions. The sequences of these enzymes are available through the Carbohydrate Active enZyme database (CAZy) that classified them into GT families on the basis of protein sequence similarity (Coutinho 2003). It is anticipated that more GTs (yet to be identified) are needed to synthesize all polysaccharides currently found in plant cell walls. Most of the GTs are predicated to be integral membrane proteins, however recent works indicate that some GTs have a cleavable signal peptide and are closely associated with other integral membranes as a complex to secure their proper subcellular localization (Zeng et al., 2010). GAX polymer, like all hemicelluloses with the exception of callose, is synthesized in the Golgi by a multi-protein complex called xylan synthase complex (XSC). Golgi apparatus is a multifunctional organelle used by a plant cell not only to synthesize plant cell wall polymers, but also to process and modify glycoproteins and sorting their destination. Thus, the proteins that populate the membranes of the Golgi are very complex in nature and content (Simon et al., 2008). The challenges in studying the Golgi-resident GTs involved in plant cell wall biosynthesis comes from the fact that they are difficult to purify in an active state and are usually considered low abundant proteins. This may explain why there are very limited reports on purification of these GTs (Perrin et al., 1999; Faik et al., 2002; Zeng et al., 2010). Instead, researchers put more efforts in developing strategies to isolate membrane preparations relatively enriched in endo-membranes or organelles (plasma membranes, endoplasmic reticulum, Golgi) along with enriched GT activities. However, these GT activities are still contaminated with other unrelated proteins. Hence, optimization of high-throughput MS methods for protein identification on these partially purified GT activities is currently lacking.

1.2 Proteomics in non-model plants

Economically important non-model crop plants such as wheat are currently lacking publicly available genomic/protein sequence information. Applying proteomics methods to these species is challenging and requires cross-species identification. Fortunately, genomes and their protein-encoding gene sequences are currently available for several plant species including *Arabidopsis thaliana, Oryza sativa, Populus trichocarpa, Sorghum bicolor, Brachypodium distachyon, Vitis vinifera, Zea mays, Medicago,* and *Glycine max* (Figure 1). These nucleotide sequences as well as the predicted amino acid sequences are available through the Phytozome website (http://www.phytozome.net/) or NCBI (http://www.ncbi.nlm.nih.gov/genomes/PLANTS/PlantList.html), both of which are considerably larger than the total protein entries available through the Uniprot database which only contains a total of 53,1192 protein entries from different plant species.

The key factor to consider in the cross-species identification is gene annotation in databanks. For example, Uniprot database has only ~29,500 protein entries that are reviewed (annotated). In the case of Arabidopsis for which the genome sequence was completed and published more than a decade ago (Arabidopsis Genome Initiative, 2000), ~10% of proteins are still not annotated and ~30% are listed with unknown biochemical function. Nevertheless, with all the wealth in genomics resources, the identification of a protein from a non-model plant is now possible through BLAST search using MS/MS spectra and *de novo* sequencing. Even if a homologous protein is not listed in protein databases, a similarity at nucleotide sequence level (~35%) would be sufficient for a reliable identification of a homolog to the peptide sequence by cross-comparison to translated available plant genes (tBLASTn).

2. Advantages and disadvantages of proteomics strategies

Proteomics methods for high-throughput identification of proteins involve three main steps: the first step is the proteolytic (mostly trypsin) digestion of the protein samples. This digestion step would need optimization as incomplete digestion may result in a loss of protein information. The second step consists of three processes: (*i*) fractionation of the resulting peptides by liquid chromatography (LC), (*ii*) ionization of the separated peptides (fractions) through a source such as Matrix-Assisted LASER Desorption/Ionization

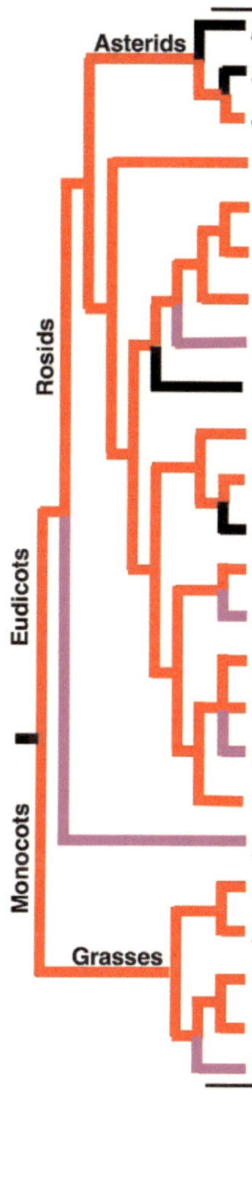

Species	Predicted Protein-coding genes	Genome size (Mb)
Mimulus guttatus (Monkey flower)	26,718	~430
Solanum lycopersicum (Tomato)	26,877	~737
Solanum tuberosum (Potato)	39,000	~840
Vitis vinifera (Grape)	26,346	~500
Arabidopsis thaliana	27,416	~120
Carica papaya (Papaya)	28,629	372
Theobroma cacao (Chocolate)	38,737	430
Citrus clementina (Clementine)	25,385	296
Eucalyptus grandis (Rose Gum)	44,974	~691
Populus trichocarpa (Poplar)	41,377	~370
Ricinus communis (Castor bean)	31,237	~320
Manihot esculenta (Cassava)	47,164	416
Glycine max (Soybean)	46,430	950
Medicago truncatula (Medicago)	>13,908	240
Malus x domestica (Apple)	57,386	~742
Fragaria vesca (Strawberry)	34,809	~240
Prunus persica (Peach)	27,852	227
Cucumis sativus (Cucumber)	26,682	~244
Aquilegia sp. (Columbine)	25,784	302
Oryza sativa (Rice)	40,577	~370
Brachypodium distachyon	25,532	272
Zea mays (Maize)	>32,000	2,300
Sorghum bicolor (Sorghum)	34,496	~700
Setaria italica (Foxtail millet)	35,471	~515

Fig. 1. Predicted number of protein-coding genes and sizes of plant genomes currently available in public databases. Red lines indicate genome sequencing completed and published; purple lines indicate unpublished genomes but publicly available; and black lines indicate plants with incomplete genomes or not fully assembled (less available). Species names are indicated along with the size of their genomes. These sequences are available through Phytozome and/or NCBI websites. Adapted from (http://synteny.cnr.berkeley.edu/wiki/index.php/Sequenced_plant_genomes)

(MALDI) or ElectroSpray Ionization (ESI), (*iii*) mass spectrometry (MS) analysis of the ionized peptides (received from the source) by tandem mass spectrometry (MS/MS), also called collision-induced dissociation (CID). This step generally delivers large information-rich files of MS/MS spectra (Bodnar et al., 2003). The third and critical step concerns the use of these MS/MS spectra in the identification of the exact proteins or at least their closest homologous proteins from the databases. Depending of the complexity of the samples, separation of the proteins prior to digestion can increase the chances of identifying these target proteins even in complex samples. Gel-based methods such as SDS-PAGE and 2-dimensional (2-D)-PAGE are widely used to separate the proteins of a sample and thus reduce its complexity. Although tremendous progress in LC and MS techniques has been made in the recent years, the application of these methods is still challenging in many complex biological processes such as plant cell wall biosynthesis (see section 1.1). There are two general proteomic strategies: gel-free and gel-based strategies.

2.1 Gel-free proteomics strategy

Direct analysis of protein composition of any biological sample (without prior separation by electrophoresis) can be achieved by a non-gel shotgun approach called MudPIT (multidimensional protein identification technology) (Washburn et al., 2001; McDonald et al., 2002). The method consists of a 2-D liquid chromatography (LC/LC) separation of the peptides (as opposed to the 2-D-PAGE of the proteins) before MS/MS analysis. The first dimension separates the peptides on the basis of their charges using a strong cation exchange (SCX) column. In the second dimension the peptides eluted from SCX column are fractionated further according to their hydrophobicity by reverse phase chromatography. Depending of the mass spectrometer used, the MudPIT strategy may have some limitations in identifying low abundant proteins. For example, there is a loading limit of protein sample onto the SCX column and it saturates easily, which can significantly impact the identification of low abundant proteins (i.e., GTs) in complex samples. Also, complex samples would require optimization of running time (retention time) for optimal protein identification (Chen et al., 2005). Lastly, although MudPIT has been successfully used to identify more than a 1000 proteins from a complex biological sample (Washburn et al., 2001), the dynamic range between the most abundant and least abundant proteins/peptides (10,000 to 1) can still limit the identification of very low abundant proteins (Wolters et al., 2001). The other key factor to consider when using gel-free proteomics methods is the precipitation step, which is used to not only concentrate the proteins, but also to remove small soluble compounds (including detergents) that are easily ionized and would mask low abundant peptide ions (Newton et al., 2004). Thus, optimizing this step is critical and may also result in a loss of some proteins from the sample. Ethanol, chloroform/methanol, cold acetone, trichloroacetic acid (TCA), or TCA/acetone are all MS-compatible reagents that can be used in proteins precipitation (Ferro et al., 2003). However, because an individual protein has specific physicochemical characteristics, any precipitation method has its own advantages and disadvantages. For example, although ethanol and TCA will precipitate around 90% of proteins, several proteins can be eliminated, reduced, or enriched in the precipitated samples (Zellner et al., 2005; Chen et al., 2010). Furthermore, ethanol is also known to precipitate salts along with the proteins, which may require dialysis of the samples before any further analysis. In our previous work, cold acetone was found to precipitate sucrose along with the proteins when sucrose amounts is >25% w/w in a fraction (Zeng et al., 2010). The increase in sucrose content in the sample decreased peptide ionization and produced low quality MS/MS spectra.

2.2 Gel-based proteomics strategy

The most direct and easy way to separate proteins in a complex sample is using a 1-D SDS-PAGE method (Laemmli, 1970). The main advantage of 1-D SDS-PAGE over the 2-D-PAGE

method is that virtually all proteins can be solubilized and separated on the gel (up to 10-15μg per sample). Loading larger amounts of protein on 1-D SDS-PAGE can be detrimental, as abundant proteins can spread over a large area of the gel and may mask low abundant proteins (Zhu et al., 2010). This can be problematic when working with proteins from photosynthesizing organs rich in ribulose-1,5-bisphosphate carboxylase/oxygenase (RuBisCO) protein, an abundant and difficult protein to eliminate (Komatsu et al., 1999). Regarding 2-D-PAGE, it is known that protein solubility is the Achilles' heel of this technique, as it does result in the loss of low abundant proteins, proteins with extreme pI, and hydrophobic proteins (integral membrane proteins) (O'Farrell, 1975; Santoni et al., 2000; Ephritikhine et al., 2004). In relation to this approach, Golgi proteins are also known to easily aggregate during sample preparation/precipitation and do not run well on 2-D-PAGE gels (Asakura et al., 2006).
Once 1-D SDS-PAGE separation is completed, the gels are usually cut into small equal slices or individual visible bands are excised. These slices/bands are subjected to trypsin digestion, and released peptides further fractionated by LC, and their structures analyzed by MS/MS (see next section).

2.3 Protein identification methods (MS/MS spectra processing)

Despite the availability of a large repository of protein-coding gene sequences from many plant species that can be used as databases for proteomic studies (Figure 1), the most daunting task in proteomics is still the identification of proteins with high sensitivity and accuracy from these databases using MS/MS spectra (Nesvizhskii et al., 2007; Patterson 2003; Service 2008). The processing of these MS/MS spectra for protein identification can be performed via two main strategies:

i. **Database-dependent strategy** in which the experimental masses of peptide ions (parent and its fragmentation product ions) from MS/MS spectra are compared to the theoretical peptide masses derived from *in silico* digestion of proteins in a database. This search would result in the identification of peptide hits. These hits are scored and ranked from best to worse matches, and any proteins assembling two or several hits are considered candidates. In the case of non-model plants, protein identification can be challenging since either the exact protein may not exist in the database or the database may not contain an evolutionary similar protein. This underscores the need for continuous effort to sequence the genomes of as many plants as possible to allow success of proteomics projects in non-model plants. Many bioinformatics algorithms such as SEQUEST, Mascot, X!Tandem, OMSSA, and Phenyx have been developed for database-dependent search using MS/MS spectra (for details about these algorithms I refer the readers to the following link: http://www.proteomesoftware.com/Proteome_software_link_software.html).
ii. **Database-independent strategy** consists of converting the fragmentation MS/MS spectra to possible *de novo* amino acid sequences. Several bioinformatics algorithms such as PepNovo, NovoHMM, Mascot, and PEAKS were designed for the heavy computation needed for the extraction of amino acid sequence information and can make MS/MS spectrum interpretable. These *de novo* peptide sequences are then used in BLAST searches of non-redundant protein database (*i.e*, NCBI, Swissprot). For non-model plants this combination (*de novo* and BLAST search) was proved to improve protein identification in terms of accuracy and rate increase (Shevchenko et al., 2001). Mascot program is a powerful search engine that can use the data from MS/MS spectra in different combinations. For example, this program can perform a search in three

modes: (*a*) peptide mass fingerprint mode using peptide mass values determined experimentally from MS/MS spectra; (*b*) sequence query mode using the combination of peptide mass data, amino acid sequence, and amino acid composition (called also sequence tag). In this situation, Mascot program extracts three or four amino acid residues of sequence data from a series of *m/z* ion peaks and build on them during the BLAST search. This search mode supports both standard and error tolerant sequence tags, and allows arbitrary combinations of fragment ion mass values, amino acid sequence data and amino acid composition data to be searched; and (*c*) the program can also use directly the raw MS/MS spectra to search a database. Because Mascot program is able to combine different types of searching modes in a single identification search, we decided to use this program for this work.

3. Application of proteomics strategies to the identification of GAX synthesizing GTs

In our previous work we demonstrated that GAX synthase is a multi-enzyme complex (XSC) formed of at least two known GTs (TaGT43-4 and TaGT47-13) and a mutase (TaGT75-4 called also reversibly glycosylated polypeptide, RGP) (Zeng et al., 2010). According to native gel electrophoresis data, this XSC has an apparent MW of ~250 kDa. In this chapter, two proteomics strategies (MudPIT and Gel-LC-MS/MS) were evaluated for efficient identification of these three proteins in membrane fractions enriched in wheat GAX synthase activity (partially purified activity). The goal is to develop a proteomics workflow for optimal identification of candidate proteins involved in plant cell wall biosynthesis in organisms with no available genome sequence information.

3.1 Experimental procedures

3.1.1 Preparation of membrane fractions enriched in GAX synthase complex

GAX synthase activity is routinely monitored as the amount of [^{14}C]glucuronic acid (GlcA) transferred from UDP-[^{14}C]GlcA into ethanol-insoluble GAX polymer in the presence or absence of UDP-xylose (UDP-Xyl) (Zeng et al., 2008, 2010). It has been shown that the rice Golgi complex has distinct compartments that could be separated by density gradient centrifugation in presence of EDTA or $MgCl_2$ (Mikami et al., 2001). Thus, we fractionated Golgi-enriched membranes prepared from etiolated wheat seedlings on a continuous 25%-40% (w/v) sucrose density gradient supplemented with 1mM EDTA and used our enzyme assay to monitor GAX synthase activity distribution (Figure 2). According to the specific activity and protein content in fraction #3 (as measured via spectrophotometer), we have achieved ~11 fold purification with a 59% yield (Table 1). However, according to SDS-PAGE analysis of this fraction, only limited number of protein bands is visible on the gel (Figure 3), rather suggesting a higher purification was achieved. Together these results indicate that fraction #3 is a good starting material for proteomics analyses for our study.

3.1.2 Nanospray tandem mass spectrometry procedures

All proteomics analyses were carried out at the Mass spectrometry and Proteomics Facility (http://www.ccic.ohio-state.edu/MS/) at Campus Chemical Instrument Center (CCIC) of The Ohio State University (Columbus, OH). Liquid chromatography-nanospray tandem mass spectrometry (nano-LC/MS/MS) of global protein identification was performed on an LTQ XL or an LTQ Orbitrap XL (Thermo Scientific) mass spectrometers using different protocols:

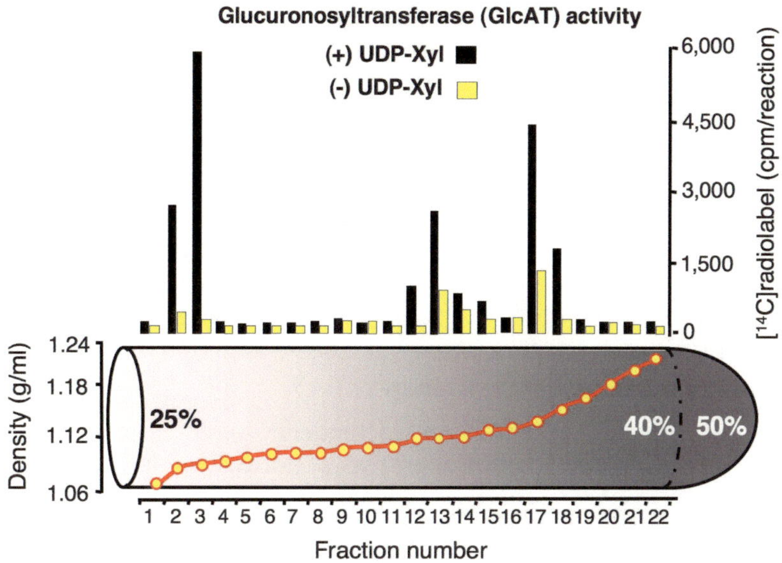

Fig. 2. Distribution of GAX synthase activity after fractionation of wheat Golgi-enriched microsomal membranes on a linear (25%-40%) sucrose density gradient. Microsomal membranes were prepared from 6-day old etiolated wheat seedlings. Fractions of 1ml were collected and each fraction was tested for GAX synthase activity ([^{14}C]GlcA incorporation in presence of UDP-xylose, as cpm/reaction) and for sucrose density (g/ml).

Fraction	Volume (ml)	Total Protein (mg)	Total Activity (pmol/h)	Specific Activity (pmol/h/mg)	Fold Purification	Yield (%)
Golgi-enriched membranes	12	60	~3060	51	1	100
Fraction #3 (after sucrose gradient)	1	0.3	~1833	~550	~11	~59

Table 1. Partial purification of wheat GAX synthase activity from Golgi-enriched microsomal membranes. The activity was measured as [^{14}C]GlcA incorporation from UDP-[^{14}C]GlcA (900cpm/pmol) into ethanol-insoluble materials and expressed as pmol GlcA incorporation per hour per milligram of protein (pmol/h/mg). Protein content was estimated using Bradford reagent.

3.1.2.1 Nano-LC/MS/MS on LTQ XL

Nano-liquid chromatography tandem mass spectrometry (Nano-LC/MS/MS) was performed on a Thermo Scientific LTQ XL mass spectrometer (Linear Quadrupole Ion Trap MSn) equipped with a nanospray source operated in positive ion mode. The LC system was an UltiMate™ 3000 system from Dionex (Sunnyvale, CA). The solvent A was water containing 50 mM acetic acid and the solvent B was acetonitrile. 5 µL of each sample was first injected on to the µ-Precolumn Cartridge (Dionex, Sunnyvale, CA), and washed with 50 mM acetic acid. The injector port was switched to inject and the peptides were eluted off of the trap onto the column. A 5 cm 75 µm ID ProteoPep II C18 column (New Objective, Inc. Woburn, MA) packed directly in the nanospray tip was used for chromatographic separations. Peptides were eluted directly off the column into the LTQ system using a gradient of 2-80%B over 45 min, with a flow rate of 300 nL/min. The total run time was 65 min. The MS/MS was acquired according to standard conditions established in the lab. Briefly, a nanospray source operated with a spray voltage of 3 KV and a capillary temperature of 200°C is used. The scan sequence of the mass spectrometer was based on the TopTen™ method; the analysis was programmed for a full scan recorded between 350 and 2000 Da, and a MS/MS scan to generate product ion spectra to determine amino acid sequence in consecutive instrument scans of the ten most abundant peaks in the spectrum. The CID fragmentation energy was set to 35%. Dynamic exclusion was enabled with a repeat count of 2 within 10 seconds, a mass list size of 200, an exclusion duration 350 seconds, the low mass width was 0.5 and the high mass width was 1.5 Da.

3.1.2.2 Capillary-LC/MS/MS on LTQ Orbitrap XL

Capillary-liquid chromatography-nanospray tandem mass spectrometry (Capillary-LC/MS/MS) of global protein identification was performed on a Thermo Scientific LTQ Orbitrap XL mass spectrometer equipped with a microspray source (Michrom Bioresources Inc, Auburn, CA) operated in positive ion mode. Samples were separated on a capillary column (0.2 X 150mm Magic C18AQ 3μ 200A, Michrom Bioresources Inc, Auburn, CA) using an UltiMate™ 3000 HPLC system from LC-Packings A Dionex Co (Sunnyvale, CA). Each sample was injected into the μ-Precolumn Cartridge (Dionex, Sunnyvale, CA) and desalted with 50 mM acetic acid for 10 min. The injector port was then switched to inject and the peptides were eluted from the trap onto the column. Mobile phase A was 0.1% formic acid in water, and 0.1% formic acid in acetonitrile was used as mobile phase B. Flow rate was set at 2 μL/min. Typically, mobile phase B was increased from 2% to 50% in 90-250 min, depending on the complexity of the sample, to separate the peptides. Mobile B was then increased from 50% to 90% in 5 min and then kept at 90% for another 5 min before being brought back quickly to 2% in 1 min. The column was equilibrated at 2% of mobile phase B (or 98% A) for 30 min before the next sample injection. MS/MS data was acquired with a spray voltage of 2 KV and a capillary temperature of 175°C is used. The scan sequence of the mass spectrometer was based on the data dependant TopTen™ method: the analysis was programmed for a full scan recorded between 300 – 2000 Da and a MS/MS scan to generate product ion spectra to determine amino acid sequence in consecutive scans of the ten most abundant peaks in the spectrum. The resolution of full scan was set at 30,000 to achieve high mass accuracy MS determination. The CID fragmentation energy was set to 35%. Dynamic exclusion is enabled with a repeat count of 30 seconds, exclusion duration of 350 seconds and a low mass width of 0.5 and high mass width of 1.5 Da. Multiple MS/MS detection of the same peptide was excluded after detecting it three times.

3.1.2.3 MS/MS data processing

Sequence information from the MS/MS spectra was processed by converting the raw data files into a merged file (.mgf) using an in-house program, RAW2MZXML_n_MGF_batch (merge.pl, a Perl script). The resulting mgf files were searched using Mascot Daemon by Matrix Science version 2.2.1 (Boston, MA) and the database searched against the full SwissProt database version 54.1 (283,454 sequences; 104,030,551 residues) or NCBI database. The mass accuracy of the precursor ions were set to 2.0 Da given that the data was acquired on an ion trap mass (LTQ) analyzer and the fragment mass accuracy was set to 0.5 Da (for analysis by high resolution Orbitrap, the mass accuracy of the precursor ions were set to 0.1 Da). Considered modifications (variable) were methionine oxidation and carbamidomethyl cysteine. Two missed cleavages for the enzyme were permitted. Peptides with a score less than 40 were filtered. Protein identifications were checked manually and proteins with a Mascot score of 50 or higher with a minimum of two unique peptides from one protein having *a* -*b* or -*y* ion sequence tag of five residues or better were accepted. Mascot also provides a "histogram of the MOWSE score distribution" for hits and a value of significance threshold (represented by a green region). A score is defined as $-10*LOG_{10}(P)$, where P is the absolute probability that the observed match is a random event. Therefore, significant matches would give scores that are higher than the significance threshold (outside of the green region). It is important to know that significance threshold values and scores of the hits are greatly affected by an increase in mass tolerance values in a MS/MS fingerprint search.

3.1.2.4 Gel-LC-MS/MS analysis

One-D SDS-PAGE was carried out according to (Laemmli, 1970). Separation gels were 10% or 12% acrylamide (150V until the loading dye reached the bottom) and molecular weight markers (Precision Plus Protein™, kaleidoscope, Bio-Rad) were used. Proteins were visualized by coomassie blue or silver staining. Briefly, the protein sample (50-60µg) was solubilized in 20 µL Invitrosol (Invitrogen) plus 1 µL of 10% amidosulphobetaine 14 (ASB-14) detergent, 2 µL of 10X SDS-PAGE loading dye, and heated at 90°C for 7 min before fractionation on 1-D SDS-PAGE acrylamide gel). The gel was rinsed with water for 10 min to remove excess SDS detergent, before staining with coomassie blue for 1 h and distained overnight. Figure 3 shows a typical silver-stained 1-D SDS-PAGE gel of fraction #3.

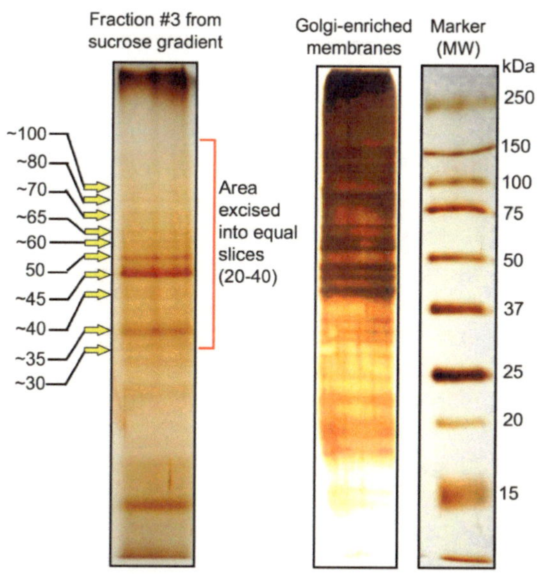

Fig. 3. SDS-PAGE (10%) and silver staining analysis of fraction #3 obtained from Figure 2. The arrows on the left indicate the position of the visible bands that were manually excised and digested with trypsin before ESI-LC-MS/MS-TRAP analysis. The area indicated on the right of the same gel corresponds to the area excised into 20-40 equal slices that were subjected to "In-gel" trypsin digestion followed by ESI-LC-MS/MS-TRAP analysis. Protein identification was carried out using Mascot program to search green plant databases at NCBI. For comparison, the original Golgi-enriched membranes were analyzed. Molecular mass markers (MW) are indicated on the right.

3.1.2.5 In gel digestion

The gels were excised in two ways (see Figure 3): (*i*) only the most visible protein bands in the gel were manually excised and individually digested with trypsin, or (*ii*) the gel area from 30 to 180 kDa was excised into 20 to 40 equal small slices and each slice was individually digested with trypsin. The resulting peptides were then analyzed by LC-MS/MS as described above.

For manual excision of visible bands, gels were digested with sequencing grade trypsin from Promega (Madison, WI) using the Multiscreen Solvinert Filter Plates from Millipore (Bedford, MA). Briefly, bands were trimmed as close as possible to minimize background polyacrylamide material. Gel pieces were then washed twice in nanopure water for 5 min each before destaining with 1:1 v/v methanol:50 mM ammonium bicarbonate for 10 min twice. The gel pieces were dehydrated with 1:1 (v/v) acetonitrile:50 mM ammonium bicarbonate, and then rehydrated by incubation with dithiothreitol (DTT) solution (25 mM in 100 mM ammonium bicarbonate) for 30 min prior to incubation in iodoacetamide solution (55 mM in 100 mM ammonium bicarbonate) for 30 min in dark. The gel bands were washed again with two cycles of water and dehydrated with 1:1 (v/v) acetonitrile:50 mM ammonium bicarobonate. Protease digestion was carried out by rehydrating gel pieces in 12 ng/mL trypsin in 0.01% ProteaseMAX Surfactant for 5 min. The gel pieces are then overlaid with 40 mL of 0.01% ProteaseMAX surfactant:50 mM ammonium bicarbonate and gently mixed on a shaker for 1 h. The digestion is stopped with addition of 0.5% TFA. The MS analysis is immediately performed to ensure high quality tryptic peptides with minimal non-specific cleavage or frozen at -80°C until samples can be analyzed.

For gel excision into 20-40 equal slices, robotic digestion was carried out using the Ettan Spot Handling Workstation (Amersham Biosciences). The slices were placed in a 96 well plate that was robotically washed, and slices digested according to the Ettan Spot Handling Workstation 2.1 User Manual. Briefly, gel slices were washed in 100 µL of 50% methanol/5% acetic acid for 30 min. This washing step was repeated 3 times and slices placed in a storage solution of 50 µL of 50% methanol/5% acetic acid until digestion. Gel slices were digested with sequencing grade trypsin from Promega (Madison, WI). Digestion was carried out by adding 100 µL water for 10 min followed by 100 µL acetonitrile for 10 min. The water and acetonitrile were removed and the gel slices were rehydrated and incubated with DTT (prepared as 32 mM in 100 mM ammonium bicarbonate) for 30 min prior incubation in iodoacetiamide solution (prepared as 80 mM in 100 mM ammonium bicarbonate solution) in dark for 30 min. The gel slices were washed again with cycles of acetonitrile and 100 mM ammonium bicarbonate in 10 and 5 min increments. The slices were dried for 10 min and then incubated in 25 µL of 50 mM ammonium bicarbonate containing trypsin (5 µg/mL) for 180 min. The peptides were extracted from the polyacrylamide with 50 µL 50% acetonitrile and 5% formic acid (3 times) and the extracts pooled together. The extracted pools were mixed with 50 µL of acetonitrile and incubated for 15 min before drying for 15 min. The MS analysis is immediately performed to ensure high quality tryptic peptides with minimal non-specific cleavage or frozen at -80°C until samples can be analyzed.

3.1.2.6 MudPIT analysis

The first MudPIT analysis on fraction #3 was carried out on a freeze-dried sample that was re-suspended in 5X Invitrosol protein solubilizer (Invitrogen) and diluted to the final 1X Invitrisol by adding 25mM ammonium bicarbonate solution. Ten microliter (10µL) of DTT (5 mg/mL solution prepared in 100mM ammonium bicarbonate) was added to the sample and incubated for 15 min. Ten microliter (10µL) of Iodoacetimide solution (15 mg/mL in 100 mM ammonium bicarobonate) was added to the sample and incubated for another 15 min in the dark. However, we noticed that this freeze-dried sample contained higher concentrations of sucrose and salts, which contributed to the production of low quality MS/MS spectra. Therefore, removal of sucrose and salts by precipitation was necessary before trypsin digestion. Two precipitation methods were evaluated: TCA/acetone or chloroform/methanol. These two methods are known to be effective in desalting and lipid removal from protein samples. In our case, TCA/acetone combination gave a better result in removing sucrose and salts from our protein sample (fraction #3), as the quality of MS/MS spectra was improved and scores were higher than the significance threshold (see section 2.3). Trypsin was prepared in 50 mM ammonium bicarobonate and added to the protein solution in trypsin:protein ratio of 1:25 (w/v) and the mixture was incubated for 60 min at 37°C (no difference was observed when incubation time was extended up to 16 h).

3.2 Results

3.2.1 Protein composition analysis of fraction #3 via MudPIT strategy

In our first MudPIT trial, the analysis was carried out on the LTQ XL (ion trap) instrument using MS/MS ion search mode with carbamidomethylation of cysteine residue as fixed modification and methionine oxidation as variable modification. Peptide mass tolerance was set to ± 2 Da, fragment mass tolerance was set to ± 0.8 Da and protein mass was unrestricted. Under these conditions, LTQ analysis of fraction #3 produced a total of 13,504 peptide queries that were used to search the green plant databases at NCBI (6,350,093 sequences). The search yielded 74 peptide matches above significance threshold (set to: Individual ions scores >59 and p<0.05) that were matched by only 501 peptide sequences (~4% of the 13,504 peptide queries). The decoy value was 41 corresponding to a false discovery rate of 11.33%. Most of the identified hits were protein entries with similar annotation (redundancy) but from different related species (orthologous proteins). Therefore, for simplicity, entries having similar annotation were eliminated, which reduced the number of total protein identified to 25 proteins (Table 2). The low protein identification rate could be due to the inhibition of peptide ionization from low abundant proteins by more abundant proteins. The other possibility could be the interference of some small molecules (i.e., sucrose, salts) present in the sample, which could produce MS/MS spectra of low quality. However, it is possible that some of these peptide queries are species-specific and do not have matches in the database. Therefore, to improve protein identification rate, we repeated the analysis of the same sample (fraction #3) using an Orbitrap high resolution mass spectrometer (LTQ Orbitrap XL). The high resolving power of the Orbitrap (determines masses with very high accuracy) has been useful in analyzing complex peptide mixtures by improving the resolution (narrow peak width), which in turn would increase the detection of more *m/z* ions. Thus, ions of similar *m/z* and proteins with multiple charges produced by ESI can be detected distinctly. Our results indicated that, even with reduced peptide mass tolerance (± 0.1 Da) and fragment mass tolerance (± 0.5 Da) values, Orbitrap analysis generated 20,123 peptide queries (6,619 more peptides compared to LTQ analysis). This analysis resulted in the identification of a total of 51 non-redundant proteins that included 37 new proteins (marked by "*" in Table 2). This increase in the rate of protein identification is the result of an increase in the number of matching peptides, 908 peptides (~408 more peptides

Accession No	Annotation	MW (Da)	emPAI	Score	Total peptides matched	Peptides with score<59
LTQ analysis						
gi|10720235	Chloroplast precursor (PCR A)	41,358	6.34	1380	145	84
gi|3318722	Leech-Derived Tryptase Inhibitor	24,142	3.35	660	134	101
gi|2493131	V-ATPase subunit B 1	54,107	0.95	685	32	18
gi|90025017	*Vacuolar proton-ATPase subunit A*	68,754	0.61	672	35	26
gi|10720236	Chloroplast precursor (PCR B)	42,350	0.45	460	25	23
gi|18274925	Vacuolar membrane proton pump	80,108	0.31	376	20	13
gi|476003	HSP70	67,146	0.11	349	12	10
gi|461465	Actin	41,940	0.19	335	13	9
gi|8928416	Tubulin beta-8 chain (Beta-8-tubulin)	50,596	0.07	232	7	6
gi|7960277	*RuBisCO activase B*	48,012	0.16	215	12	9
gi|461988	Elongation factor 1-alpha	49,489	0.16	202	19	15
gi|90903441	*Lipoxygenase 2*	17,788	ND	110	4	4
gi|125548120	Hypothetical protein OsI_015175	92,128	0.04	154	4	3
gi|118498764	Glyceraldehyde-3-phosphate dehydrogenase	30,091	0.13	136	6	5
gi|7579064	*Cytosolic glyceraldehyde-3-phosphate dehydrogenase*	25,448	0.15	135	5	4
gi|13925731	*Cyclophilin A-1*	18,721	0.21	100	3	3
gi|20302473	*Ferredoxin-NADP(H) oxidoreductase*	40,491	0.08	102	2	1
gi|914031	*WBP2=lipid transfer protein homolog*	3,003	1.56	98	4	4
gi|6911146	Glycine-rich RNA-binding protein 2	16,312	0.24	95	8	5
gi|303844	Eukaryotic initiation factor 4A	47,187	0.08	93	2	1
gi|944842	*ATP/ADP carrier protein*	36,013	0.10	90	5	3
gi|56606827	*Calreticulin-like protein*	47,404	0.14	82	2	1
gi|56713236	*Non-specific lipid transfer protein*	12,807	ND	66	2	2
gi|4099408	*Tonoplast intrinsic protein*	25,431	0.15	65	3	2
gi|115349890	*Fasciclin-like protein FLA3*	41,600	ND	59	1	0
Total peptides					505	
Orbitrap analysis (Hits marked by "*" are unique hits to from Orbitrap analysis)						
gi|10720235	Chloroplast precursor (PCR A)	41,358	24.30	3667	142	14
gi|90025017	*Vacuolar proton-ATPase subunit A*	68,754	2.48	2670	88	8
gi|10720236	Chloroplast precursor (PCR B)	42,350	1.60	1111	48	6
*gi|157339402	*ATPase subunit A*	69,767	0.51	961	29	2
gi|2493131	V-ATPase subunit B 1	54,107	1.28	906	35	4
gi|32481063	Rubisco activase	47,371	0.82	871	28	3
gi|7960277	*RuBisCO activase B*	48,012	0.97	759	28	5
gi|19880505	*Aquaporin PIP1*	31,083	1.26	648	27	20
gi|18274925	Vacuolar membrane proton pump	80,108	0.43	637	37	15
*gi|162457723	Binding protein homolog1	73,211	0.32	522	20	11
gi|56606827	*Calreticulin-like protein*	47,404	0.41	486	12	6
*gi|115383189	*Plasma membrane intrinsic protein*	30,046	1.85	471	21	14
*gi|34922469	Lipoxygenase 2.1	106,254	0.31	409	24	18
*gi|6525065	Chloroplast TE factor Tu	50,551	0.38	385	9	2
gi|303844	Eukaryotic initiation factor 4A	47,187	0.29	381	12	5
*gi|2827002	*HSP70*	71,385	0.58	380	23	10
*gi|26986186	Hexose transporter	79,631	0.17	348	13	7
gi|461465	Actin	41,940	0.62	323	21	15
*gi|125535913	Hypothetical protein OsI_036360	17,064	1.00	318	19	15
*gi|8928423	Tubulin beta-4 chain	50,791	0.49	316	12	6
*gi|115874	Serine carboxypeptidase III	55,869	0.16	304	10	4

Accession No	Annotation	MW (Da)	emPAI	Score	Total peptides matched	Peptides with score<59
***gi|14017584**	*Cytochrome f*	35,425	0.77	276	15	12
***gi|14017579**	*ATP synthase CF1 beta subunit*	53,881	0.83	265	33	29
***gi|739292**	Oxygen-evolving complex protein 1	26,603	0.57	261	18	15
***gi|7452979**	Allene oxide synthase	53,616	0.16	253	7	1
*gi|5123910	*HSP80-2*	80,533	0.22	252	12	7
***gi|122022**	Histone H2B.1	16,423	0.62	248	8	3
gi|90903441	*Lipoxygenase 2*	17,788	2.05	241	13	12
*gi|127519390	*Lipid transfer protein*	11,629	0.45	197	6	4
*gi|493591	*Disulfide isomerase precursor*	56,845	0.44	195	11	10
*gi|1658313	Os07g0683900	39,146	0.23	183	10	9
*gi|73912433	*Aspartic proteinase*	54,965	0.25	179	9	8
*gi|4158232	*Reversibly glycosylated polypeptide*	41,985	0.28	168	9	6
*gi|111073715	*Triticain alpha*	51,572	0.17	160	7	4
*gi|15289940	Putative actin	41,929	0.64	149	17	15
***gi|1556446**	Alpha tubulin	50,266	0.38	140	10	9
*gi|32400836	*Elongation factor*	18,638	0.24	139	4	1
*gi|72256525	*Geranylgeranyl hydrogenase*	50,966	0.17	139	7	6
*gi|6013196	*Heat shock protein 101*	100,949	0.13	128	4	2
gi|944842	*ATP/ADP carrier protein*	36,013	0.12	108	5	4
*gi|3646373	RGP1 [Oryza sativa (indica cultivar-group)]	40,079	0.17	108	4	3
***gi|4158221**	Reversibly glycosylated polypeptide [Oryza sativa (indica cultivar-group)]	41,861	0.17	108	4	3
***gi|108709682**	Alpha-1,4-glucan-protein synthase [Oryza sativa (japonica cultivar-group)]	30,549	0.17	108	4	3
*gi|3334138	Calnexin homolog precursor	62,270	0.07	100	4	4
*gi|12585487	V-ATPase subunit C	40,131	0.22	91	3	2
*gi|18650668	*Temperature stress-induced lipocalin*	21,809	0.15	90	2	1
***gi|399414**	Elongation factor 1-alpha	49,480	0.28	75	12	12
***gi|90652740**	*Beta-glucosidase*	64,980	0.06	75	5	5
gi|115349890	*Fasciclin-like protein FLA3*	44,024	0.07	69	2	2
***gi|13027362**	Unknown protein (rice)	30,690	0.11	69	2	2
*gi|58339283	*Leucine zipper protein zip1*	48,111	0.29	57	3	3
Total peptides					908	

Table 2. List of proteins identified in fraction #3 by MudPIT using LTQ or Orbitrap instruments. Fraction #3 was obtained from sucrose density gradient in Figure 2. Hits corresponding to wheat proteins are in italic. Unique hits identified in the Orbitrap analysis are marked by "*".

compared to LTQ) and generation of new MS/MS spectra by the higher resolution of the Orbitrap. In addition, the Orbitrap analysis improved the scores (i.e., better MOWSE scores) of the identified proteins. Interestingly, some of the proteins identified in LTQ analysis were absent (or have lower scores) in Orbitrap analysis, which may indicate a higher rate of false positive in LTQ analysis. In addition, the decoy value for Orbitrap analysis was 54 corresponding to ~7% false discovery rate (compared to 11.33% in LTQ analysis). It is worth mentioning that we were able to improve protein identification by reducing false positive rate to ~3% after cleaning our samples from residual sucrose, lipids and slats without major increase in protein identification rate.

In the context of GAX biosynthesis, only the Orbitrap analysis resulted in the identification of several hits corresponding to reversibly glycosylated polypeptides (RGPs, GT75 family) from rice (gi|3646373, gi|4158221, gi|108709682) and wheat (gi|4158232). These proteins were identified by the following peptides (all with scores higher than 100): GIFWQEDIIPFFQNATIPK, NLDFLEMWRPFFQPYHLIIVQDGDPSK, YVDAVLTIPK, TGLPYLWHSK, VPEGFDYELYNR, NLLSPSTPFFFNTLYDPYR, and EGAPTAVSHGLWLNIPDYDAPTQMVKPR. However, both LTQ and Orbitrap analyses failed to identify TaGT43-4 and TaGT47-13 or any members of the GT43 or GT47 families. These results underscore the limitations of MudPIT in identifying some GTs involved in plant cell wall biosynthesis.

3.2.2 Gel-LC-MS/MS strategy to analyze protein content of fraction #3

3.2.2.1 Analysis of individual visible bands on 1-D SDS-PAGE

When individual visible bands on the gel (according to coomassie blue staining) were trypsin-digested and the resulting peptides analyzed by LTQ, a total of 169 proteins were matched by 2,106 peptide sequences (~5% of the total peptide sequences, Table 3). Among these 169 proteins, only three proteins (gi|4158232, gi|2218152, and Os01g0926600) were identified as GTs involved in GAX biosynthesis, and one of these three GTs (Os01g0926600)

Approximate MW of the bands (kDa)	Total No of peptide queries	No of peptide sequences with hits	No of hits identified	Top hit [No of peptides matched]	GTs involved in GAX biosynthesis [No of peptides matched]
30	4,356	42	14	gi|10720235 [15]	None
35	4,766	116	17	gi|10720235 [48]	None
40	4,790	122	21	gi|10720235 [21]	gi|4158232, gi|2218152, GT75 [3]
48	3,972	36	10	gi|10720235 [11]	None
50	4,124	127	19	gi|5917747 [15]	Os01g0926600, GT47 [1]
59	4,526	790	29	gi|2493131 [83]	None
64	4,275	376	20	gi|14017569 [62]	None
70	4,739	267	9	gi|90025017 [114]	None
80	4,430	169	17	gi|90025017 [28]	None
100	3,306	61	13	gi|544242 [3]	None
Total	43,284	2,106	169		

Table 3. Proteins identified from Gel-LC-MS/MS analysis of visible bands on SDS-PAGE gel of fraction #3 (Figure 3) involved in GAX biosynthesis. Fraction #3 was from EDTA-supplemented sucrose gradient (see Figure 2). The approximate MW of gel bands is indicated along with the number of proteins identified under the same band.

was newly identified by this analysis (not present in MudPIT analysis). Table 4 lists non-redundant proteins identified by each analysis of individual proteins band. There were 57 new proteins identified by this strategy (not present in MudPIT analysis). Among these new hits, Os01g0926600 (MW 47,271) was identified from the analysis of gel band around 50 kDa by the following peptide **IEGSAGDVLEDDPVGR** (score 79). This rice protein has an exostosin domain belonging to the family GT47 that also contains wheat members known to be involved in GAX synthesis (Zeng et al., 2010). Again, this strategy failed in identifying any members of the GT43 family.

3.2.2.2 Analysis of equal slices covering 30-180 kDa area of the 1-D SDS-PAGE

When the gel area between 30 and 180 kDa was sliced into 20-40 equal slices and each slice subjected to trypsin digestion, a total of 233 proteins were identified through LC-MS/MS using LTQ analysis. These hits were matched by 1,283 peptide sequences, a ~0.9% of the total peptide sequences (Table 5). Table 5 lists the total peptide queries resulted from each slice along with the number of hits identified in NCBI databases, and the number of peptides that matched these hits. This table also lists the top hit from the analysis of each slice (often the top hit is similar in many slices). Table 6 lists the 41 identified proteins

Accession No	Annotation	Score	Peptides matched
Band around 25 kDa			
gi|22204124	Putative 40S ribosomal protein S3 [Triticum aestivum]	117	2
gi|115458216	Os04g0405100 [Oryza sativa (japonica cultivar-group)]	116	2
gi|115475680	Os08g0277900 [Oryza sativa (japonica cultivar-group)]	103	2
gi|41529149	Putative acid phosphatase [Hordeum vulgare subsp. vulgare]	97	4
gi|22022400	Glutathione-S-transferase 19E50 [Triticum aestivum]	72	2
gi|47607439	Mitochondrial ATP synthase precursor [Triticum aestivum]	61	3
	Band around 30 kDa		
gi|162459667	Annexin p33 [Zea mays]	188	4
gi|115436780	Os01g0501800 [Oryza sativa (japonica cultivar-group)]	171	4
gi|162462814	Toc34-2 protein [Zea mays]	170	8
gi|82502214	Vacuolar proton ATPase subunit E [Triticum aestivum]	166	13
gi|2586127	β-keto acyl reductase [Hordeum vulgare]	124	4
gi|146231063	Hypersensitive response protein [Triticum aestivum]	97	4
gi|32308080	α-SNAP [Hordeum vulgare subsp. vulgare]	70	2
gi|15226197	Leucine-rich repeat transmembrane protein kinase [Arabidopsis thaliana]	58	3
	Band around 35 kDa		
gi|1658313	osr40g2 [Oryza sativa Indica Group]	163	5
gi|20302471	Ferredoxin-NADP(H) oxidoreductase [Triticum aestivum]	162	7
gi|115474135	Os07g0683600 [Oryza sativa (japonica cultivar-group)]	141	5
gi|21952858	Putative 60S ribosomal protein L5 [Oryza sativa Japonica Group]	132	3
gi|115434012	Os01g0104400 [Oryza sativa (japonica cultivar-group)]	118	6
gi|218155	Chloroplastic aldolase [Oryza sativa Japonica Group]	97	4
	Band around 40 kDa		
gi|146552329	Glyceraldehyde 3-phosphate dehydrogenase [Zehneria baueriana]	155	6
gi|120670	Glyceraldehyde-3-phosphate dehydrogenase [Zea mays]	130	5
gi|125553548	Hypothetical protein OsI_020490 [Oryza sativa (indica cultivar-group)]	120	9
gi|115437984	Os01g0587000 [Oryza sativa (japonica cultivar-group)]	119	2
gi|110278822	ATP synthase subunit gamma [Zea mays]	111	6

Accession No	Annotation	Score	Peptides matched
gi|157328342	Unnamed protein product [Vitis vinifera]	97	6
gi|2218152	Type IIIa membrane protein cp-wap13 [Vigna unguiculata]	86	3
gi|218157	Cytoplasmic aldolase [Oryza sativa Japonica Group]	63	2
gi|115473689	Os07g0643700 [Oryza sativa (japonica cultivar-group)]	62	3
gi|53791484	DUF642 containing protein, unknown protein [Oryza sativa Japonica Group]	58	2
	Band around 45 kDa		
gi|46911561	Putative vacuolar ATPase subunit H [Populus deltoides x maximowiczii]	140	2
gi|6015084	Elongation factor Tu [Pisum sativum]	137	3
gi|75243541	Probable V-type proton ATPase subunit H [Oryza sativa Japonica Group]	111	2
gi|50251779	Citrate synthase, glyoxysomal precursor [Oryza sativa Japonica Group]	78	2
	Band around 50 kDa		
gi|5917747	Elongation factor-1 alpha 3 [Lilium longiflorum]	241	15
gi|7452981	Allene oxide synthase [Hordeum vulgare subsp. vulgare]	125	6
gi|115465711	Os05g0585500 [Oryza sativa (japonica cultivar-group)]	91	2
gi|115441967	Os01g0926600 [Oryza sativa (japonica cultivar-group)] GT47	79	1
gi|37361675	RubisCO large subunit [Aloe niebuhriana]	64	5
gi|90289596	Alpha tubulin-2A [Triticum aestivum]	58	1
gi|2274988	Unnamed protein product [Hordeum vulgare subsp. vulgare]	57	2
	Band around 59 kDa		
gi|15233891	Vacuolar ATP synthase B2 [Arabidopsis thaliana]	1999	75
gi|115589736	Serine hydroxymethyltransferase [Triticum monococcum	551	21
gi|162462751	Mitochondrial F-1-ATPase subunit 2 [Zea mays]	528	19
gi|47522360	Putative calcium-dependent protein kinase [Triticum aestivum]	167	3
gi|81176509	atp1 [Triticum aestivum]	82	7
gi|77554943	Aminoacyl-tRNA synthetase [Oryza sativa (japonica cultivar-group)]	63	2
	Band around 64 kDa		
gi|14017569	ATP synthase CF1 alpha subunit [Triticum aestivum]	1363	62
Band around 70 kDa			
gi|1181331	Calnexin [Zea mays]	95	3
gi|51592190	Nucleotide pyrophosphatase/phosphodiesterase [Hordeum vulgare]	57	2
	Band around 80 kDa		
gi|50897038	Methionine synthase [Hordeum vulgare subsp. vulgare]	821	28
gi|8134568	5-methyltetrahydropteroyltriglutamate--homocysteine methyltransferase [common iceplant]	243	13
gi|82582811	Heat shock protein 90 [Triticum aestivum]	198	8
gi|32492578	RNA binding protein Rp120 [Oryza sativa (japonica cultivar-group)]	83	2
gi|115452177	Os03g0271200 [Oryza sativa (japonica cultivar-group)]	64	2
	Band around 100 kDa		
gi|544242	Endoplasmin homolog [Hordeum vulgare]	93	3
gi|2429087	Lipoxygenase 2 [Hordeum vulgare subsp. vulgare]	63	2

Table 4. Newly identified proteins by LC-MS/MS analysis from individual visible bands on SDS-PAGE gels (not identified in MudPIT analysis). Only proteins with scores >55 and/or two peptide matches are listed.

Gel area covered MW (kDa)	Slice #	Total No of peptide queries	No of sequences with hits	No of hits identified	Top hit [No of peptides matched]	GTs involved in GAX biosynthesis
~70 - ~180	1	4,683	88	13	gi|6715512	None
	2	4,845	141	11	gi|90025017	None
	3	4,761	41	7	gi|90025017	None
	4	5,262	137	23	gi|10720235	None
	5	5,229	37	5	gi|10720235	None
	6	4,971	17	5	gi|10720235	None
	7	4,866	23	7	gi|6715512	None
	8	5,009	23	6	gi|10720235	None
	9	4,249	3	1	gi|10720235	None
	10	4,804	12	1	gi|10720235	None
	11	4,897	23	4	gi|10720235	None
	12	4,623	160	15	gi|2493132	None
	13	4,283	38	11	gi|10720235	None
	16	3,937	10	3	gi|13375563	None
	17	3,793	3	2	gi|129707	None
~30 to ~70	1	5,236	21	5	gi|10720235	None
	2	4,722	16	4	gi|739292	None
	3	4,552	13	4	gi|10720235	None
	4	4,564	2	1	gi|10720235	None
	5	4,623	49	4	gi|14017569	None
	6	4,243	30	5	gi|6715512	None
	7	3,950	4	2	gi|20322	None
	8	4,165	4	2	gi|439586	None
	9	4,596	40	8	gi|10720235	gi|159470791, gi|159471277, GT47 family
	10	3,834	18	5	gi|57471704	None
	11	3,481	5	2	gi|10720235	None
	12	4,171	68	9	gi|90025017	None
	13	4,852	17	3	gi|129708	None
	14	3,880	21	5	gi|10720235	None
	15	3,764	283	14	gi|2493131	None
	16	1,007	26	4	gi|10720235	gi|2218152, gi|4158232, GT75 family
	17	1,994	37	6	gi|10720235	None
	18	2,212	40	8	gi|10720235	gi|2218152, gi|4158232, GT75 family
	19	4,766	116	17	gi|10720235	None
	20	4,790	122	21	gi|10720235	None
Total	37	149,614	1283	233		

Table 5. Proteins involved in GAX biosynthesis identified in fraction #3 by Gel-LC-MS/MS and LTQ strategy. The gel area between 30 and 180 kDa of SDS-PAGE was sliced into 20-40 slices (see Figure 3) and each slice was trypsin-digested and analyzed.

(among the 233 hits) that were unique to this strategy (not in previous analyses). Among the unique hits identified by this strategy, two Chlamydomonas reinhardtii GTs (gi|159470791, and gi|159471277) belonging to the GT47 family (both annotated as exostosin-like glycosyltransferase) were identified by the following peptide RVAEADIPRL (score 56). This

Accession No	Annotation	Score	Peptides matched
gi|6715512	V-type H+ ATPase B subunit [Nicotiana tabacum]	117	18
gi|2493650	RuBisCO large subunit-binding protein subunit beta	77	2
gi|475600	BiP isoform B [Glycine max]	102	5
gi|123656	Heat shock-related protein [Spinacia oleracea]	94	7
gi|115458184	Calreticulin family, Os04g0402100	62	2
gi|42541152	Delta tonoplast intrinsic protein TIP2;2 [Triticum aestivum]	120	3
gi|1709846	Photosystem II 22 kDa protein [Lycopersicon esculentum]	109	2
gi|4099406	Camma-type tonoplast intrinsic protein [Triticum aestivum]	95	3
gi|28569578	Allene oxide synthase [Triticum aestivum]	59	3
gi|1709358	Nucleoside-triphosphatase [Pisum sativum]	81	2
gi|904147	Adenosine triphosphatase [Sinofranchetia chinensis]	237	11
gi|15010616	AT4g38510/F20M13_70 [Arabidopsis thaliana]	235	8
gi|24496452	Actin [Hordeum vulgare]	172	5
gi|115589744	S-adenosylmethionine synthetase 1 [Triticum monococcum]	74	2
gi|13375563	Lipid transfer protein precursor [Triticum aestivum]	154	4
gi|14017578	ATP synthase CF1 epsilon subunit [Triticum aestivum]	99	2
gi|16225	Calmodulin [Arabidopsis thaliana]	69	2
gi|75108545	Peroxiredoxin Q, chloroplastic [Triticum aestivum]	95	2
gi|464517	50S ribosomal protein L12-1 [Secale cereale]	67	4
gi|115442509	Cyt-b5 family, Os01g0971500 [Oryza sativa (japonica cultivar-group)]	65	3
gi|42565453	Cyclophilin [Hyacinthus orientalis]	59	2
gi|68566191	Cytochrome b6-f complex [Triticum aestivum]	129	5
gi|118104	Peptidyl-prolyl cis-trans isomerase [Zea mays]	89	2
gi|154761388	Cyclophilin [Triticum aestivum]	89	2
gi|231496	Actin-58 [Solanum tuberosum]	166	8
gi|115467154	Os06g0221200 annexin family [Oryza sativa (japonica cultivar-group)]	86	2
gi|52548250	ADP-ribosylation factor [Triticum aestivum]	162	6
gi|74048999	Eukaryotic translation initiation factor 5A1 [Triticum aestivum]	71	2
gi|57471704	Ribosomal protein L11 [Triticum aestivum]	186	8
gi|432607	Ras-related GTP binding protein possessing GTPase activity [Oryza sativa]	119	4
gi|115441299	Os01g0869800, PsbS subunit [Oryza sativa (japonica cultivar-group)]	73	2
gi|115444503	Os02g0171100 [Oryza sativa (japonica cultivar-group)]	283	4
gi|16304127	Glyceraldehyde 3-phosphate dehydrogenase 1 [Fragaria x ananassa]	132	3
gi|18071421	Putative dehydrogenase [Oryza sativa (japonica cultivar-group)]	125	4
gi|166627	Nucleotide-binding subunit of vacuolar ATPase [Arabidopsis thaliana]	115	2
gi|8272480	Fructose 1,6-bisphosphate aldolase precursor [Avena sativa]	92	5
gi|159470791	Exostosin-like glycosyltransferase [Chlamydomonas reinhardtii]	56	2
gi|159471277	Exostosin-like glycosyltransferase [Chlamydomonas reinhardtii]	56	1
gi|115451383	Os03g0200800 [Oryza sativa (japonica cultivar-group)]	202	4
gi|147858623	Hypothetical protein [Vitis vinifera]	65	2

Table 6. List of unique hits identified through Gel-LC-MS/MS and LTQ analysis of 20-40 slices covering 30-180 kDa area of the SDS-PAGE. Only hits with scores >55 and/or two peptide matches are listed.

strategy, however, identified the exact wheat RGP protein (TaGT75-4, gi|4158232) with the following peptides VPEGFDYELYNR and YVDAVLTIPK (both with score 59). Therefore, this strategy successfully identified TaGT75-4 protein and homolog to TaGT47-13 but failed to identify the exact TaGT47-13 protein or any homolog to TaGT43-4 protein.

4. Discussion

Hemicellulosic polymers such as GAX represent up to 40% (w/w) of grass cell walls (in particular from growing tissues). In sharp contrast with the abundance of these polymers, the GTs that synthesize these compounds are present in low amounts in Golgi membranes of the plant cell. This observation suggests that these enzymes are highly active and may not be required in large quantities in the plant cell. This low abundance of GTs has been the main limiting factor in applying proteomics approaches to plant cell wall biosynthesis. To further complicate the issue, isolation of GTs from Golgi membranes (or simply disrupting these membranes) generally results in a drastic reduction or loss of transferase activity *in vitro*. To detect this weak transferase activity *in vitro*, it is necessary to use very sensitive biochemical assays (i.e., [^{14}C]radiolabeled sugars-based assay). Since the loss of transfer activity is GT-dependent, the biochemical assays are not the best way to estimate the abundance of these enzymes in a particular protein preparation. Therefore, when working with plant cell wall GTs, all these factors should be taken in consideration. In this work, such *in vitro* assay was used to monitor the distribution of GAX synthase activity (from Golgi-enriched membranes) on a linear sucrose density gradient supplemented with EDTA as described earlier (Zeng et al., 2010). According to our *in vitro* assay, fraction #3 was substantially enriched in GAX synthase activity (Figure 2), and it can be assumed that this fraction is also enriched in TaGT43-4, TaGT47-13, and TaGT75-4 proteins. Therefore, fraction #3 is an excellent starting material to evaluate proteomics strategies in identifying these three GTs among a mixture of proteins. Furthermore, because genome and protein sequence information from five grass species are currently publicly available (Figure 1), it can be expected that proteomics analysis on wheat would be successful.

Our analyses indicated that gel-based proteomics approach (gel-LC-MS/MS) has a superior result compared to gel-free approach (i.e., MudPIT). In the MudPIT strategy, LTQ and Orbitrap analyses identified a total of 83 non-redundant proteins, but only 14 of these proteins where in common (Figure 4). However, the Orbitrap gave higher scores and protein identification rates. On the other hand, the Gel-LC-MS/MS strategy resulted in the identification of a total of 180 non-redundant proteins, among which 83 proteins were in common with MudPIT analyses (97 new proteins) (Figure 4).

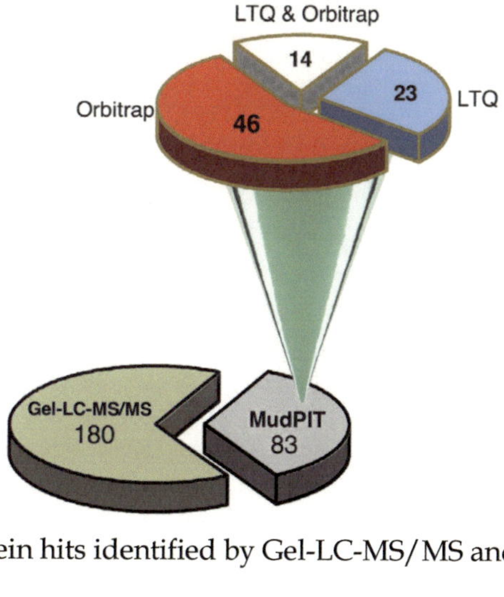

Fig. 4. Distribution of protein hits identified by Gel-LC-MS/MS and MudPIT strategies.

Regarding the ability to identify GTs, the Gel-LC-MS/MS strategy identified most of the GTs associated with GAX biosynthesis. Intriguingly, all the strategies used failed to identify TaGT43-4 or any closest homolog from the NCBI database. Three possibilities could explain this result: (*i*) the TaGT43-4 protein may be lost during the precipitation step (preparation of the sample); (*ii*) TaGT43-4 is a very active enzyme and is present in only small amounts in fraction #3, which may not be detectable by the LC-MS/MS methods used in this work, or (*iii*) TaGT43-4 protein is somehow resistant to trypsin digestion.

Our hypothesis is that most of the TaGT43-4 protein was lost during the precipitation step, as Golgi proteins are known to easily aggregated during precipitation and are very difficult to re-solubilize in a buffer containing detergent. Although it has been shown that ASB-14 and SDS detergents are suited for solubilizing hydrophobic proteins (Herbert, 1999), their use in this study may not be efficient in re-solubilizing freeze-dried or TCA/acetone precipitated wheat Golgi proteins. In support of this hypothesis, fraction #3 should be enriched in Golgi proteins (Zeng et al., 2010), but our analysis indicates that fraction #3 was actually enriched in endoplasmic reticulum (14%), tonoplast (17%), and plastid (28%) proteins, and Golgi proteins represented only 2% of the total hits (according to NCBI annotation of possible subcellular localizations) (Figure 5). Therefore, a reliable 'precipitation-re-solubilization' strategy appears to present a crucial step that must be optimized for minimal protein loss. Alternatively, improving enrichment strategies to overcome protein loss during the `precipitation-re-solubilization` step should be developed.

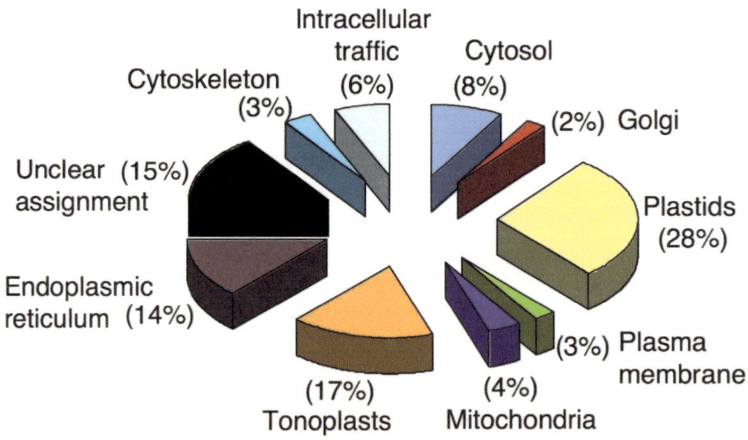

Fig. 5. Classification of proteins identified in fraction #3 according to NCBI annotation of their possible sub-localization.

Although all proteomics strategies employed here have failed to reveal the exact identity of some GTs associated with GAX biosynthesis, proteomics is still a powerful tool, as many low abundant GTs (among the 2% proteins from the Golgi) could be identified. Furthermore, this work demonstrated that working with a non-model species without a fully sequenced genome such as wheat did not seem to be an issue, as most (40-60%) of the proteins identified were either from wheat sequences available in the databases, or were closest homologs to the anticipated wheat proteins from grass species (rice, barley, maize, or sorghum). The other limitation in applying proteomics to plant cell wall biosynthesis is the capacity of a mass spectrometer analyzer to extract as many MS/MS spectra as possible to increase the detection rate of proteins. To overcome all these issues and depending of the complexity of the protein sample, we are proposing a workflow to carry out a successful proteomics analysis (Figure 6). In this workflow, the first step is to assess the quality of the sample by optimizing the precipitation step (removal of salts and contaminants) without any protein loss. Our work demonstrated that "precipitation-re-solubilization" step is crucial in a successful proteomics analysis of Golgi membrane proteins. Depending of the complexity of the samples, the simplest proteomics strategy to try is the combination of MudPIT fractionation with LTQ analysis. If the sample contains more than 500 proteins, the use of high resolution mass spectrometry (e.g. Orbitrap) in combination with MudPIT could

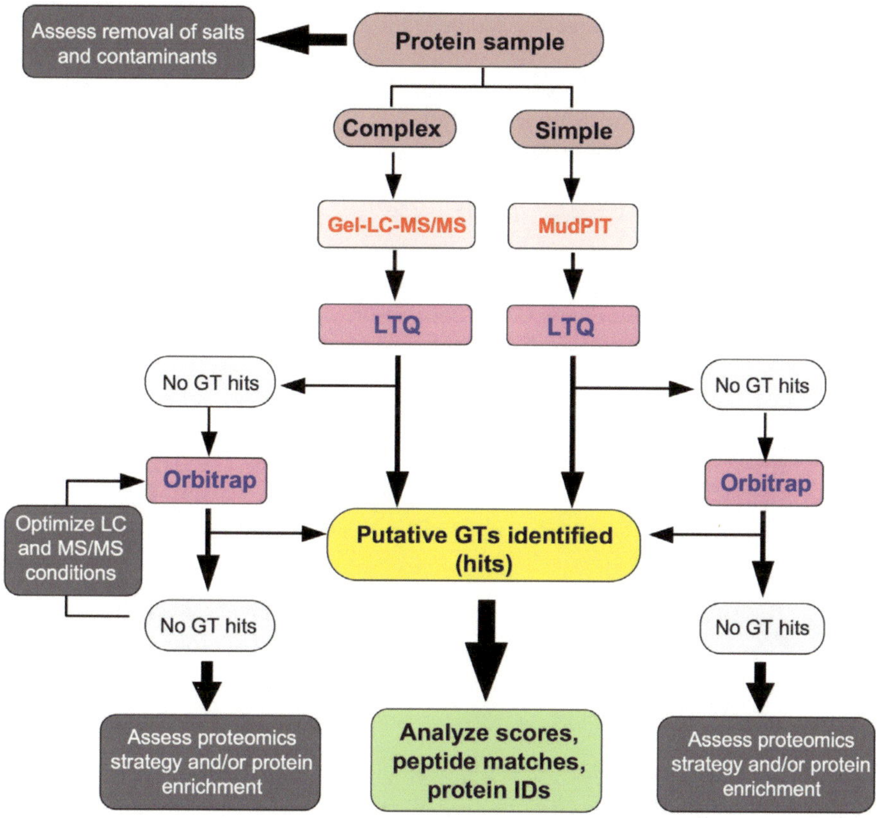

Fig. 6. A proteomics workflow pipeline for efficient protein identification from unknown samples.

be the easiest strategy to test. For more complex protein samples (more than 1000 proteins) it may be necessary to combine 1-D SDS-PAGE fractionation, LC-MS/MS and the high resolving power of the Orbitrap analyzer for optimal protein identification (Figure 6).
The distribution of GAX synthase over sucrose density gradient was intriguing. In the absence of EDTA, all GAX synthase activity stabilized at the expected density of ~1.16g/mL (fractions 17 and 18 in Figure 2) along with the Golgi marker activity IDPase (Zeng et al., 2010). The inclusion of EDTA in the sucrose gradient resulted in splitting of the activity into three density areas, namely around density 1.09g/mL (fractions 2 and 3 in Figure 2), around density 1.14g/mL (fractions 12 and 13 in Figure 2), and around density 1.16g/mL (fractions 17 and 18 in Figure 2). Fraction #3 contained the highest GAX synthase activity. This shift in the density

may suggest that GAX synthase activity is associated with various Golgi compartments. The presence of such Golgi compartments was reported earlier by Mikami et al. (2001) in rice but not in tobacco cells. They showed that rice Golgi complex fractionated into several compartments by simple centrifugation on density gradient in presence of EDTA or $MgCl_2$. Recently, Asakura et al., (2006) used this strategy to isolate (and analyze by proteomics) rice cis-Golgi membranes labeled with green fluorescent protein (GFP) fused to a cis-Golgi marker SYP31 (which belongs to a family of SNARE proteins; soluble N-ethyl-melaeimide sensitive factor attachment protein receptor). Interestingly, their proteomics results gave very similar protein composition to our data, except that no members of the GT43, 47, and 75 were identified in their study (Asakura et al., 2006). Taking together, these results suggest that the cis-Golgi is less tightly attached to the medial and trans-Golgi compartments in grasses. Therefore, we are tempted to propose a possible explanation of the effect of EDTA on the dissociation of these Golgi compartments (Figure 7). Our hypothesis is that some ions (*i.e.*, Ca^{2+}) are involved in linking cis-Golgi (and probably the trans Golgi network [TGN]) to the medial and trans-Golgi cisternae. In the absence of EDTA, the whole Golgi complex would stabilize at an apparent high density (~1.16g/mL, fraction 17 and 18 in Figure 2). Upon addition of EDTA into sucrose gradient, the ions are chelated leaving different Golgi compartments stabilized at their corresponding densities (1.09, 1.12, and 1.16g/mL in Figure 2). In any case, the fractionation of the Golgi complex from grasses in the presence of EDTA is an excellent tool for isolating different Golgi compartments.

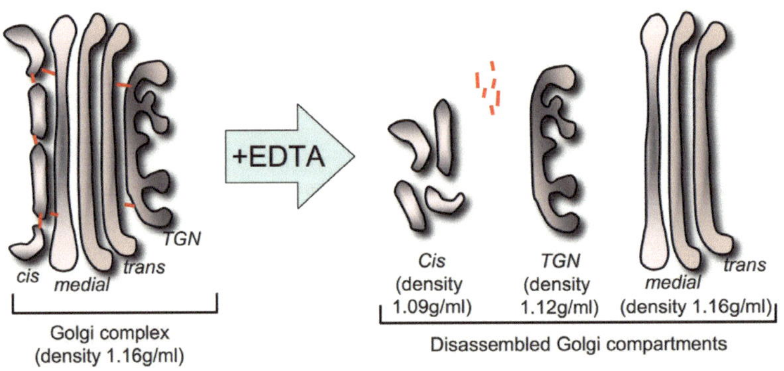

Fig. 7. A diagrammatic representation of the possible effect of EDTA on the dissociation of Golgi compartments, cis-, medial-, trans-Golgi, and trans Golgi network (TGN). EDTA would chelate metal ions that mediate the attachment (red lines) of cis-Golgi compartment to the medial and trans-Golgi cisternae.

5. Conclusion

We evaluated two proteomics strategies for MS/MS identification of GTs associated with GAX synthase complex in wheat for which little genome sequence is available. The evaluation of these strategies is based on their capacity to identify the exact wheat proteins TaGT43-4, TaGT47-13, and TaGT75-4, or at least their closest homologous proteins from other grass species such as rice, barley, maize, or sorghum. Therefore, these strategies are MS/MS spectra quality-dependent and cross-species-dependent using error tolerant BLAST search and *de novo* peptide sequences generated from the MS/MS spectra. Our data indicated that the highest number of unique hits identified (180 proteins) was obtained through Gel-LC-MS/MS strategy using the Orbitrap analyzer, but the fact that TaGT43-4 and/or its homologous proteins were not identified by this method underscores the importance of optimizing sample preparation step (precipitation-re-solubilization). Based on our results and interpretations, we have proposed a workflow chart that includes routinely used proteomics methods and optimization steps to help increase the detection rate of proteins of plant cell wall GTs from non-model plants.

6. Acknowledgment

The author would like to thank Dr. Green-Church for her valuable comments and editing of the experimental procedures dealing with proteomics analysis. Thanks to Wei Zeng and Nan Jiang for the excellent work preparing protein samples for proteomics. This work was supported by the National Science Foundation (grant no. IOS–0724135 to A.F.)

7. References

Asakura, T., Hirose, S., Katamine, H., Kitajima, A., Hori, H., Sato, M.H., Fujiwara, M., Shimamoto, K. & Mitsui, T. (2006) Isolation and proteomic analysis of rice Golgi membranes: cis-Golgi membranes labeled with GFP-SYP31. *Plant Biotechnology* 23: 475-485

Arabidopsis Genome Initiative (2000) Analysis of the genome sequence of the flowering plant *Arabidopsis thaliana*. *Nature* 408: 796–815

Bodnar, W.M., Blackburn, R.K., Krise, J.M. & Moseley, M.A. (2003) Exploiting the complementary nature of LC/MALDI/MS/MS and LC/ESI/MS/MS for increased proteome coverage. *J. Am. Soc. Mass Spectrom.* 14: 971-979

Carpita, N.C. (2011) Update on Mechanisms of Plant Cell Wall Biosynthesis: How Plants Make Cellulose and Other (1-4)-β-D-Glycans. *Plant Physiol.*, 155: 171-184

Chen, H.S., Rejtar, T., Andreev, V., Moskovets, E. & Karger, B.L. (2005) Enhanced characterization of complex proteomic samples using LC-MALDI MS/MS: exclusion of redundant peptides from MS/MS analysis in replicate runs. *Anal. Chem.*, 77: 7816-7825

Chen, Y.Y., Lin, S.Y., Yeh, Y.Y., Hsiao, H-H., Wu, C-Y., Chen, S-T. & Wang A. H-J. (2005) A modified protein precipitation procedure for efficient removal of albumin from serum. *Electrophoresis*, 26: 2117-27

Coutinho, P.M., Deleury, E., Davies, G.J. & Henrissat, B. (2003) An evolving hierarchical family classification for glycosyltransferases. *J. Mol. Biol.* 328: 307-317

Ephritikhine, G., Ferro, M. and Rolland, N. (2004) Plant membranes proteomics: *Plant Physiol. Biochem.*, 42: 943-962

Faik, A. (2010) xylan biosynthesis: news from the grass. *Plant Physiol.*, 153: 396-40

Faik, A., Bar-Peled, M., DeRocher, A.E., Zeng, W., Perrin, R.M., Wilkerson, C., Raikhel, N.V. & Keegstra K. (2000) Biochemical characterization and molecular cloning of an alpha-1,2-fucosyltransferase that catalyzes the last step of cell wall xyloglucan biosynthesis in pea. *J. Biol. Chem.*, 275: 15082-15089.

Ferro, M., Salvi, D., Brugiere, S., Miras, S., Kowalski, S., Louwagie, M., Garin, J., Joyard, J. & Rolland, N. (2003) Proteomics of the chloroplast envelop membranes from *Arabidopsis thaliana*. *Mol. Cell Proteomics*, 2: 325-345

Herbert, B. (1999) Advances in protein solubilization for two-dimenstional electrophoresis. *Electrophoresis*, 20: 660-663

Komatsu, S., Muhammad, A. & Rakwal, R. (1999) Separation and characterization of proteins from green and etiolated shoots of rice (*Ozyza sativa* L.): towards a rice proteome. *Electrophoresis*, 20: 630-636

Laemmli, U.K. (1970) Cleavage of structural proteins during the assembly of the head of bacteriophage T4. *Nature*, 227: 680-685

McDonald, W.H., Ohi, R., Miyamoto, D.T., Mitchison, T.J. & Yates III, J.R. (2002) Comparison of three directly coupled HPLC MS/MS, 2-phase MudPIT, and 3-phase MudPIT. *Int. J. Mass Spectrom.* 219: 245-251

Mikami, S., Hori, H. & Mitsui, T. (2001) Separation of distinct compartments of rice Golgi complex by sucrose density gradient centrifugation. *Plant Sci.*, 161: 665-675

Newton, R.P., Breton, A.G., Smith, C.J. & Dudley, E. (2004) Plant proteome analysis by mass spectrometry: principales, problems, pitfalls and recent developments. *Phytochemistry*, 65: 1449-1485

Nesvizhskii, A.I., Vitek, O. & Aebersold, R. (2007) Analysis and validation of proteomic data generated by tandem mass spectrometry. *Nat. Methods* 4: 787-797

O'Farrell, P.H. (1975) High resolution two-dimensional electrophoresis of proteins. *J. Biol. Chem.*, 250: 4007-4021

Patterson, S.D. (2003) Data analysis-the Achilles heel of proteomics. *Nat. Biotechnol.*, 21: 221-222

Perrin, R.M., DeRocher, A.E., Bar-Peled, M., Zeng, W., Norambuena, L., Orellana, A., Raikhel, N.V. & Keegstra, K. (1999) Xyloglucan fucosyltransferase, an enzyme involved in plant cell wall biosynthesis. *Science* 284: 1976–1979

Santoni, V., Kieffer, S., Desclaux, D., Masson, F. & Rabilloud, T. (2000) Membrane proteomics: use of additive main effects with multiplicative interaction model to classify plasma membrane proteins according to their solubility and electrophoretic properties. *Electrophoresis*, 21: 3329-3344

Service, R.F. (2008) Proteomics. Proteomics ponders prime time. *Science* 321: 1758-1761

Shevchenko, A., Sunyaev, S., Loboda, A., Shevehenko, A., Bork, P., Ens, W. & Standing, K.G. (2001) Charting the proteomes of organisms with unsequenced genomes by MALDI-quadrupole time of flight mass spectrometry and Blast homology searching. *Anal. Chem.*, 73: 1917-1926

Simon, W.J., Maltman, D.J. & Slabas, A.R. (2008) Isolation and fractionation of the endoplasmic reticulum from castor bean (Ricinus communis) endosperm for proteomic analysis. *Methods Mol. Biol.*, 425: 203-215

Washburn, M.P., Wolters, D. & Yates III, J.R. (2001) Large-scale analysis of the yeast proteome by multidimensional protein identification technology. *Nat. Biotechnol.*, 19: 242-247

Wolters, D.A., Washburn, M.P. & Yates III, J.R. (2001) An automated multidimentional protein identification technology for shotgun proteomics. *Anal Chem.*, 73: 5683-5690

Zellner, M., Winkler, W., Hayden, H., Diestinger. M., Eliasen, M., Gesslbauer, B., Miller, I., Chang, M., Kungl, A., Roth, E. & Oehler, R. (2005) Quantitative validation of different protein precipitation methods in proteome analysis of blood platelets *Electrophoresis*,26: 2481-2489

Zhu, W., Smith, J.W. & Huang C.M. (2010) Mass spectrometry-based label-free quantitative proteomics. *J. Biomed. Biotechnol.*, 840518. Doi:10.1155/2010/840518

Zeng, W., Jiang, N., Nadella, R., Killen, T.L., Nadella, V. & Faik, A. (2010) A glucurono(arabino)xylan synthase complex from wheat contains members of the GT43, GT47, and GT75 families and functions cooperatively. *Plant Physiol.*, 154: 78-97

8

The Current State of the Golgi Proteomes

Harriet T. Parsons[1], Jun Ito[1], Eunsook Park[2], Andrew W. Carroll[1], Hiren J. Joshi[1], Christopher J. Petzold[1], Georgia Drakakaki[2] and Joshua L. Heazlewood[1]

[1]Joint BioEnergy Institute and Physical Biosciences Division, Lawrence Berkeley National Laboratory
[2]Department of Plant Sciences, University of California, Davis
USA

1. Introduction

The Golgi apparatus plays a central role in the eukaryotic secretory pathway shuttling products between a variety of destinations throughout the cell. It is a major site for the post translational modification and processing of proteins as well as having a significant role in the synthesis of complex carbohydrates. The Golgi apparatus exists as a contiguous component of the endomembrane system which encompasses the endoplasmic reticulum (ER), plasma membrane, vacuoles, endosomes and lysozymes (Morre & Mollenhauer, 2009). The Golgi apparatus is tightly linked to numerous signaling processes through membrane and vesicular trafficking throughout the endomembrane. This interconnection provides communication and recycling networks between the Golgi apparatus, the plasma membrane, vacuoles and lysosomes. Thus, the Golgi apparatus represents significant structure within the eukaryotic cell by regulating an array of complex biosynthetic processes.

The Golgi apparatus was first described at the end of the 19th century by Camillo Golgi using light microscopy on nerve tissue samples (Golgi, 1898; Dröscher, 1998). A half century passed and with the development of the electron microscope a more detailed picture emerged highlighting the extreme complexity and heterogeneity of the organelle in the eukaryotic cell (Dalton & Felix, 1953). The classic structure of the Golgi apparatus is that of a distinct membranous stack disassociated within the cytosol (Fig.1). This familiar image conceals the underlying complexity and interconnected nature of this organelle within the cell. With the development in the last decade of routine mass spectrometry-based proteomics, applying these methods to functionally characterize biological systems has been a major focus. The complexity and integrated structure of the Golgi apparatus and associated membrane systems makes analysis of this organelle one of the most complicated subcellular compartments to address with modern proteomics techniques. This chapter will highlight recent advances in our knowledge about the Golgi apparatus in eukaryotic systems that have been largely driven by the development of isolation procedures and subsequent proteomic analysis.

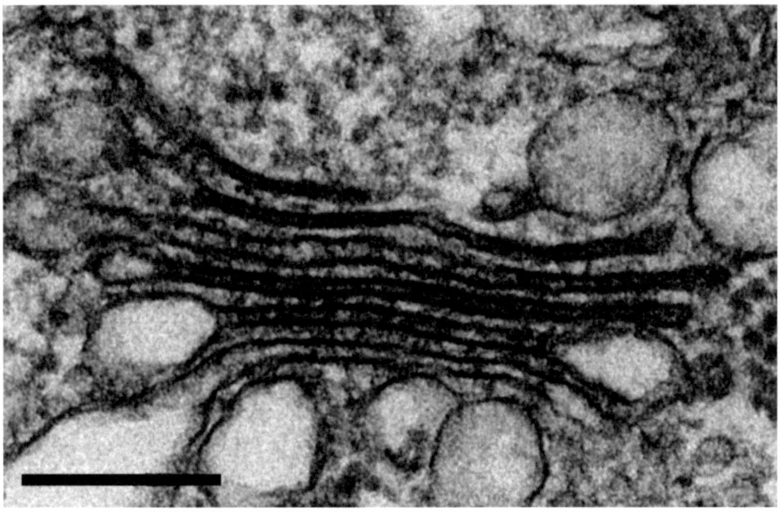

Fig. 1. Electron micrograph of a Golgi stack from the model plant *Arabidopsis thaliana* highlighting the integrative structure and membrane organization. Scale bar = 200nm.

2. Differential density enrichment of Golgi

The basic technique of differential density enrichment of Golgi membranes represents the most common isolation and enrichment process for downstream proteomic analyses. The technique was well-established in most eukaryotic systems prior to the development of mass spectrometry-based identification techniques. Consequently, analysis of the enriched Golgi apparatus fraction using this method was a logical approach, although one limited by contaminating organellar membranes and low gradient resolution. Yet, insights into membrane systems associated with trafficking, post-translational modifications and complex carbohydrate biosynthesis have been revealed.

Many of the initial approaches used to characterize Golgi associated proteins by differential centrifugation employed SDS-PAGE arraying techniques prior to identification of proteins by mass spectrometry. Nearly all of the early proteomic studies on enriched Golgi fractions are from easily accessible samples such as rat livers, likely reflecting the need for the development of purification techniques. The earliest 'proteomic' analyses of the Golgi employed two-dimensional gel electrophoresis (2-DE) to array enriched stacked Golgi fractions from rat livers (Taylor et al., 1997b). The study used cyclohexamide in an effort to clear transitory proteins from the secretory pathway and reduce non-specific Golgi proteins. While only a handful of proteins were identified by cross comparing reference maps and immunoblotting the study demonstrated the validity of a proteomic approach to analyze the Golgi and could discern resident proteins from cargo and cytosolic proteins (Taylor et al., 1997b). Significantly, the study employed a recently developed sequential sucrose gradient enrichment method (Taylor et al., 1997a) and enabled the reliable visualization of Golgi proteins by 2-DE with reduced contaminants. The sequential sucrose enrichment of stacked Golgi from liver samples involved initially loading a clarified homogenate (post-nuclear supernatant) between 0.86 M and 0.25 M sucrose steps followed by centrifugation. The resultant 0.5/0.86M interface (Int-2) was removed, adjusted to 1.15M sucrose and overlaid with 1.0M, 0.86M and 0.25M sucrose and centrifuged. The resultant 0.25/0.86 interface represented the enriched stacked Golgi fraction (Fig. 2). This fraction was estimated to represent 200 to 400- fold enrichment over the post-nuclear fraction (Taylor et al., 1997a). With the development of reliable protein identification through tandem mass spectrometry this Golgi enrichment technique could be more readily exploited for functional proteomic studies. This was undertaken through 2-DE arraying of Golgi samples using the sequential sucrose technique on rat liver and mammary epithelial samples (Taylor et al., 2000; Wu et al., 2000). Both studies employed early liquid chromatography-tandem mass spectrometry (LC-MS/MS) methods using iontrap MS and resulted in the identification of 71 proteins from 588 unique 2-DE spots of rat liver (Taylor et al., 2000) and over 30 distinct proteins from mammary epithelial cells (Wu et al., 2000). This later work outlined a comparative study of Golgi isolated from cells transiting from basal (steady-state) to maximal secretion. Proteins identified by this study were upregulated during this transition and comprised a series of Rab proteins, structural components such as microtubule motor proteins and membrane fusion proteins e.g. Annexins (Wu et al., 2000). Another early attempt at characterizing Golgi associated proteins by proteomic methods employed a previously developed strategy to enrich WNG fractions (nuclear associated Golgi fractions) from rat liver homogenates (Dominguez et al., 1999). The technique employs multiple rounds of low speed centrifugation steps in conjunction with a 0.25 to 1.10 M sucrose gradient. For analysis by mass spectrometry, proteins from the resultant rat liver WNG fraction were partitioned into a detergent fraction using Triton X-114 to enrich for membrane proteins and arrayed by SDS-PAGE or 1-DE (Bell et al., 2001). A total of 81 proteins were identified from 1-DE arrayed samples using MALDI-MS, nano-MS/MS or Edman sequencing and included a range of well-characterized Golgi proteins including lectins, KDEL receptor, glycosyltransferases and trafficking proteins (Rabs, SNAREs, SCAMPs). While the results demonstrated the applicability of the overall approach, a significant number of contaminants were also identified in the samples (Bell et al., 2001).

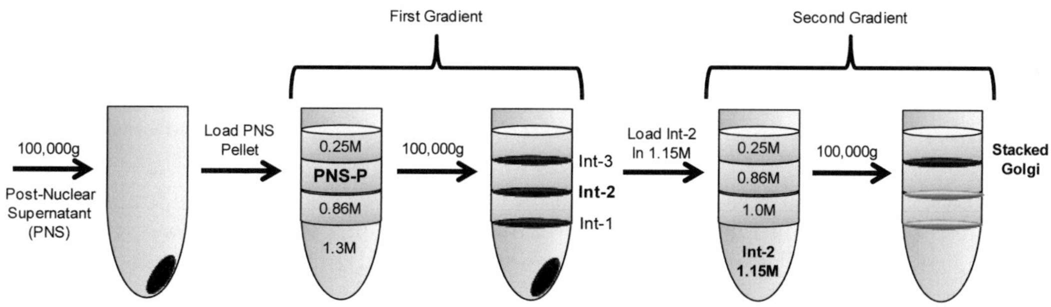

Fig. 2. Schematic outline of the sequential sucrose density gradient procedure developed for the isolation and purification of Golgi stacks from rat liver (Taylor et al., 1997a). The values within the tubes indicate concentration of sucrose (M: mol/L).

With the development of advanced gel-free approaches like multidimensional protein identification technology (MudPIT) for the analysis of complex lysates by mass spectrometry (Wu et al., 2003), a number of previously established enrichment strategies were again analyzed. The two independent and sequential sucrose step gradient technique (outlined above) was employed to isolate Golgi membranes from the livers of rats treated with cyclohexamide (Wu et al., 2004). These fractions were analyzed directly by MudPIT, using a 12-step strong cation exchange (SCX) fractionation coupled to reverse phase LC-MS/MS by iontrap MS. From this approach, 421 proteins were identified from Golgi fractions isolated from rat liver homogenates (Wu et al., 2004). About 26% of the proteins (110) could be confidently allocated to Golgi-associated functions with about 10% allocated as unknowns. A large proportion of the identifications represented proteins from contaminating organelles including ER (23%), plasma membrane (9%) and cytosol (9%). Nonetheless, a large number of proteins were identified from major Golgi functional processes including transporters, SNARES, glycosyltransferases, G-proteins and clathrin proteins (Wu et al., 2004). Importantly, the gel-free approach enabled the detection of protein modifications with a total of 10 Golgi proteins identified with arginine dimethylation, a potential regulator of protein function (Wu et al., 2004). Stacked Golgi membranes isolated from rat liver homogenates were again analyzed by MudPIT (using a Q-TOF) in a separate study. After initially enriching the membranes by a discontinuous sucrose gradient, repeated centrifugations through 1.3M sucrose were used to clarify and purify the Golgi stacks (Takatalo et al., 2006). Replicate experiments were fractionated offline by SCX prior to analysis by LC-MS/MS. A total of 1125 proteins were identified in the two experiments by combining results from two separate MS/MS search algorithms (Takatalo et al., 2006). Analysis of these protein identifications found that 35% contained at least one predicted transmembrane domain and 201 proteins could be assigned as Golgi or transport related based on functional annotations. When compared to the similar approach of Wu et al., (2004) a total of 399 proteins were identified by both studies. The multiple isolation and analysis strategies enabled the number of unknown proteins from the rat liver Golgi fraction to be significantly extended by 89 proteins. Overall, proteins associated with glycosylation, Golgi matrix complexes, enzymes involved in isomerase and transferase activities and protein trafficking components were all identified (Takatalo et al., 2006). Nonetheless, limitations in the purification procedure are highlighted by positive allocations of this proteome to the cytosol (15%), contaminating membranes (20%) and the cryptically named 'undetermined location' (20%) (Takatalo et al., 2006).

In an effort to address contamination issues and to better tease apart and define the ER, quantitative proteomic strategies were subsequently employed on rat liver Golgi enriched by sucrose density centrifugation (Gilchrist et al., 2006). Previous methods used to isolate rat Golgi fractions (Bell et al., 2001) were used as well as rough and smooth microsomal preparations (Paiement et al., 2005). This was followed by 1-DE, band isolation and analysis by nanoLC-MS/MS (Q-TOF MS). Based on annotation information, Golgi fractions were estimated to comprise around 10% mitochondrial protein, although there was a distinction in proteins identified between the Golgi and microsomal fractions. Multiple biological replicates for the Golgi and microsomes were undertaken and proteins quantified by redundant spectral counting techniques (Blondeau et al., 2004). Hierarchical clustering with Pearson correlation was used to visualize protein abundance measures and this analysis identified four major groups which co-clustered with organelle markers sec61 (rough microsome), calnexin (ER), p97 (smooth microsomes) and mannosidase II (Golgi). This co-clustering with defined markers indicated that quantifying distinct fractions could be used to tease apart the secretory pathway (Gilchrist et al., 2006). Nonetheless only 43 additional proteins could be confidently assigned to the Golgi compared to a previous analysis that had identified 81 proteins (Bell et al., 2001). Consequently COP1 vesicle enrichment from the Golgi fraction was also undertaken to extend the proteome. Overall 1430 proteins were identified from the rat liver secretory system with quantitative proteomics enabling the confident assignment of 193 proteins to Golgi/COP1 vesicles, with 405 proteins shared between the Golgi and ER fractions. The work identified a host of secretory proteins including Rabs, SNARES, Sec proteins and unknowns (Gilchrist et al., 2006). A similar MudPIT approach in mouse was undertaken by employing spectral counting for quantification of proteins from multiple subcellular fractions (cytosol, microsomes, mitochondria and nuclei) to determine distinct proteomes (Kislinger et al., 2006). This study also isolated compartments from different organs and undertook organ specific transcript profiles to develop machine-learning techniques to classify subcellular localizations. The enrichment method for the microsomal fraction (Golgi) resulted from high speed centrifugation of a post-nuclear supernatant and thus represents a fairly crude membrane preparation. Nonetheless, by employing samples from multiple organs, transcript profiling and modeling techniques a significant number of secretory proteins were identified and expressions profiled (Kislinger et al., 2006).

The focus on cellular secretory system as an important component in agricultural milk production was explored with the analysis of microsomal fractions from bovine mammary glands (Peng et al., 2008). The study used a 100,000g microsomal pellet from homogenized mammary tissue clarified at 10,000g to isolate enriched microsomal fractions. Samples were arrayed by 1-DE and analyzed by nanoLC-MS/MS, by linear iontrap (LTQ). Of the 703 proteins identified from these fractions nearly half were designated as likely secretory components with a total of 48 (~ 7%) allocated as originating from the Golgi apparatus. These included ADP-ribosylation factors, coatomer protein complex components, transporters, Rab and Sec proteins (Peng et al., 2008).

Surprisingly, only one study has used basic density centrifugation techniques and proteomics to examine the Golgi apparatus of yeast, the most widely studied eukaryotic system (Forsmark et al., 2011). The *sec6-4* yeast mutant was identified several decades ago as a temperature sensitive mutant that accumulates post-Golgi secretory vesicles (Walworth & Novick, 1987) although until recently no proteomic analysis had been undertaken. Recently the *sec6-4* mutant was used to undertake comparative proteomics against the *sec23-1*, a yeast mutant that is depleted of vesicles (Kaiser & Schekman, 1990). Comparative proteomics was undertaken between these two yeast mutants using iTRAQ on fractionated vesicles purified on 9-step linear sorbitol gradients. After density centrifugation, fractions showing the highest signal for SNARE proteins from the two mutant lines were pooled and used for analysis by nanoLC-MS/MS by Orbitrap MS. A total of 242 proteins were identified from the pooled fractions with 91 having greater abundance in the *sec6-4* mutant lines. Many of the most highly differentially expressed proteins were cell wall associated and plasma membrane transporters all of which were likely cargo proteins. The analysis also identified many vesicle proteins which included SNAREs, GTPases, protein glycosylation components and V-ATPase complex components (Forsmark et al., 2011).

While there has been significant interest in plant cell wall biosynthesis for the past decade, only two studies have used proteomics to explore density enriched Golgi fractions from plants. The absence of many studies likely reflects the reported difficulties in isolating Golgi membranes from plants. With this problem in mind, one of the first attempts employed a rice suspension cell culture expressing a *cis*-Golgi marker (*35S::GFP-SYP31* construct) in an attempt to characterize enriched fractions from discontinuous sucrose density gradient (Asakura et al., 2006). The enrichment procedure utilized sequential discontinuous sucrose gradients on a microsomal fraction to greatly enrich membranes containing the *35S::GFP-SYP31* construct. The technique also employed the addition of $MgCl_2$ which had previously been shown to separate distinct Golgi compartments (Mikami et al., 2001). Twice purified Golgi membranes were arrayed by both 1-DE and 2-DE, but the 2-DE arrays yielded few identifications. Protein bands from 1-DE were excised and analyzed by MALDI-TOF MS. A total of 63 proteins were identified with 70% containing predicted transmembrane domains. A variety of Golgi proteins were identified including COP complex components, GTPases, Rab proteins, an EMP70 and phospholipase D (Asakura et al., 2006). Very little is known about the role of the Golgi apparatus in secondary cell wall formation in plants due to considerable technical issues in sample preparation. Recently, an analysis of membrane enriched fractions from developing compression wood of *Pinus radiata* has attempted to address this shortage in knowledge (Mast et al., 2010). Microsomal fractions were enriched from compression wood homogenates using discontinuous gradients with maximal Golgi marker activity found at the 8/27% interface. Fractions were further extracted using a TX-114 Triton phase separation technique in order to remove contaminating proteins (*e.g.*, actin). Samples were analyzed using nanoLC-MS/MS by linear iontrap (LTQ). A total of 175 proteins were identified, 66 in the aqueous phase and 103 in the detergent phase. Only a handful of proteins were confidently allocated to the Golgi apparatus after functional annotations, further highlighting the inherent difficulties in working with this tissue. These included laccases, cellulose synthases and a xylosyltransferase (Mast et al., 2010).

3. Immunoaffinity purification of Golgi

The immunoaffinity purification (IP) of organelles has been a widely used technique for decades. Since the approach relies on the presence of differential epitopes rather than differences in physical parameters it is highly amenable for isolating functionally distinct subcellular organelles (Richardson & Luzio, 1986). This approach is often combined with sucrose gradient fractionation. The use of IP and proteomics to isolate and characterize subcellular compartments of the secretory system has been most effectively applied in animal systems. Early approaches investigating Golgi immunoisolation employed baby hamster kidney (BHK) cells infected with mutant vesicular stomatitis virus (VSV). After temperature induction the viral G protein accumulates at the *trans*-Golgi network (TGN) serving as bait for the immunoisolation of the corresponding compartment. Immunoisolated TGN was visualized with TEM confirming its structural integrity. Several polypeptides were enriched in the isolated fraction (de Curtis et al., 1988).

An early study which highlighted future approaches, Hobman et al., (1998) successfully used IP to identify ER exit sites in BHK cells. This study used the Rubella virus E1 glycoprotein, which normally localizes in a subset of Smooth ER (SER) and does not overlap with COPI coated ER. Tubular networks of SER were separated by cell fractionation, followed by immunoisolation using Dynabeads coated with an antibody against epitope tagged E1. Using electron microscopy, the authors demonstrated that the isolated membrane structures were distinguished from COPI coated ER. Western blot analysis against known markers identified the presence of proteins in this compartment and included ER-Golgi intermediate and transitional ER markers. These results indicated that this tubular distinct subdomain of SER provides the site for COPII vesicle biogenesis. A further early study demonstrating the viability of the approach used antibody-conjugated magnetic beads facilitating the isolation of peroxisomes from rat liver (Kikuchi et al., 2004). Peroxisomes separated by cell fractionation were further isolated with an antibody against PMP70 (70-kDa peroxisomal membrane protein). Proteomic analysis was carried out, using in gel digestion and LC-MS/MS was accommodated in a hybrid type-Q-TOF mass spectrometer. The proteome contained 34 known peroxisomal proteins and a minor number of mitochondrial proteins. Two unknown proteins were added to the peroxisomal proteome by this study; a Lon protease and a bi-functional protein consisting of an aminoglycoside phosphotransferase-domain.

The application of IP followed by protein identification through mass spectrometry addressing the proteome of the secretory components was first undertaken to examine the ER-Golgi intermediate compartments or ERGIC of humans (Breuza et al., 2004). The approach used the drug Brefeldin A to cause an over accumulation of ERGIC clusters in human HepG2 cells. The post nuclear supernatant was loaded onto a linear 13-29% Nycodenz gradient and fractions enriched for the ERGIC-53 marker pooled. Dynabeads coupled to KDEL receptor monoclonal antibodies were used to IP the ERGIC membranes. Samples were arrayed by 1-DE and protein bands analyzed by MALDI-TOF MS. A total of 19 proteins were identified including ERGIC-53, the KDEL receptor, SEC22b, cargo receptors and membrane trafficking components.

The use of IP on recombinant lines expressing specific epitopes was first undertaken in yeast (Inadome et al., 2005). Strains expressing recombinant SNARE proteins Myc_6-sed5 and Myc_6-Tlg2 were employed to isolate vesicles associated with early (sed5) and late (Tlg2) Golgi compartments. Since recombinant Myc tagged proteins are employed the IP does not require protein specific antibodies. The approach utilized the supernatant from a 100,000g centrifugation step to purify both Myc_6-sed5 and Myc_6-Tlg2 vesicles using an anti-Myc monoclonal antibody. Vesicles were affinity purified from each strain using Protein A-Sepharaose beads and resultant proteins arrayed by 1-DE. Protein bands were excised and analyzed by MALDI-TOF MS. A total of 29 proteins were identified from the Sed5 vesicles and 32 proteins from the Tlg2 vesicles. Both proteomes contained large proportions of known Golgi proteins including Rab GTPases, SNAREs, mannosyltransferases, V-ATPase proteins (Inadome et al., 2005). Interestingly a number of proteins were identified exclusively in the either the early Sed5 vesicles, namely COPII components involved in transport from ER to Golgi and mannosyltransferases involved in protein glycosylation. This fraction included a protein of unknown function (svp26) that was shown to be involved in retention of membrane proteins in early Golgi compartments. Similarly, in the late vesicle Tgl2 proteome, a significant number of proteins with no known function were identified.

Isolation of synaptic vesicles by immunopurification from mammalian systems is a well-developed procedure (Morciano et al., 2005; Burre et al., 2007). Due to the dynamic nature of synaptic vesicle trafficking, it has been challenging to isolate synaptic vesicles from their corresponding plasma membrane with conventional vesicle isolation techniques. Consequently an IP approach was employed on samples from rat brains and enabled the identification of several proteins that might be involved in the regulation of neurotransmitter release and the structural dynamics of the nerve terminal (Morciano et al., 2005). Synaptic vesicles were isolated from two synaptosomal enriched sucrose fractions, followed by IP with Dynabeads covered with antibodies against synaptic vesicle protein 2 (SV2). The isolated proteins were separated by BAC/SDS-PAGE and identified by MALDI-TOF MS. The free vesicle fraction contained 72 proteins while 81 were identified in the plasma membrane containing, denser fraction. Although many proteins, involved in vesicle trafficking and tethering were identified in both fractions, several proteins were specific to one fraction. For example, several isoforms of Rab2, Rab3, Rab11, Rab14 were only found in the free vesicles. The PM associated fractions contained several proteins that could modulate presynaptic function, including GTPases and Na+/K+-ATPases, which are potentially associated with the plasma membrane. Several physicochemical conditions were evaluated to enrich low abundant proteins (Burre et al., 2007), among these the addition of SB/DTT or NP40 showed that it can improve the efficacy of protein elution from the magnetic beads. Moreover, the use of phase separation can divide proteins according to their hydrophobicity, resulting in the isolation of more integral membrane proteins.

The selection of bait proteins might be a crucial prerequisite for specific vesicle immunoisolation. Motor proteins could be excellent baits for dissecting a subset of post-Golgi membrane compartments. Recently, a kinesin motor, calsyntenin-1, has been successfully used to identify its cargo vesicles from neuronal axon in mouse brain (Steuble et al., 2010). This study involved IP from two distinct subfractionated organelle populations, using Dynabeads coupled with anti-calsyntenin antibodies. Solubilized proteins were analyzed by LTQ-ICR-FT MS. This approach allowed the identification of endosomes that contained calsyntenin-1 and identification of their specific resident proteins. A combination of biochemical analysis and immunocytochemistry lead to a model of two distinct, non-overlapping endosomal populations of calsyntenin-1. An early endosome population, containing β-amyloid precursor protein (APP) and second, a APP negative, recycling endosomal population.

In contrast with the studies in animal and microbial systems only one pioneering study has been undertaken in plants (Drakakaki et al., 2011). The SYP61 TGN and early endosome compartment was isolated, employing an IP approach from SYP61-CFP transgenic Arabidopsis plants (Fig. 3). This involved a two-step procedure, comprising sucrose gradient fractionation followed by IP using agarose beads coupled with antibodies against GFP, facilitating the SYP61 compartment isolation. In total, 145 proteins were identified by MudPIT nano-LC MS/MS. These include the SYP61 SNARE complex and its regulatory proteins (SYP41, VTI12, and VPS45), GTPases and proteins involved in vesicle trafficking.

Other SYP4 members such as SYP43 were also found in the SYP61 vesicle proteome, establishing new protein association with SYP61. Several proteins of unknown function,

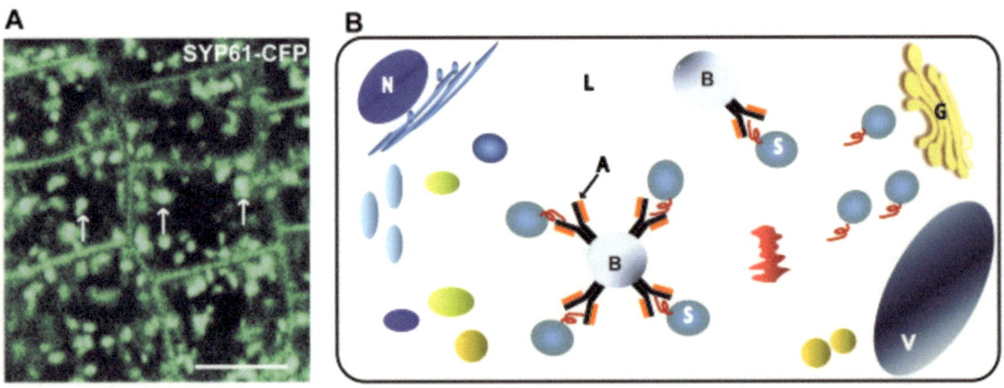

Fig. 3. Schematic representation of vesicle immunoisolation. A). Subcellular localization of SYP61-CFP in transgenic Arabidopsis root cells. Arrows indicate SYP61-CFP vesicles. Scale bar = 10μm. B). Diagram of vesicle IP. Vesicles or compartments are isolated from cell lysate with the aid of an antibody against a target protein. Compartments are not to scale. B: Protein A-Agarose bead; A: GFP antibody; G: Golgi apparatus; S: SYP61 containing vesicles; L: cell lysates, V:vacuole.

such as ECHIDNA were included providing the opportunity of new cargo identification. Plasma membrane associated proteins such as the SYP121-complex and members of the cellulose synthase family were also present. These findings suggested a role of SYP61 in exocytic trafficking and possibly in transporting of cell wall components.

Immunoisolation of TGN vesicles can not only facilitate the identification of its proteome but also other components. A recent study in yeast has analyzed the lipid profile of TGN vesicles transporting a transmembrane raft protein. It demonstrated that the isolated compartment was selectively enriched with ergosterol and sphingolipid, thus providing evidence that TGN can sort membrane lipids (Klemm et al., 2009). As more different subcompartments of Golgi are purified, we will gain greater insights into the nature of the sorted components and their transported cargo.

4. Subcellular correlation analysis of Golgi from complex lysates

For proteomic analysis of an isolated organelle, it is critical that the sample is of high purity to minimize the inclusion of contaminant proteins. If the organelle sample is only partially enriched, far greater scrutiny is required to confidently differentiate between proteins that genuinely reside in the organelle and contaminants. Highly purified samples are relatively straightforward to obtain for proteomic analyses of mitochondria (Sickmann et al., 2003; Taylor et al., 2003; Heazlewood et al., 2004), chloroplasts (Friso et al., 2004; Kleffmann et al., 2004) and nuclei (Bae et al., 2003; Turck et al., 2004; Mosley et al., 2009), mainly through the use of differential centrifugation techniques. However, obtaining pure organelles of the endomembrane system such as the Golgi, endoplasmic reticulum (ER) and plasma membrane is difficult as they are highly interconnected and are of similar sizes and densities (Wu et al., 2004; Hanton et al., 2005). To avoid the task of having to isolate pure endomembrane system organelles, several groups have instead developed techniques to analyze multiple subcellular compartments simultaneously with crude organelle samples. One such technique is the localization of organelle proteins by isotope tagging (LOPIT), where organelles are partially separated by equilibration centrifugation in a self-forming iodixanol density gradient (Dunkley et al., 2004; Sadowski et al., 2006; Lilley & Dunkley, 2008). Protein distribution patterns are quantified by differential isotope tagging of proteins across organelle fractions with LC-MS/MS on a Q-TOF instrument. Finally, multivariate analysis of isotopically-tagged peptide fragment ion data is used to correlate proteins with similar density gradient ion distributions with that of known organelle marker proteins to determine their subcellular location.

The first LOPIT study was performed on twelve crude membrane fractions from Arabidopsis callus culture that were selectively tagged with light or heavy isotope-coded affinity tags (ICATs) (Dunkley et al., 2004). ICATs are chemical probes containing 1) a reactive group targeting free cysteine residues, 2) an isotopically coded linker to distinguish between heavy (deuterium or ^{13}C) and light tags (protons or ^{12}C) and 3) an affinity tag such as biotin or avidin to capture the labeled peptide or protein (Gygi et al., 1999). The twelve membrane fractions were organized into six pair-wise comparisons of ICAT light- and heavy-tagged fractions. The six ICAT protein sample pairs were pooled, digested with trypsin and the ICAT-labeled peptides were avidin-affinity purified and analyzed by LC-MS/MS (Q-TOF). Multivariate analysis of LC-MS/MS data determined the relative abundances of 170 identified Arabidopsis proteins. Of these, a subset of 28 known or predicted Arabidopsis organelle marker proteins were used by multivariate analysis to highlight the clear separation between the known and predicted Golgi- and ER-localized protein clusters. Significantly a number of cell wall biosynthetic enzymes were identified including a number of glycosyltransferases. This confirmed LOPIT as a valid method for discriminating between Golgi- and ER-localized proteins from Arabidopsis crude membrane fractions (Dunkley et al., 2004). Further development of the LOPIT technique replaced ICAT with isotope tagging of Arabidopsis membrane peptide fractions for both relative and absolute protein quantitation (iTRAQ) (Dunkley et al., 2006) (Fig. 4). The iTRAQ method is a progression of ICAT by labeling the free primary amines of peptides with four different iTRAQ reporter tags (114, 115, 116 and 117 m/z). They are detectable by MS/MS, which allows for simultaneous quantification analysis of up to four peptide samples (Wiese et al., 2007). Arabidopsis membrane peptide fractions were differentially tagged with the four iTRAQ reporters, fractionated and analyzed by MudPIT and Q-TOF MS. The addition of SCX to RP LC-MS/MS provided superior peptide separation and identification, resulting in 689 Arabidopsis protein identifications. Multivariate analysis of iTRAQ-labeled MS/MS data revealed 89 proteins in the Golgi density gradient cluster. This more extensive analysis further validated the approach as further cell wall biosynthetic enzymes such as glycosyltransferases and sugar interconverting enzymes were identified as well as transporters, V-ATPase components and a variety of proteins with likely Golgi functions (Dunkley et al., 2006). This was a significant improvement on the initial LOPIT set of ten Arabidopsis Golgi-localized proteins by ICAT and LC-MS/MS (Dunkley et al., 2004).

To test its robustness in other biological system, LOPIT was used to investigate the subcellular distribution of proteins from Drosophila embryos. A total of 329 Drosophila proteins were identified and localized to three subcellular locations; the plasma membrane (94), mitochondria (67) and the ER/Golgi (168) (Tan et al., 2009). The lack of distinction between ER- and Golgi-residing Drosophila proteins by LOPIT underscored the significant challenges faced when dissecting complex and heterogeneous biological samples, as opposed to a simplified system of crude membranes from a relatively homogenous Arabidopsis cell culture.

A similar strategy to LOPIT but employing label-free quantitation techniques is protein correlation profiling (PCP). PCP uses quantitation of unmodified peptide ions by MS to bypass the chemical modification step in ICAT and iTRAQ, which results in less complicated MS/MS spectra and higher confidence in peptide identifications (Andersen et al., 2003; Foster et al., 2006). However, it is heavily reliant on invariable conditions in 2D LC-MS/MS for reproducible quantitation between samples. Proof of concept for PCP was first demonstrated with purified human centrosomes (Andersen et al., 2003) and in the cellular context with sucrose density gradient separations of mouse liver homogenate (Foster et al., 2006). A total of 1,404 mouse liver proteins were identified by 2D LC-MS/MS (LTQ-FT) and their MS ion distribution profiles were mapped by PCP to ten different subcellular locations. These results were corroborated with MS ion distribution profiles and enzymatic assays of known organelle marker proteins and immunofluorescence staining of mouse liver cells for visual confirmation of select proteins with overlapping or non-overlapping PCPs. While this study reported rates of 61 to 93% overlap from comparing its mitochondrial-localized protein set with previous human and mouse mitochondrial proteomes, the rates of overlap were considerably lower for proteins localized to the plasma membrane (49%) and Golgi (36%). Nonetheless, they made significant inroads in characterizing the mouse Golgi proteome and identified a series of Rab proteins, mannosyltransferases, COP components, transporters and a diverse range of transferases (Foster et al., 2006).

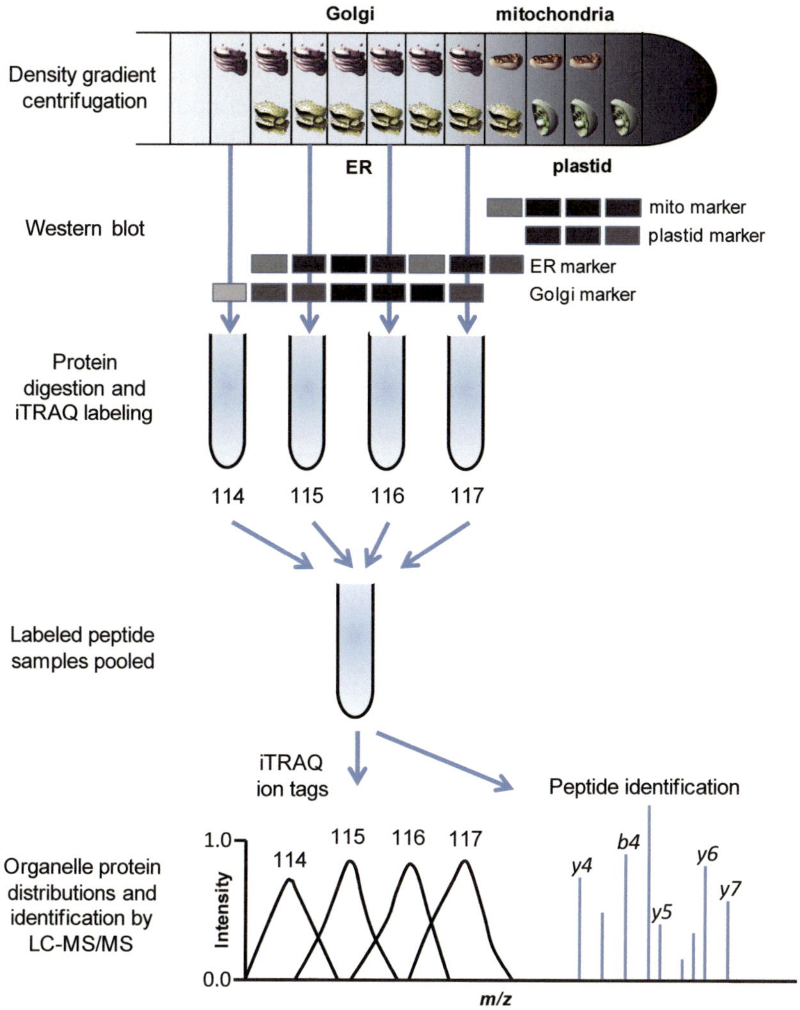

Fig. 4. Outline of the LOPIT technique using crude cellular extracts. LOPIT employs centrifugation of a self-forming iodixanol density gradient to partially resolve organelle fractions. Western blotting of the fractions for known Golgi and ER marker proteins show that in most cases, there is overlap between them. A series of four protein fractions are digested with trypsin and treated with iTRAQ reagents containing the labels 114, 115, 116 or 117m/z and pooled for LC-MS/MS analysis. Ion intensity measurements of the iTRAQ reporter ion fragments 114 to 117 m/z providing the basis of protein quantitation with simultaneous analysis of the major b, y and other fragment ions for protein identification.

The introductions of LOPIT and related organelle purification-free methods were intended to address the issue of separating Golgi from other endomembrane system components, but this still remains rather difficult to achieve with complex biological systems. Refining these methods by optimizing density gradient conditions to enhance the resolution of Golgi, along with continuing development of multivariate techniques are seen as pivotal to expand the set of genuine Golgi-residing proteins in semi-purified samples (Foster et al., 2006; Trotter et al., 2010).

5. Free flow electrophoresis (FFE) purification of Golgi

Free Flow Electrophoresis, though 50 years old has adapted well to contemporary research fields, recently filling a particular niche in subcellular proteomics, in combination with mass spectrometry. This section explores the role of FFE in isolation of the Golgi apparatus from plant and mammalian tissues. Essentially, an electric field is applied perpendicular to a sample as it moves up a separation chamber in a liquid medium. Subcellular components are therefore separated according to surface charge and organelle streams collected as 96 fractions (Fig. 5). Hydrodynamic stability of the liquid is crucial; convection currents arising from localized joule heating can disrupt organelle streams. Apparatus design has consistently advanced along with the fields to which FFE has been applied. MicroFFE apparatus designs (Turgeon & Bowser, 2009) have overcome some of the imperfections inherent in the technique. Entirely liquid phase and continuous, FFE is appropriate for large scale, preparative fractionation of cells, organelles, proteins and peptides. The apparatus can be operated in two modes: zonal electrophoresis (ZE), or isoelectric focusing (IEF) mode. ZE-FFE is becoming recognized for its impressive separation and purification capacity of plant, mammalian and yeast organelles (reviewed by Islinger et al., 2010).
The first use of FFE for Golgi was applied to mammalian Golgi membranes and lead to separation of sub-Golgi compartments, demonstrated by a series of enzyme assays (Hartelschenk et al., 1991). However, this was prior to the proteomic era and was never

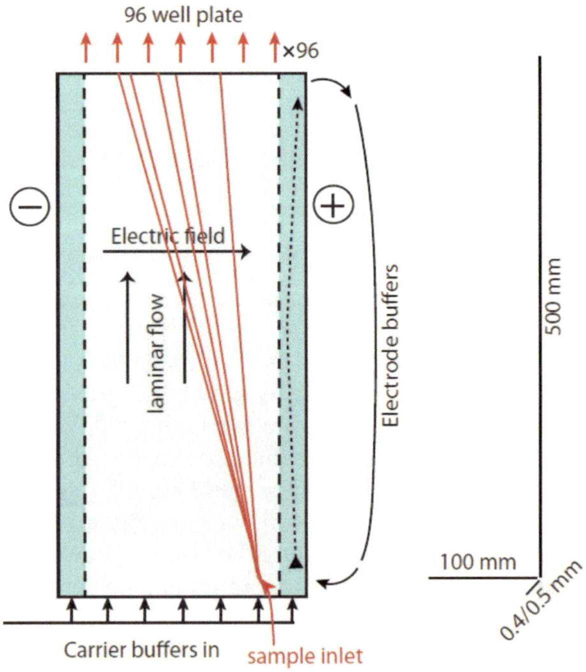

Fig. 5. A schematic diagram of a large scale FFE setup with dimensions shown on the right. The diagram outlines a late commercially model available through BD Diagnostics, with counter flow at sample outlets and stabilization buffers at the extreme anodic and cathodic carrier buffer inlets (Islinger et al., 2010). MicroFFE apparatus are similar with 56.5 mm ×35 mm × 30 mm dimensions (Turgeon & Bowser, 2009).

revisited with modern mass spectrometry tools. Plant homogenates were first subjected to FFE some decades ago (Kappler et al., 1986; Sandelius et al., 1986; Bardy et al., 1998) but these first forays demonstrated little potential for Golgi isolation. With plant Golgi antibodies then, as now, commercially unavailable, enzyme assays were the primary means of determining fraction composition. Profiling by enzyme assays was not sufficiently precise or efficient for tracking lower-abundance Golgi proteins amidst a relatively complex background of contaminants, although the distribution of enzyme activities reported by (Sandelius et al., 1986) are broadly consistent with later proteomic analyses.

The first isolation of plant Golgi membranes has depended on both FFE and proteomic advances (Parsons and Heazlewood, unpublished data). Semi-high throughput mass spectrometry was used to track the electrophoretic migration of Golgi membranes. The proteins identified in individual fractions were matched against markers protein lists for each subcellular location, including the cytosol, compiled from SUBA, the SUBcellular Arabidopsis database (Heazlewood et al., 2007). This allowed simultaneous monitoring of over 50 proteins in most fractions without recourse to antibodies or enzyme assays. Overlaid on the total protein output for all 96 fractions, marker lists revealed a detailed picture of organelle migration (Fig. 6). Once the shoulder peak corresponding to the purest Golgi fractions had been identified, parameters could be fine tuned, exploiting the electronegativity of Golgi vesicles and enhancing the cathodic migration of this area relative to the main protein peak. Total protein output from this targeted Golgi purification study showed a broader main protein peak and a prominent shoulder on the cathodic edge when compared to earlier studies on plant homogenates (Kappler et al., 1986; Bardy et al., 1998). Careful balancing of the carrier buffer flow rate to voltage ratio maximized the separation range of organelles whilst organelle streams remained focussed. Cathodic migration increased with voltage but was limited by increasing the flow rate as exposure time to the electric field was shorter. Lateral diffusion of organelle streams dictated the lower flow rate

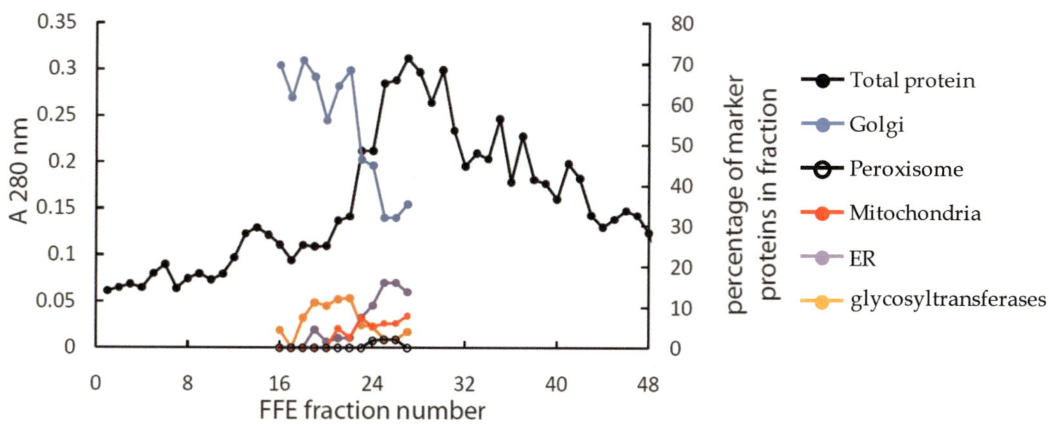

Fig. 6. Golgi membrane migration profile after FFE separation. A portion of the total protein output, measured at 280 nm (fractions 1 to 48) is shown. Around 50 proteins were identified in each fraction scanned using semi-high throughput LC-MS/MS. Overlaid are matches from marker protein lists compiled from the SUBA subcellular database and the ~50 identified proteins from each fraction. Many glycosyltransferases are located in the Golgi and were used as a further guide for Golgi membrane migration.

limit. Golgi fractions with minimal contamination were identified through continued monitoring and selected for detailed proteomic characterization (Parsons and Heazlewood, unpublished data).

The application of FFE, mass spectrometry and proteomic data as tools for Golgi isolation and characterization marked a precedent for plant Golgi proteomics. Previously, relatively few plant Golgi proteins had been identified by proteomic techniques (Dunkley et al., 2006). The application of FFE to isolate high purity Golgi fractions resulted in a Golgi proteome of 425 proteins identified in at least two of three biological replicates. This included over 50 glycosyltransferases, 25 transporters, the entire V-ATPase complex, a variety of trafficking components, methyltransferases and acetyltransferases (Parsons and Heazlewood, unpublished data). While proteins identified in a single preparation were excluded from the final proteome, they nevertheless present a useful resource for functional analysis of the plant Golgi apparatus. With so little Golgi proteomic data resources, common contaminants originating from the Golgi in other proteomes were difficult to identify. This therefore represents both significant progress in our potential to understand Golgi processes and consolidation of the current state of subcellular protein localization in plants. As an example the ectoapyrase protein APY1 is currently classified as a plasma membrane protein involved in extracellular signaling through the hydrolysis of phosphate from ATP (Wu et al., 2007). The APY1 protein was identified in all three replicates and YFP tagging confirmed its Golgi localization. Heterologous expression of this protein in the yeast nucleoside diphosphatase (NDPase) mutant *gda1,* rescued the glycosylation phenotype in this mutant, thus functionally characterizing the APY1 protein as a Golgi-resident NDPase (Parsons and Heazlewood, unpublished data). Since most glycosylation occurs in the Golgi, the APY1 protein represents a resident and functional Golgi protein, rather than a transitory plasma membrane localized protein. Furthermore, plasma membrane and Golgi compartments are easily separated using FFE (Bardy et al., 1998) with Golgi and ER compartments partially separated (Fig. 6). Thus, selectively pre-enriching organelles and tailoring FFE parameters for maximal separation has considerable potential in distinguishing between resident and transitory proteins in the secretory system. Some proteins observed after FFE purification of the plasma membrane were present in all three Golgi preparations and can be readily classified as 'transient proteins' rather than contaminants (Parsons and Heazlewood, unpublished data).

Given the success achieving high purity fractions (Taylor et al., 1997a) and sub-compartmental resolution of Golgi structures (Hartelschenk et al., 1991), it is surprising that a corresponding proteomic study has not been undertaken in rats. FFE was foremost amongst techniques compared for purification of mouse mitochondria (Hartwig et al., 2009) whilst impressive results were achieved after separating populations of PM vesicles (Cutillas et al., 2005), suggests that FFE still has much to contribute to both Golgi and other subcellular proteomes. In Arabidopsis, the Golgi proteome was characterized from only two to three fractions out of approximately 15 fractions over which Golgi proteins were detected. Further studies suggested this reflects medial to *trans*-Golgi separation (Parsons and Heazlewood, unpublished data). Could FFE separate the remainder of the Golgi from contaminating membranes or even Golgi sub-compartments? Chemical modification of Golgi compartments holds some promise; addition of ATP was found to enhance migration of membrane compartments towards the cathode (Barkla et al., 2007). Unfortunately no mass spectrometry was undertaken in this study. A low ionic strength two-component buffer system permits separation at lower currents, reducing convection from joule heating, as could the use of microFFE setups, enhancing sub-compartment separation. FFE has already enhanced our knowledge of Golgi proteomics but its role is clearly far from over and there is much potential for further advances using FFE.

6. Comparative analysis of the Golgi proteomes

The characterization of the Golgi apparatus and associated secretory components by mass spectrometry has been undertaken on a range of species. While most of these organisms represent model systems with extensive genetic resources and well annotated genomes, analyses have been undertaken in less tractable systems, namely pine trees (Mast et al., 2010). Nonetheless, with the exception of work undertaken in rat, only a handful of analyses have focused on the proteomic characterization of the Golgi and its associated membranes from model systems. This is in contrast to the extensive series of proteomic studies undertaken on organelles from many of these systems. For example, in the model plant Arabidopsis over ten separate proteomic analyses have been undertaken on plasma membrane fractions, six studies on mitochondria and eight analyses of the plastid (Heazlewood et al., 2007). These facts further highlight the technical challenges when attempting to isolate high purity Golgi fractions and associated structures, even from well studied model systems. Overall, searches of the literature were able to identify over twenty separate studies that have employed proteomics techniques to address the characterization of the Golgi apparatus and associated secretory components. These studies have been undertaken using a diverse collection of isolation and enrichment techniques over the past decade and have employed a range of proteomics approaches including 2-DE (Taylor et al., 1997b; Morciano et al., 2005), 1-DE (Peng et al., 2008), iTRAQ (Dunkley et al., 2006), spectral counting (Foster et al., 2006) and MudPIT (Wu et al., 2004). These studies also covered the range of protein identification methods namely Peptide Mass Fingerprinting (Morciano et al., 2005), Edman degradation (Bell et al., 2001) and MS/MS (Gilchrist et al., 2006).

The protein identifications outlined in these works were extracted from the published manuscripts and online supplementary material to produce a collection of proteins identified in each study. Protein sequences were obtained from GenBank or UniProt for each accession and consolidated at the species level using BLAST analysis tool against minimally redundant protein sets where available. These comprised the International Protein Index (Kersey et al., 2004) for human, mouse, rat and bovine, The Arabidopsis Information Resource (Swarbreck et al., 2008) for Arabidopsis, the Saccharomyces Genome Database (Cherry et al., 1997) for yeast, FlyBase (Tweedie et al., 2009) for Drosophila and the Rice Genome Annotation Project (Ouyang et al., 2007) for rice. This enabled the classification of the total number of proteins identified from the Golgi apparatus and associated membranes based on each isolation method and by each species (Table 1). Finally, the total number of non-redundant proteins currently assigned to the Golgi apparatus and associated membrane components for each species could also be ascertained (Table 1). Where possible, we relied on annotation information and classifications outlined in each manuscript to determine whether a protein should be included in the final lists. This included early endosome, secretory and unknowns (when efforts to classify contaminants had been undertaken). The largest number of proteins assigned to the Golgi of any one species is that of rat. This reflects the number of individual studies and the fact that this represented the major system used to study the Golgi proteome for a number of years.

Species	Density centrifugation	Immuno-affinity	Free Flow Electrophoresis	Correlation Analysis	Total
Pine	10				10
Human		24			18
Rice	49				43
Drosophila				168	168
Bovine	252				238
Yeast	241	52			276
Mouse	2711[a]	56		490	428
Arabidopsis		145	425	92	534
Rat	1117	57			996

Table 1. The total number of proteins, by species and technique, currently identified by proteomic approaches from the Golgi apparatus and associated membrane systems. [a]The analysis of mouse microsomes by density centrifugation (Kislinger et al., 2006) has not been included in the final total for this species as it represents a crude microsomal fraction.

The set of non-redundant protein sequences compiled from the proteomic analyses of the Golgi were assembled for cross species orthology analysis. In order to remove identical genes and splice variants, these sequences were first clustered at 95% sequence identity and only one representative from each cluster carried over for subsequent analysis. Following this, the sequences were clustered at 30% identity. All clustering was performed with the program uCLUST (Edgar, 2010). A protein was mapped to an ortholog of another species if at least one representative of that species was present in the same cluster. Proteins were considered paralogs when two or more sequences from the same species were present in a cluster in which sequences from no other species were present (Fig. 7)

After homology matching, a number of gene families were found across the Golgi proteomes of most species. These included Rab GTPases, heat-shock proteins, alpha-mannosidases, thioredoxins, and cyclophilins. Apart from the Rab GTPases, which mediate vesicle trafficking, the other families are involved in protein folding and protein glycosylation. There were a number of large clusters containing only Arabidopsis genes and these clusters were contained glycosyltransferases associated with synthesis of the plant cell wall (Scheller & Ulvskov, 2010). In addition, there was a cluster of pine sequences containing laccases, which may be associated with the synthesis of lignin in woody tissue (Ranocha et al., 2002). In general, when only a few proteins had been reported in a species, those proteins were more likely to have orthologs in the other species in the set. This suggests that the most easily detected proteins in proteomics studies are abundant proteins involved in core Golgi-related functions that have not diverged as greatly over evolutionary history as the less abundant and harder to find proteins.

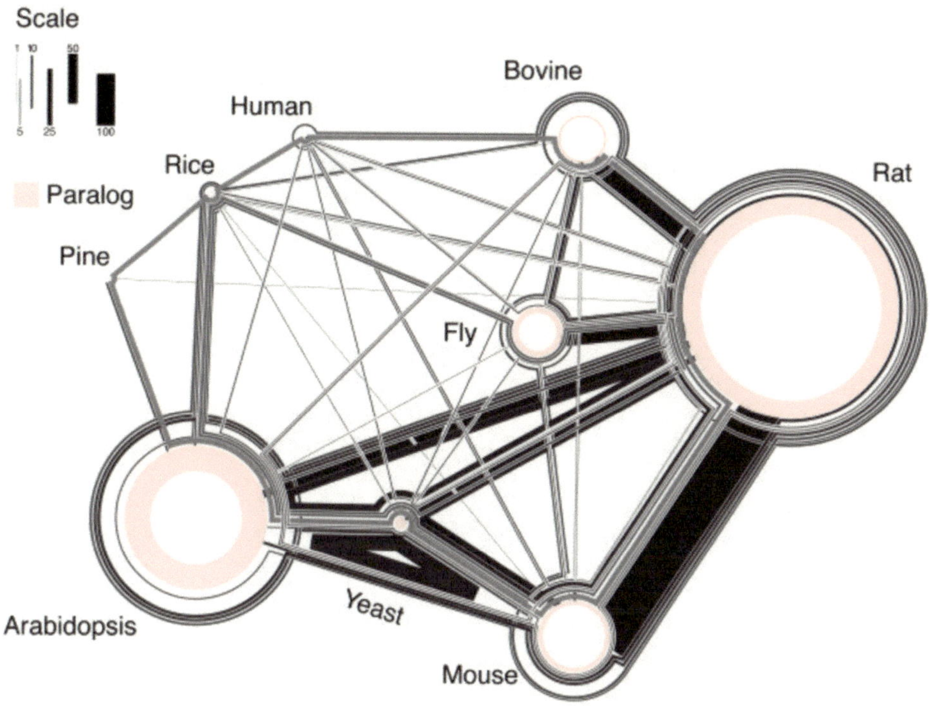

Fig. 7. Orthology interaction map of the non-redundant Golgi proteome sets. The size of the species circle indicates the number of proteins identified in the Golgi proteome of that species. The pink shading indicates the number of paralogs for a given species. The lines indicate orthology connections between the species with the thickness indicating the number of proteins. The Scale refers to the number of proteins represented by the thickness of the line.

7. Conclusion

The characterization of the Golgi proteome from various systems represents an important technical and biological achievement. Its central role within the cell in functions ranging from cell wall biosynthesis to protein glycosylation to secretion is of significant importance. Knowledge about these functions contributes to both our fundamental understanding of complex eukaryotic systems to their exploitation in areas of biofuels (cell wall manipulation) and agriculture (milk production). While there is clearly more basic knowledge required to understand the functionally complex roles of the Golgi apparatus, advances made by work outlined in this chapter demonstrate that the first decade of proteomics has been fruitful and improvements to isolation and analysis methods are promising for the field going forward.

8. Acknowledgment

The work conducted by the Joint BioEnergy Institute was supported by the Office of Science, Office of Biological and Environmental Research, of the U.S. Department of Energy under Contract No. DE-AC02-05CH11231. GD and EP were supported by start-up funds from the University of California, Davis.

9. References

Andersen, J.S.; Wilkinson, C.J.; Mayor, T.; Mortensen, P.; Nigg, E.A. & Mann, M. (2003). Proteomic characterization of the human centrosome by protein correlation profiling. *Nature*, Vol.426, No.6966, pp. 570-574

Asakura, T.; Hirose, S.; Katamine, H.; Sato, M.; Hujiwara, M.; Shimamoto, K.; Hori, F. & Mitsui, T. (2006). Rice Golgi proteome: Analysis of GFP-syp31 labeled cis Golgi membrane. *Plant and Cell Physiology*, Vol.47, pp. S26-S26

Bae, M.S.; Cho, E.J.; Choi, E.Y. & Park, O.K. (2003). Analysis of the Arabidopsis nuclear proteome and its response to cold stress. *Plant Journal*, Vol.36, No.5, pp. 652-663

Bardy, N.; Carrasco, A.; Galaud, J.P.; Pont-Lezica, R. & Canut, H. (1998). Free-flow electrophoresis for fractionation of *Arabidopsis thaliana* membranes. *Electrophoresis*, Vol.19, No.7, pp. 1145-1153

Barkla, B.J.; Vera-Estrella, R. & Pantoja, O. (2007). Enhanced separation of membranes during free flow zonal electrophoresis in plants. *Analytical Chemistry*, Vol.79, No.14, pp. 5181-5187

Bell, A.W.; Ward, M.A.; Blackstock, W.P.; Freeman, H.N.; Choudhary, J.S.; Lewis, A.P.; Chotai, D.; Fazel, A.; Gushue, J.N.; Paiement, J.; Palcy, S.; Chevet, E.; Lafreniere-Roula, M.; Solari, R.; Thomas, D.Y.; Rowley, A. & Bergeron, J.J. (2001). Proteomics characterization of abundant Golgi membrane proteins. *Journal of Biological Chemistry*, Vol.276, No.7, pp. 5152-5165

Blondeau, F.; Ritter, B.; Allaire, P.D.; Wasiak, S.; Girard, M.; Hussain, N.K.; Angers, A.; Legendre-Guillemin, V.; Roy, L.; Boismenu, D.; Kearney, R.E.; Bell, A.W.; Bergeron, J.J. & McPherson, P.S. (2004). Tandem MS analysis of brain clathrin-coated vesicles reveals their critical involvement in synaptic vesicle recycling. *Proceedings of the National Academy of Sciences of the United States of America*, Vol.101, No.11, pp. 3833-3838

Breuza, L.; Halbeisen, R.; Jeno, P.; Otte, S.; Barlowe, C.; Hong, W. & Hauri, H.P. (2004). Proteomics of endoplasmic reticulum-Golgi intermediate compartment (ERGIC) membranes from brefeldin A-treated HepG2 cells identifies ERGIC-32, a new cycling protein that interacts with human Erv46. *Journal of Biological Chemistry*, Vol.279, No.45, pp. 47242-47253

Burre, J.; Zimmermann, H. & Volknandt, W. (2007). Immunoisolation and subfractionation of synaptic vesicle proteins. *Analytical Biochemistry*, Vol.362, No.2, pp. 172-181

Cherry, J.M.; Ball, C.; Weng, S.; Juvik, G.; Schmidt, R.; Adler, C.; Dunn, B.; Dwight, S.; Riles, L.; Mortimer, R.K. & Botstein, D. (1997). Genetic and physical maps of Saccharomyces cerevisiae. *Nature*, Vol.387, No.6632, pp. 67-73

Cutillas, P.R.; Biber, J.; Marks, J.; Jacob, R.; Stieger, B.; Cramer, R.; Waterfield, M.; Burlingame, A.L. & Unwin, R.J. (2005). Proteomic analysis of plasma membrane vesicles isolated from the rat renal cortex. *Proteomics*, Vol.5, No.1, pp. 101-112

Dalton, A.J. & Felix, M.D. (1953). Studies on the Golgi Substance of the Epithelial Cells of the Epididymis and Duodenum of the Mouse. *American Journal of Anatomy*, Vol.92, No.2, pp. 277-305

de Curtis, I.; Howell, K.E. & Simons, K. (1988). Isolation of a fraction enriched in the *trans*-Golgi network from baby hamster-kidney cells. *Experimental Cell Research*, Vol.175, No.2, pp. 248-265

Dominguez, M.; Fazel, A.; Dahan, S.; Lovell, J.; Hermo, L.; Claude, A.; Melançon, P. & Bergeron, J.J.M. (1999). Fusogenic domains of golgi membranes are sequestered into specialized regions of the stack that can be released by mechanical fragmentation. *The Journal of Cell Biology*, Vol.145, No.4, pp. 673-688

Drakakaki, G.; van de Ven, W.; Pan, S.; Miao, Y.; Wang, J.; Keinath, N.K.; Weatherly, B.; Jiang, L.; Schumacher, K.; Hicks, G. & Raikhel, N. (2011). Isolation and proteomic analysis of the SYP61 compartment reveal its role in exocytic trafficking in Arabidopsis. *Cell Research*, doi: 10.1038/cr.2011.1129

Dröscher, A. (1998). Camillo Golgi and the discovery of the Golgi apparatus. *Histochemistry and Cell Biology*, Vol.109, No.5, pp. 425-430

Dunkley, T.P.; Watson, R.; Griffin, J.L.; Dupree, P. & Lilley, K.S. (2004). Localization of organelle proteins by isotope tagging (LOPIT). *Molecular & Cellular Proteomics*, Vol.3, No.11, pp. 1128-1134

Dunkley, T.P.J.; Hester, S.; Shadforth, I.P.; Runions, J.; Weimar, T.; Hanton, S.L.; Griffin, J.L.; Bessant, C.; Brandizzi, F.; Hawes, C.; Watson, R.B.; Dupree, P. & Lilley, K.S. (2006). Mapping the Arabidopsis organelle proteome. *Proceedings of the National Academy of Sciences of the United States of America*, Vol.103, No.17, pp. 6518-6523

Edgar, R.C. (2010). Search and clustering orders of magnitude faster than BLAST. *Bioinformatics*, Vol.26, No.19, pp. 2460-2461

Forsmark, A.; Rossi, G.; Wadskog, I.; Brennwald, P.; Warringer, J. & Adler, L. (2011). Quantitative proteomics of yeast post-Golgi vesicles reveals a discriminating role for Sro7p in protein secretion. *Traffic*, Vol.12, No.6, pp. 740-753

Foster, L.J.; de Hoog, C.L.; Zhang, Y.; Xie, X.; Mootha, V.K. & Mann, M. (2006). A mammalian organelle map by protein correlation profiling. *Cell*, Vol.125, No.1, pp. 187-199

Friso, G.; Giacomelli, L.; Ytterberg, A.J.; Peltier, J.B.; Rudella, A.; Sun, Q. & Wijk, K.J. (2004). In-depth analysis of the thylakoid membrane proteome of *Arabidopsis thaliana* chloroplasts: new proteins, new functions, and a plastid proteome database. *Plant Cell*, Vol.16, No.2, pp. 478-499

Gilchrist, A.; Au, C.E.; Hiding, J.; Bell, A.W.; Fernandez-Rodriguez, J.; Lesimple, S.; Nagaya, H.; Roy, L.; Gosline, S.J.; Hallett, M.; Paiement, J.; Kearney, R.E.; Nilsson, T. & Bergeron, J.J. (2006). Quantitative proteomics analysis of the secretory pathway. *Cell*, Vol.127, No.6, pp. 1265-1281

Golgi, C. (1898). Intorno alla struttura della cellula nervosa. *Archives Italiennes de Biologie*, Vol.30, pp. 60-71

Gygi, S.P.; Rist, B.; Gerber, S.A.; Turecek, F.; Gelb, M.H. & Aebersold, R. (1999). Quantitative analysis of complex protein mixtures using isotope-coded affinity tags. *Nature Biotechnology*, Vol.17, No.10, pp. 994-999

Hanton, S.L.; Bortolotti, L.E.; Renna, L.; Stefano, G. & Brandizzi, F. (2005). Crossing the divide--transport between the endoplasmic reticulum and Golgi apparatus in plants. *Traffic*, Vol.6, No.4, pp. 267-277

Hartelschenk, S.; Minnifield, N.; Reutter, W.; Hanski, C.; Bauer, C. & Morre, D.J. (1991). Distribution of glycosyltransferases among Golgi-apparatus subfractions from liver and hepatomas of the rat. *Biochimica et Biophysica Acta*, Vol.1115, No.2, pp. 108-122

Hartwig, S.; Feckler, C.; Lehr, S.; Wallbrecht, K.; Wolgast, H.; Muller-Wieland, D. & Kotzka, J. (2009). A critical comparison between two classical and a kit-based method for mitochondria isolation. *Proteomics*, Vol.9, No.11, pp. 3209-3214

Heazlewood, J.L.; Verboom, R.E.; Tonti-Filippini, J.; Small, I. & Millar, A.H. (2007). SUBA: The Arabidopsis subcellular database. *Nucleic Acids Research*, Vol.35, pp. D213-D218

Heazlewood, J.L.; Tonti-Filippini, J.S.; Gout, A.M.; Day, D.A.; Whelan, J. & Millar, A.H. (2004). Experimental analysis of the Arabidopsis mitochondrial proteome highlights signaling and regulatory components, provides assessment of targeting prediction programs, and indicates plant-specific mitochondrial proteins. *Plant Cell*, Vol.16, No.1, pp. 241-256

Hobman, T.C.; Zhao, B.; Chan, H. & Farquhar, M.G. (1998). Immunoisolation and characterization of a subdomain of the endoplasmic reticulum that concentrates proteins involved in COPII vesicle biogenesis. *Molecular Biology of the Cell*, Vol.9, No.6, pp. 1265-1278

Inadome, H.; Noda, Y.; Adachi, H. & Yoda, K. (2005). Immunoisolaton of the yeast Golgi subcompartments and characterization of a novel membrane protein, Svp26, discovered in the Sed5-containing compartments. *Molecular & Cellular Biology*, Vol.25, No.17, pp. 7696-7710

Islinger, M.; Eckerskorn, C. & Volkl, A. (2010). Free-flow electrophoresis in the proteomic era: A technique in flux. *Electrophoresis*, Vol.31, No.11, pp. 1754-1763

Kaiser, C.A. & Schekman, R. (1990). Distinct sets of SEC genes govern transport vesicle formation and fusion early in the secretory pathway. *Cell*, Vol.61, No.4, pp. 723-733

Kappler, R.; Kristen, U. & Morre, D.J. (1986). Membrane flow in plants: Fractionation of growing pollen tubes of tobacco by preparative free-flow electrophoresis and kinetics of labeling of endoplasmic reticulum and Golgi apparatus with [3H]leucine. *Protoplasma*, Vol.132, No.1-2, pp. 38-50

Kersey, P.J.; Duarte, J.; Williams, A.; Karavidopoulou, Y.; Birney, E. & Apweiler, R. (2004). The International Protein Index: an integrated database for proteomics experiments. *Proteomics*, Vol.4, No.7, pp. 1985-1988

Kikuchi, M.; Hatano, N.; Yokota, S.; Shimozawa, N.; Imanaka, T. & Taniguchi, H. (2004). Proteomic analysis of rat liver peroxisome: presence of peroxisome-specific isozyme of Lon protease. *Journal of Biological Chemistry*, Vol.279, No.1, pp. 421-428

Kislinger, T.; Cox, B.; Kannan, A.; Chung, C.; Hu, P.; Ignatchenko, A.; Scott, M.S.; Gramolini, A.O.; Morris, Q.; Hallett, M.T.; Rossant, J.; Hughes, T.R.; Frey, B. & Emili, A. (2006). Global survey of organ and organelle protein expression in mouse: combined proteomic and transcriptomic profiling. *Cell*, Vol.125, No.1, pp. 173-186

Kleffmann, T.; Russenberger, D.; von Zychlinski, A.; Christopher, W.; Sjolander, K.; Gruissem, W. & Baginsky, S. (2004). The *Arabidopsis thaliana* chloroplast proteome reveals pathway abundance and novel protein functions. *Current Biology*, Vol.14, No.5, pp. 354-362

Klemm, R.W.; Ejsing, C.S.; Surma, M.A.; Kaiser, H.J.; Gerl, M.J.; Sampaio, J.L.; de Robillard, Q.; Ferguson, C.; Proszynski, T.J.; Shevchenko, A. & Simons, K. (2009). Segregation of sphingolipids and sterols during formation of secretory vesicles at the *trans*-Golgi network. *Journal of Cell Biology*, Vol.185, No.4, pp. 601-612

Lilley, K.S. & Dunkley, T.P. (2008). Determination of genuine residents of plant endomembrane organelles using isotope tagging and multivariate statistics. *Methods in Molecular Biology*, Vol.432, pp. 373-387

Mast, S.; Peng, L.; Jordan, T.W.; Flint, H.; Phillips, L.; Donaldson, L.; Strabala, T.J. & Wagner, A. (2010). Proteomic analysis of membrane preparations from developing Pinus radiata compression wood. *Tree Physiology*, Vol.30, No.11, pp. 1456-1468

Mikami, S.; Hori, H. & Mitsui, T. (2001). Separation of distinct compartments of rice Golgi complex by sucrose density gradient centrifugation. *Plant Science*, Vol.161, No.4, pp. 665-675

Morciano, M.; Burre, J.; Corvey, C.; Karas, M.; Zimmermann, H. & Volknandt, W. (2005). Immunoisolation of two synaptic vesicle pools from synaptosomes: a proteomics analysis. *J Neurochem*, Vol.95, No.6, pp. 1732-1745

Morre, J. & Mollenhauer, H.H. (2009). The Golgi apparatus: The first 100 years. Springer, ISBN 978-0-387-74346-2, New York, USA

Mosley, A.L.; Florens, L.; Wen, Z. & Washburn, M.P. (2009). A label free quantitative proteomic analysis of the Saccharomyces cerevisiae nucleus. *Journal of Proteomics*, Vol.72, No.1, pp. 110-120

Ouyang, S.; Zhu, W.; Hamilton, J.; Lin, H.; Campbell, M.; Childs, K.; Thibaud-Nissen, F.; Malek, R.L.; Lee, Y.; Zheng, L.; Orvis, J.; Haas, B.; Wortman, J. & Buell, C.R. (2007). The TIGR Rice Genome Annotation Resource: improvements and new features. *Nucleic Acids Research*, Vol.35, pp. D883-887

Paiement, J.; Young, R.; Roy, L. & Bergeron, J.J. (2005). Isolation of rough and smooth membrane domains of the endoplasmic reticulum from rat liver, In: *Cell Biology: A Laboratory Handbook*, J. Celis, N. Carter, K. Simons, V. Small, T. Hunter, & D.M. Shotton, (Eds.) 41-44, Elsevier Academic Press, ISBN 978-0-12-164730-8, Burlington, MA, USA

Peng, L.; Rawson, P.; McLauchlan, D.; Lehnert, K.; Snell, R. & Jordan, T.W. (2008). Proteomic analysis of microsomes from lactating bovine mammary gland. *Journal of Proteome Research*, Vol.7, No.4, pp. 1427-1432

Ranocha, P.; Chabannes, M.; Chamayou, S.; Danoun, S.; Jauneau, A.; Boudet, A.M. & Goffner, D. (2002). Laccase down-regulation causes alterations in phenolic metabolism and cell wall structure in poplar. *Plant Physiology*, Vol.129, No.1, pp. 145-155

Richardson, P.J. & Luzio, J.P. (1986). Immunoaffinity purification of subcellular particles and organelles. *Applied Biochemistry and Biotechnology*, Vol.13, No.2, pp. 133-145

Sadowski, P.G.; Dunkley, T.P.; Shadforth, I.P.; Dupree, P.; Bessant, C.; Griffin, J.L. & Lilley, K.S. (2006). Quantitative proteomic approach to study subcellular localization of membrane proteins. *Nature Protocols*, Vol.1, No.4, pp. 1778-1789

Sandelius, A.S.; Penel, C.; Auderset, G.; Brightman, A.; Millard, M. & Morre, D.J. (1986). Isolation and highly purified fractions of plasma-membrane and tonoplast from the same homogenate of soybean hypocotyls by Free-Flow Electrophoresis. *Plant Physiology*, Vol.81, No.1, pp. 177-185

Scheller, H.V. & Ulvskov, P. (2010). Hemicelluloses. *Annual Review of Plant Biology*, Vol.61, pp. 263-289

Sickmann, A.; Reinders, J.; Wagner, Y.; Joppich, C.; Zahedi, R.; Meyer, H.E.; Schonfisch, B.; Perschil, I.; Chacinska, A.; Guiard, B.; Rehling, P.; Pfanner, N. & Meisinger, C. (2003). The proteome of Saccharomyces cerevisiae mitochondria. *Proceedings of the National Academy of Sciences of the United States of America*, Vol.100, No.23, pp. 13207-13212

Steuble, M.; Gerrits, B.; Ludwig, A.; Mateos, J.M.; Diep, T.M.; Tagaya, M.; Stephan, A.; Schatzle, P.; Kunz, B.; Streit, P. & Sonderegger, P. (2010). Molecular characterization of a trafficking organelle: dissecting the axonal paths of calsyntenin-1 transport vesicles. *Proteomics*, Vol.10, No.21, pp. 3775-3788

Swarbreck, D.; Wilks, C.; Lamesch, P.; Berardini, T.Z.; Garcia-Hernandez, M.; Foerster, H.; Li, D.; Meyer, T.; Muller, R.; Ploetz, L.; Radenbaugh, A.; Singh, S.; Swing, V.; Tissier, C.; Zhang, P. & Huala, E. (2008). The Arabidopsis Information Resource (TAIR): gene structure and function annotation. *Nucleic Acids Research*, Vol.36, pp. D1009-1014

Takatalo, M.S.; Kouvonen, P.; Corthals, G.; Nyman, T.A. & Rönnholm, R.H. (2006). Identification of new Golgi complex specific proteins by direct organelle proteomic analysis. *Proteomics*, Vol.6, pp. 3502-3508

Tan, D.J.; Dvinge, H.; Christoforou, A.; Bertone, P.; Martinez Arias, A. & Lilley, K.S. (2009). Mapping organelle proteins and protein complexes in Drosophila melanogaster. *Journal of Proteome Research*, Vol.8, No.6, pp. 2667-2678

Taylor, R.S.; Jones, S.M.; Dahl, R.H.; Nordeen, M.H. & Howell, K.E. (1997a). Characterization of the Golgi complex cleared of proteins in transit and examination of calcium uptake activities. *Molecular Biology of the Cell*, Vol.8, No.10, pp. 1911-1931

Taylor, R.S.; Fialka, I.; Jones, S.M.; Huber, L.A. & Howell, K.E. (1997b). Two-dimensional mapping of the endogenous proteins of the rat hepatocyte Golgi complex cleared of proteins in transit. *Electrophoresis*, Vol.18, No.14, pp. 2601-2612

Taylor, R.S.; Wu, C.C.; Hays, L.G.; Eng, J.K.; Yates, J.R., 3rd & Howell, K.E. (2000). Proteomics of rat liver Golgi complex: minor proteins are identified through sequential fractionation. *Electrophoresis*, Vol.21, No.16, pp. 3441-3459

Taylor, S.W.; Fahy, E.; Zhang, B.; Glenn, G.M.; Warnock, D.E.; Wiley, S.; Murphy, A.N.; Gaucher, S.P.; Capaldi, R.A.; Gibson, B.W. & Ghosh, S.S. (2003). Characterization of the human heart mitochondrial proteome. *Nature Biotechnology*, Vol.21, No.3, pp. 281-286

Trotter, M.W.; Sadowski, P.G.; Dunkley, T.P.; Groen, A.J. & Lilley, K.S. (2010). Improved sub-cellular resolution via simultaneous analysis of organelle proteomics data across varied experimental conditions. *Proteomics*, Vol.10, No.23, pp. 4213-4219

Turck, N.; Richert, S.; Gendry, P.; Stutzmann, J.; Kedinger, M.; Leize, E.; Simon-Assmann, P.; Van Dorsselaer, A. & Launay, J.F. (2004). Proteomic analysis of nuclear proteins from proliferative and differentiated human colonic intestinal epithelial cells. *Proteomics*, Vol.4, No.1, pp. 93-105

Turgeon, R.T. & Bowser, M.T. (2009). Micro free-flow electrophoresis: theory and applications. *Analytical and Bioanalytical Chemistry*, Vol.394, No.1, pp. 187-198

Tweedie, S.; Ashburner, M.; Falls, K.; Leyland, P.; McQuilton, P.; Marygold, S.; Millburn, G.; Osumi-Sutherland, D.; Schroeder, A.; Seal, R. & Zhang, H. (2009). FlyBase: enhancing Drosophila Gene Ontology annotations. *Nucleic Acids Research*, Vol.37, pp. D555-559

Walworth, N.C. & Novick, P.J. (1987). Purification and characterization of constitutive secretory vesicles from yeast. *Journal of Cell Biology*, Vol.105, No.1, pp. 163-174

Wiese, S.; Reidegeld, K.A.; Meyer, H.E. & Warscheid, B. (2007). Protein labeling by iTRAQ: a new tool for quantitative mass spectrometry in proteome research. *Proteomics*, Vol.7, No.3, pp. 340-350

Wu, C.C.; Yates, J.R., 3rd; Neville, M.C. & Howell, K.E. (2000). Proteomic analysis of two functional states of the Golgi complex in mammary epithelial cells. *Traffic*, Vol.1, No.10, pp. 769-782

Wu, C.C.; MacCoss, M.J.; Howell, K.E. & Yates, J.R., 3rd. (2003). A method for the comprehensive proteomic analysis of membrane proteins. *Nature Biotechnology*, Vol.21, No.5, pp. 532-538

Wu, C.C.; MacCoss, M.J.; Mardones, G.; Finnigan, C.; Mogelsvang, S.; Yates, J.R., 3rd & Howell, K.E. (2004). Organellar proteomics reveals Golgi arginine dimethylation. *Molecular Biology of the Cell*, Vol.15, No.6, pp. 2907-2919

Wu, J.; Steinebrunner, I.; Sun, Y.; Butterfield, T.; Torres, J.; Arnold, D.; Gonzalez, A.; Jacob, F.; Reichler, S. & Roux, S.J. (2007). Apyrases (nucleoside triphosphate-diphosphohydrolases) play a key role in growth control in Arabidopsis. *Plant Physiology*, Vol.144, No.2, pp. 961-975

Part 4

Comparative Approaches in Biology

9

Differentiation of Four Tuna Species by Two-Dimensional Electrophoresis and Mass Spectrometric Analysis

Tiziana Pepe[1], Marina Ceruso[1], Andrea Carpentieri[2], Iole Ventrone[1], Angela Amoresano[2], Aniello Anastasio[1] and Maria Luisa Cortesi[1]
[1]*Dipartimento di Scienze Zootecniche e Ispezione degli Alimenti – Università di Napoli*
[2]*Dipartimento di Chimica Organica e Biochimica – Università di Napoli*
Italy

1. Introduction

Species belonging to the genus *Thunnus* are pelagic predator fishes, commonly known as tuna. The species within this genus are of commercial value, and six of them are considered the most valued in world trade (D.M., MIPAAF, 31 Gennaio 2008). *Thunnus* species originate from a variety of geographic areas, and for this reason the different species can be characterized by the presence of different biological contaminants and sensory characteristics. The species *Thunnus thynnus* has a higher quality and commercial value due to its excellent organoleptic features.

Tuna species are usually consumed as fillets or processed products. The loss of the external anatomical and morphological features makes the authentication of a fish species difficult or impossible and enables fraudulent substitutions (Marko et al., 2004). Species substitution is very common in fish products, due to the profits resulting from the use of less expensive species. For species of tuna, substitutions have both commercial and health implications (Agusa et al., 2005; Besada et al., 2006; Storelli et al., 2010), thus, analytical techniques to differentiate fish species are essential. The development of suitable analytical methods for fish species identification in prepared and transformed fish products is of great interest to enforcement agencies involved with labelling regulations and the authentication of fish in various products to prevent the substitution of fish species (Mackie et al., 2000; Meyer et al., 1995).

Several biochemical techniques enable the study and identification of fillet or minced fish species. Among these methods, isoelectric focusing (IEF) (Etienne et al., 2000; Rehbein et al., 2000; Renon et al., 2001;), capillary zone electrophoresis (Acuña et al., 2008), and amplification of selected DNA sequences by the polymerase chain reaction (PCR) have been used for the identification of certain groups of fish species (Espiñeira et al., 2008; Hubalkova et al., 2008; Pepe et al., 2005, 2007; Trotta et al., 2005).

Presently, PCR is the most frequently used technique, as DNA is heat-stable and resistant to heat treatments that may be applied to the tuna during processing. However, obtaining an accurate species identification is very difficult if the species show a high degree of homology as *Thunnus* does (Chow & Kishino, 1995; Lopez & Pardo, 2005; Michelini et al., 2007; Pardo

& Begoña, 2004; Terio et al., 2010; Viñas & Tudela, 2009). The sequences usually used as species molecular markers are the DNA mitochondrial fragments especially *cytochome b* (*cyt b*) genes and the ribosomal 16S and 12S subunits (Kochzius et al., 2010; Russo et al., 1996; Zehner et al., 1998). Previous studies demonstrated that these molecular markers are not discriminating for *Thunnus* species, because they have few polymorphisms expressed by point mutations (Bottero et al., 2007).

EU Commission Regulation no. 2065/2001 of 22 October 2001 has established detailed rules for consumer information to be included on labels regarding fish species. Accordingly, it is also necessary to develop new methods to prevent illegal species substitutions in seafood products (EC No 2065/2001). Proteins are playing an increasing role in the international scientific community and proteomics, the large-scale analysis of proteins expressed by a cell or a tissue contributes greatly to the study of gene function (Pandey & Mann, 2000). Recently, proteomics has been applied in the fishing industry with several aims, e.g., to examine the water-soluble muscle proteins from farm and wild fish to show aquaculture effects on seafood quality (Monti et al., 2005) or to elucidate the influence of internal organ colonization by *Moraxella* sp. in internal organs of *Sparus aurata* (Addis et al., 2010). Proteomics has also been considered as a tool for species identification in seafood products with interesting results (Carrera et al., 2006, 2007; Chen et al., 2004; López et al., 2002; Piñeiro et al., 1999, 2001).

The aim of this chapter is to examine the potential of proteomics to identify four tuna species through characterisation of specific sarcoplasmic proteins. We investigated *T. albacares, T. alalunga,* and *T. obesus* two dimensional gel electrophesis (2-DE) patterns and also verified the presence of specie-specific proteins for these tuna species. Muscle extracts from four tuna species of the genus *Thunnus* (*T. thynnus, T. alalunga, T. albacares, T. obesus*) were evaluated by both mono and 2-DE and mass spectrometric techniques. In preliminary results (Pepe et al., 2010), proteomics was applied for the identification of a species-specific protein in *T. thynnus* by 2-DE profiles. The analysis of two dimensional gels by ImageMaster™ 2D Platinum software revealed the presence of a protein with a molecular weight of approximately 70 kDa in the *T. thynnus*' 2-DE pattern, which was absent in the other species. This protein, identified as Trioso fosfato isomerasi (gi46909469) through mass spectrometric techniques might be considered a specific marker. The aim of this chapter was to investigate *T. albacares, T. alalunga,* and *T. obesus* 2- DE patterns and verify the presence of species-specific proteins for these tuna species.

2. Materials and methods

2.1 Fish samples

In this study, a total of four different tuna species were tested, with three specimens from each species. The whole tuna specimens were identified, according to their anatomical and morphological features, as belonging to *T. thynnus, T. alalunga, T. albacores,* and *T. obesus* species at the Department of Animal Science and Food Inspection, University of Naples, "Federico II". *T. thynnus* and *T. alalunga* specimens were fished in the Mediterranean Sea and supplied by "Pozzuoli fish market", *T. albacares* specimens were fished in the Indian Ocean and supplied by Salerno P.I.F. (Posto di Ispezione Frontaliera), and *T. obesus* specimens were fished in the South East Atlantic Ocean and were obtained from Philadelphia, Pennsylvania, United States. Fish were frozen on board at - 20 ° C and shipped in insulated boxes to the laboratory. Tuna muscle samples were taken and stored at -80 °C for further analysis.

2.2 Extraction of sarcoplasmic proteins

Raw muscle tissue (3 g) was dipped in 6 mL of 10 mM Tris-HCl buffer at 4 ° C, pH 7.2, supplemented with 5 mM PMSF (phenylmethanesulfonylfluoride). Samples were minced with an "Ultra-turrax" at 4 ° C, for 30 s at 15,000 g to obtain a homogeneus sample of water-soluble proteins. Minced tissues were centrifuged at 15,000 g at 4 ° C for 20 min. The supernatants were then recovered and filtered using Ultrafree CL (0.22 μm) filters, and stored at -20 ° C until analysis by electrophoresis (Carrera et al., 2007). The efficacy and the reproducibility of the extraction protocol of sarcoplasmic proteins was evaluated using *T. alalunga*. The extraction protocol was carried out in triplicate and further checked for quality and quantity by SDS-PAGE.

2.3 SDS polyacrylamide gel electrophoresis (SDS-PAGE)

Protein concentration was measured by the Bradford method (Bradford, 1976) using bovine serum albumin as the standard. Proteins (50 μg) were separated on a 12.5% (w/w) polyacrylamide gel at 25 mA/gel constant current. Gels were stained for 50 min with Coomassie Brilliant Blue R-250 and destained with MilliQ grade water.

2.4 Two dimensional electrophoresis (2-DE)

The first dimensional electrophoresis (isoelectric focusing, IEF) was carried out on non-linear wide-range immobilized pH gradients (pH 3-10; 7 cm long IPG strips; GE Healthcare, Uppsala, Sweden) using the Ettan IPGphor system (GE Healthcare, Uppsala, Sweden). Analytical-run IPG-strips were rehydrated with 50 μg of total proteins in 125 μl of rehydratation buffer and 0.2% (v/v) carrier ampholyte for 12h, at 50 mA, at 20° C. The strips were then focused according to the following electrical conditions at 20°C: 500 V for 30 min, 1000 V for 30 min, 5000 V for 10h, until a total of 15000 V was reached. For preparative gels 100 μg of total proteins were used. After focusing, analytical and preparative IPG strips were equilibrated for 15 min in 6 M urea, 30% (V/V) glycerol, 2% (w/V) SDS, 0.05 M Tris-HCl, pH 6.8, 1% (w/V) DTT, and subsequently for 15 min in the same urea/SDS/Tris buffer solution but substituting the 1% (w/V) DTT with 2.5% (w/V) iodoacetamide. The second dimension was carried out on 12.5% (w/w) polyacrylamide gels (10 cm x 8 cm x 0.75 mm) at 25 mA/gel constant current and 10°C until the dye front reached the bottom of the gel, according to (Hochstrasser et al., 1988) MS-preparative gels were stained for 50 min with Coomassie Brilliant Blue R-250 and destained with MilliQ grade water. The software ImageMaster™ 2D Platinum was used for the analysis of the two dimensional gel images.

2.5 Image analysis

Gels images were acquired with an Epson expression 1680 PRO scanner. Computer-aided 2-D image analysis was carried out using the ImageMaster™ 2D Platinum software. Relative spot volumes (%V) (V=integration of OD over the spot area; %V = V single spot/V total spot) were used for quantitative analysis in order to decrease experimental errors. The normalized intensity of spots on three replicate 2-D gels was averaged and standard deviation was calculated for each condition.

A few initial reference points (landmarks) were affixed for gels alignment, the first step of the image analysis. Landmarks are positions in one gel that correspond to the same position in the other gels. Then, the software automatically detects spots, which represent the proteins on the gels. The software "matches" the gels, and the corresponding spots are paired. The pair is the association between spots that represent the same protein in different gels. Pairs are automatically determined using ImageMaster powerful gel matching algorithm. The different 2DE images can be compared by synchronized 3-D spots view.

2.6 Protein identification by mass spectrometry

2.6.1 In situ digestion

The analysis was performed on the Comassie blue-stained spots excised from gels. The excised spots were washed first with acetonitrile and then with 0.1M ammonium bicarbonate. Enzymatic digestion was carried out with trypsin (10 ng/µl) in 10mM ammonium bicarbonate pH 8.5 at 4° C for 2 h. The buffer solution was then removed and a new aliquot of the enzyme/buffer solution was added for 16 h at 37° C. A minimum reaction volume, enough for the complete rehydratation of the gel was used. Peptides were then extracted washing the gel particles with 1% formic acid and ACN at room temperature.

2.6.2 MALDI-TOF mass spectrometry

Positive Reflectron MALDI spectra were recorded on a Voyager DE STR instrument (Applied Biosystems, Framingham, MA). The MALDI matrix was prepared by dissolving 10 mg of alpha-cyano-4-hydroxycinnamic acid in 1 mL of acetonitrile / water (90:10 v/v). Typically, 1 µl of matrix was applied to the metallic sample plate, and 1 µl of analyte was then added. Acceleration and reflector voltages were set up as follows: target voltage at 20 kV, first grid at 95% of target voltage, delayed extraction at 600 ns to obtain the best signal-to-noise ratios and the best possible isotopic resolution with multipoint external calibration using a peptide mixture purchased from Applied Biosystems. Each spectrum represents the sum of 1500 laser pulses from randomly chosen spots per sample position. Raw data were analysed using the computer software provided by the manufacturers and are reported as monoisotopic masses. Spectra were manually interpreted, there was no need of any De-isotopic or other post acquisition processing due to the good signal to noise ratio. Peak lists were generated manually and used for proteins identification.

2.6.3 LC-MS/MS analysis

A mixture of peptide solution was analysed by LC-MS/MS analysis using a 4000Q-Trap (Applied Biosystems) coupled to an 1100 nano HPLC system (Agilent Technologies) and Agilent HPLC-Chip/MS. The mixture was loaded on an Agilent reverse-phase pre-column cartridge (Zorbax 300 SB-C18, 5x0.3 mm, 5 µm) at 10 µl/min (A solvent 0.1% formic acid, loading time 5 min). Peptides were separated on a Agilent reverse-phase column (Zorbax 300 SB-C18, 150 mm X 75µm, 3.5 µm), at a flow rate of 0.3 µl/min with a 0% to 65% linear gradient in 60 min (A solvent 0.1% formic acid, 2% ACN in MQ water; B solvent 0.1% formic acid, 2% MQ water in ACN). Nanospray source was used at 2.5 kV with liquid coupling, with a declustering potential of 20 V, using an uncoated silica tip from NewObjectives (O.D. 150 µm, I.D. 20 µm, T.D. 10 µm). Data were acquired in information-dependent acquisition (IDA) mode, in which a full scan mass spectrum was followed by MS/MS of the 5 most abundant ions (2 s each). In particular, spectra acquisition of MS-MS analysis was based on a survey Enhanced MS Scan (EMS) from 400 m/z to 1400 m/z at 4000 amu/sec. This scan mode was followed by an Enhanced Resolution experiment (ER) for the five most intense ions and then MS^2 spectra (EPI) were acquired using the best collision energy calculated on the bases of m/z values and charge state (rolling collision energy) from 100 m/z to 1400 m/z at 4000 amu/sec. Data were acquired and processed using Analyst software (Applied Biosystems).

2.6.4 MASCOT analysis

The mass spectra obtained were then used for protein identification using the MASCOT software that compares peptide masses obtained by MALDI-TOF MS and LC-MS/MS of each spot with the theoretical peptide masses from all the proteins accessible in the databases (Peptide Mass Fingerprinting, PMF). Spectral data were analyzed using Analyst software (version 1.4.1) and MS-MS centroid peak lists were generated using the MASCOT.dll script (version 1.6b9). MS/MS centroid peaks were threshold at 0.1% of the base peak. MS/MS spectra having less than 10 peaks were rejected. MS/MS spectra were searched against NBCI (*National Center for Biotechnology Information*) database, (2006.10.17 version) using the licensed version of Mascot 2.1 version (Matrix Science), after converting the acquired MS/MS spectra in mascot generic file format. The Mascot search parameters were: taxonomy: Animalia; significance threshold: higher than 50 (according to Mascot scoring system, Pappin et al., 1993), allowed number of missed cleavages 3; enzyme trypsin; variable post-translational modifications, methionine oxidation, pyro-glu N-term Q; peptide tolerance 100ppm and MS/MS tolerance 0.5 Da; peptide charge, from +2 to +3 and top 20 protein entries. Spectra with a MASCOT score <25 having low quality were rejected. The score used to evaluate quality of matches for MS/MS data was higher than 30. However, spectral data were manually validated and contained sufficient information to assign peptide sequence.
Little genomic information is available for *Thunnus* genus, so protein identification is limited to a scarce number of tuna sequences deposited in the database. Therefore, once a significant protein match was made, protein sequence data were used for BLAST homology searches against other species in the NCBI database.

3. Results

3.1 SDS-PAGE

The protein extraction protocol developed for *T. alalunga* was used for all examined samples and showed high reproducibility; the extracted proteins were of both good quality and quantity (Figure 1). Protein samples of *T. thynnus, T. albacares, T. alalunga,* and *T. obesus* were fractionated by SDS gel electrophoresis as shown in Figure 2. After SDS-PAGE fractionation, some differences could be observed between the different tuna species. SDS-PAGE protein bands, in fact, showed inter-species differences, in particular for proteins with molecular weights lower than 25 kDa.

3.2 Analysis by 2-DE

In order to better elucidate the protein maps of different tuna species, the four samples were subjected to 2D fractionation. Deep analysis of the muscle proteome from the tuna species was undertaken by 2-DE image analysis using the ImageMaster™ 2D Platinum software. The tuna 2-DE images were aligned choosing four landmarks (L1, L2, L3 and L4) in each gel (Figure 3).

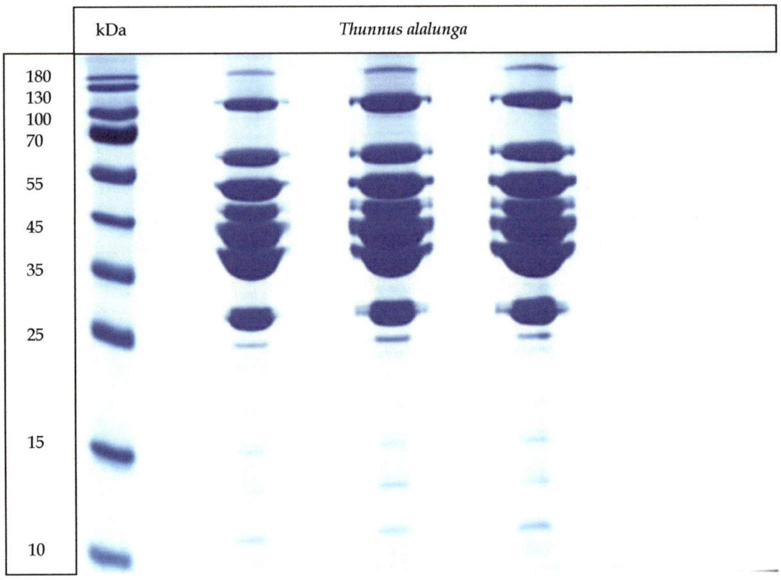

Fig. 1. *T. alalunga* SDS-PAGE. Three different protein samples were compared to verify the reproducibility of the extraction protocol.

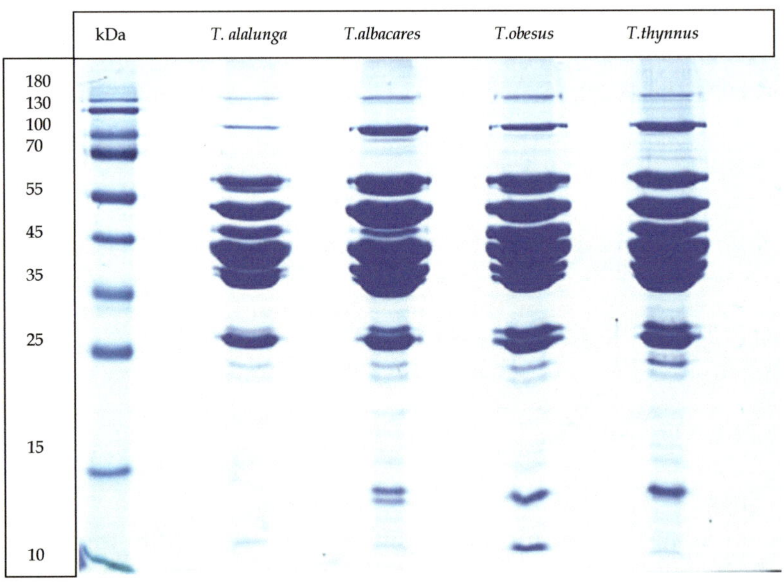

Fig. 2. *T. alalunga, T. albacares, T. obesus,* and *T. thynnus* SDS-PAGE. Proteins with molecular weight lower than 25 kDa are different among species.

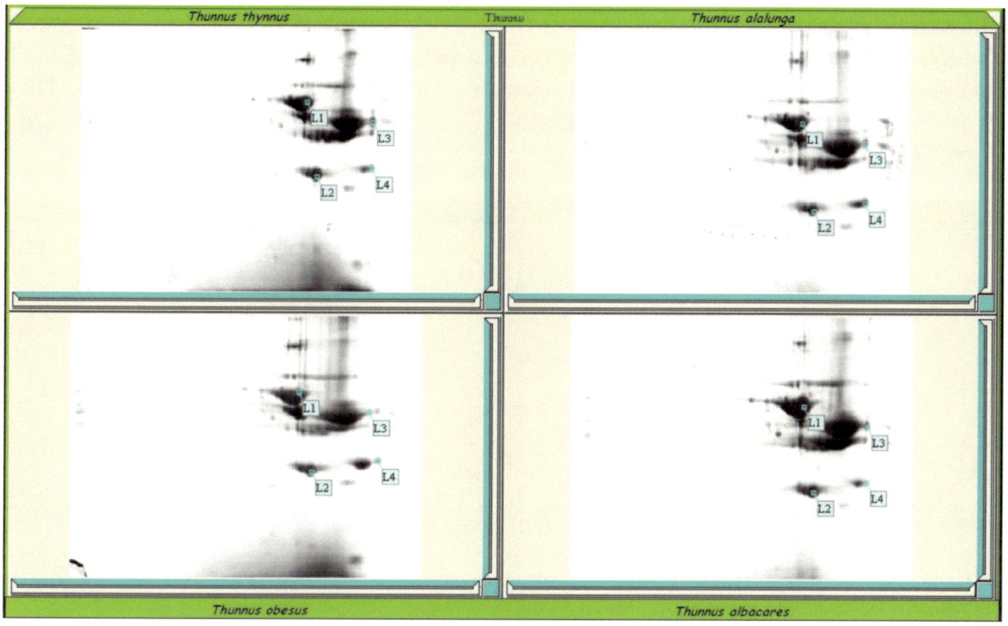

Fig. 3. 2DE gel images alignment: landmarks affixing (L1, L2, L3 and L4).

The software correctly detected and aligned spots between the four tuna 2-DE gel images, as reported in Figure 4. The ImageMaster™ 2D Platinum found: 107 total spots on *T. thynnus* 2-DE gel, 93 total spots on the 2-DE gel of *T. alalunga*, 115 total spots on *T. albacares* 2-DE gel and 123 total spots on the 2-DE gel of *T. obesus*.

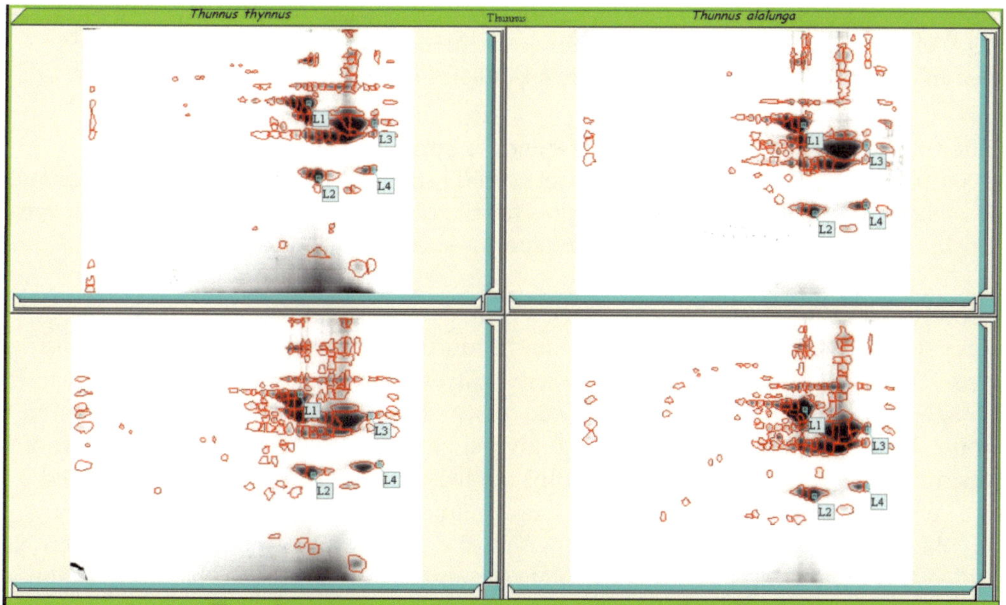

Fig. 4. Spot detection. Spots from the 2-DE arrayed samples representing proteins are circled in red.

Gel matching of tuna 2-DE images indicated the presence of spots that were both common to the four species, and the presence of spots that were specific for each species (Fig. 5). The software detected 28 specific spots on *T. thynnus* 2-DE gel, 48 specific spots on the 2-DE gel of *T. alalunga* , 65 specific spots on *T. albacares* 2-DE gel and 60 specific spots on the 2-DE gel of *T. obesus*.

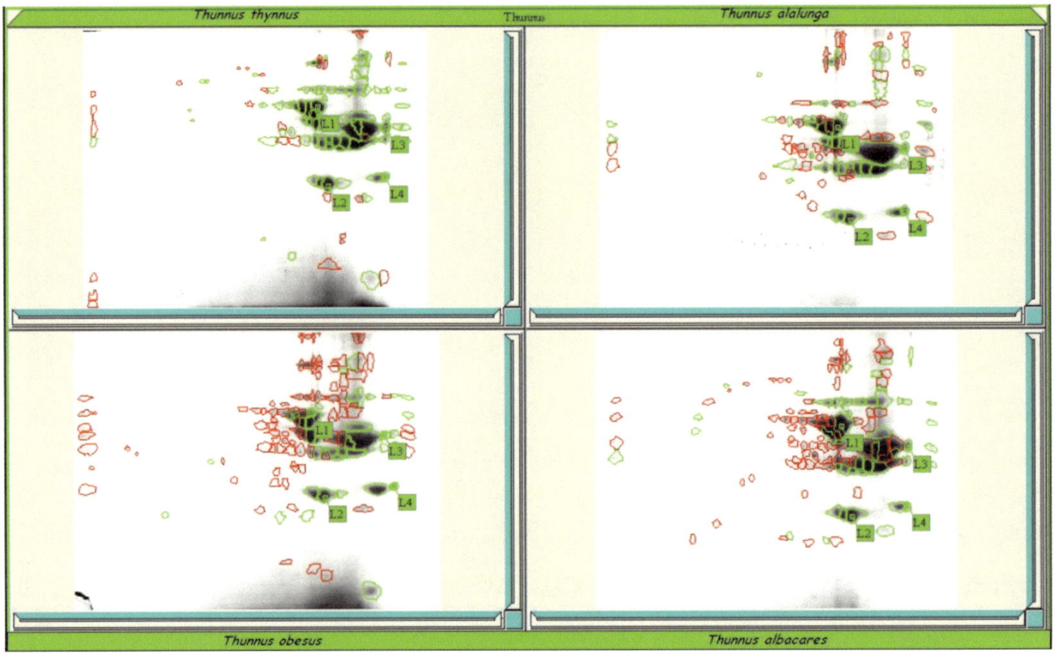

Fig. 5. Gel matching: spots circled in green are common to the four tuna species, spots circled in red are not paired and therefore specific for each species.

3.3 Identification of non-paired/specie-specific spots

The comparison of the 3-D view of the "not paired" spots in the four 2-DE gel images makes it possible to find the most interesting spots for the characterization of the four tuna species (Fig 6-13). These proteins were considered species-specific markers.

3.4 Protein identification

Protein spots were excised from the gel and reduced, alkylated, and in-gel digested with trypsin. The resulting peptide mixtures were analyzed directly by MALDI-TOF MS and/or LC MS/MS. The MS/MS spectra were used to search for a non-redundant match using the in-house MASCOT software, thus taking advantage of the specificity of trypsin and of the taxonomic category of the samples. NCBInr database updates are regularly uploaded to in house version of MASCOT. We filtered identifications restricting to Animalia taxonomy.
Molecular weights values that matched within the given mass accuracy of 100 ppm were recorded and the proteins that had the highest number of peptide matches were examined. Protein identification is limited to a scarce number of tuna sequences deposited in the database. Therefore, for proteins identified with low MASCOT score, protein sequence data were used for BLAST homology searches against other species in the NCBI database. The

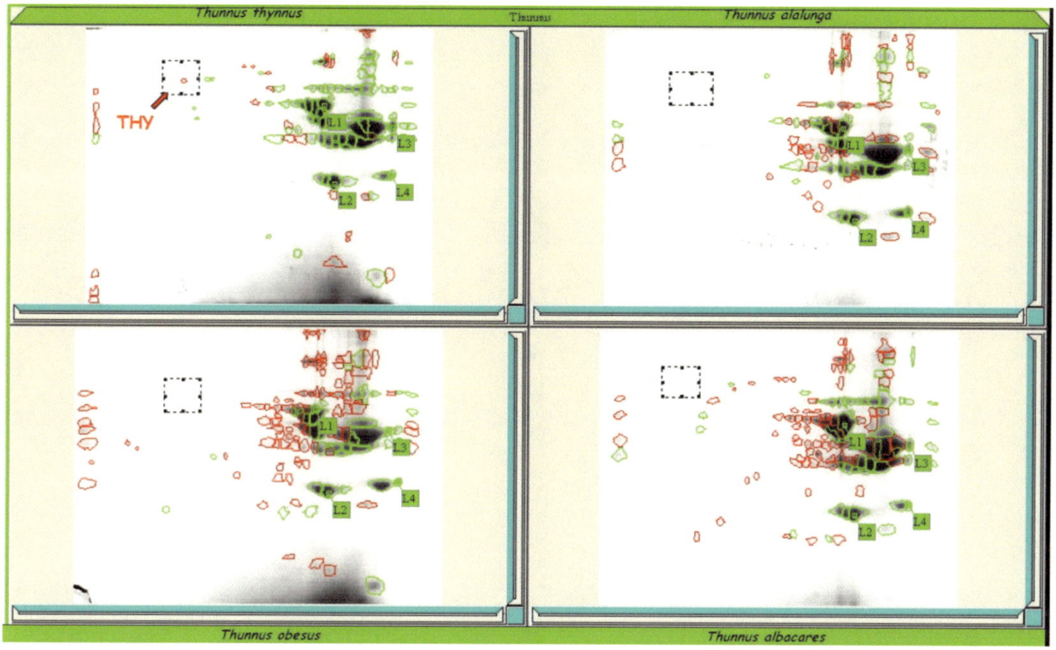

Fig. 6. An example of a *T. thynnus* spot that might be a specific marker (labeled THY). Equivalent areas on all gels highlighted with a box.

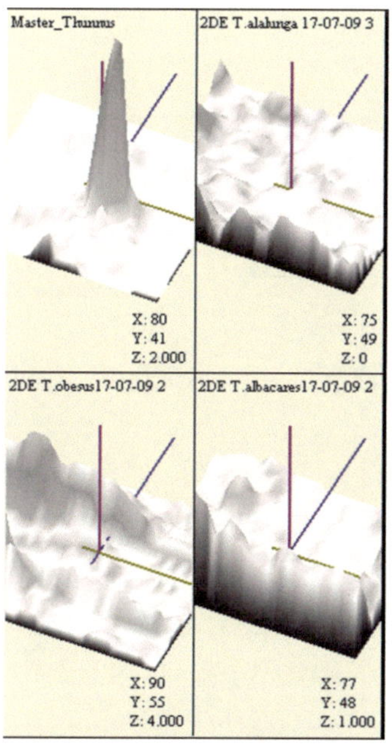

Fig. 7. 3-D view of the *T. thynnus* species-specific spots and 3-D view of the same area from the other species gels. No spot/protein is present in the gel arrayed samples from the other species.

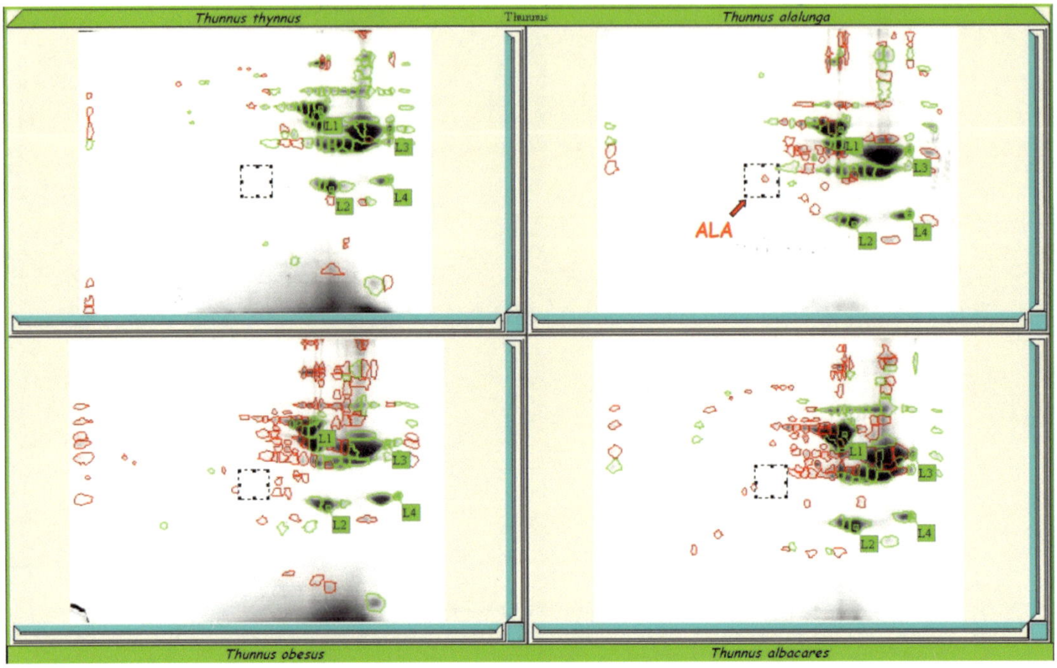

Fig. 8. An example of a *T. alalunga* spot that might be a specific marker (labeled ALA). Equivalent areas on all gels highlighted with a box.

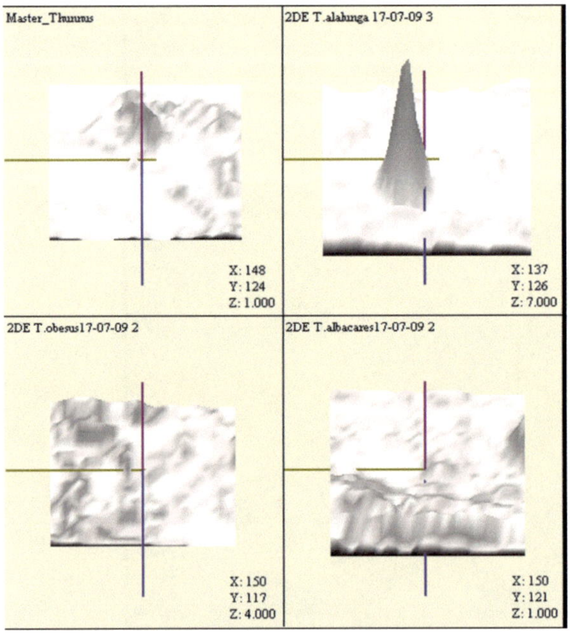

Fig. 9. 3-D view of the *T. alalunga* species-specific spots and 3-D view of the same area from the other species gels. No spot/protein is present in the gel arrayed with samples from the other species.

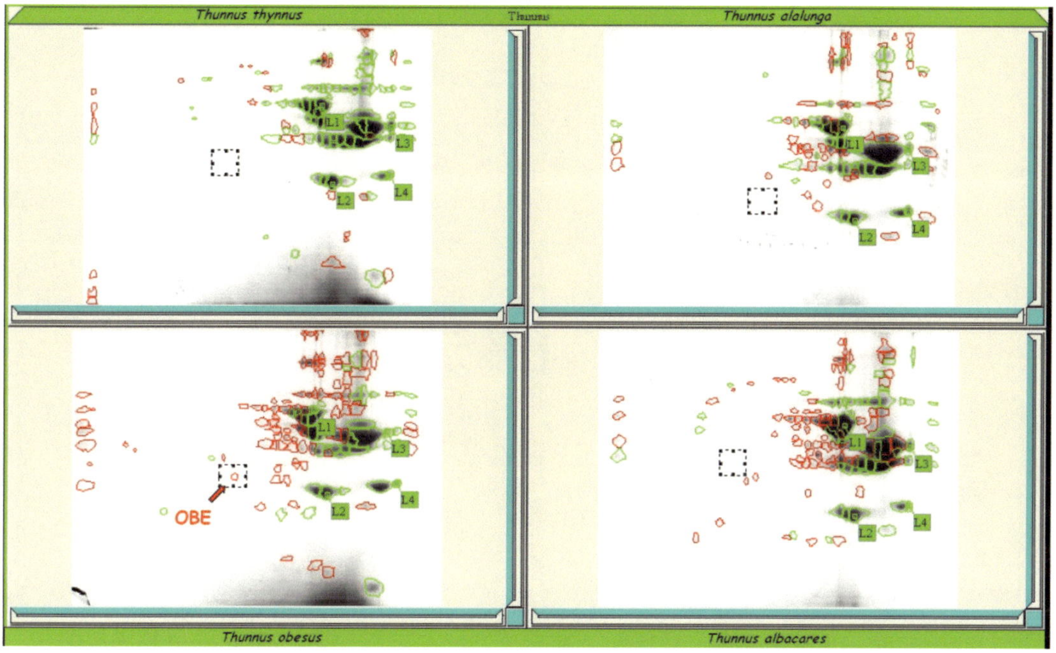

Fig. 10. An example of a *T. obesus* spot that might be a specific marker (labeled OBE). Equivalent areas on all gels highlighted with a box.

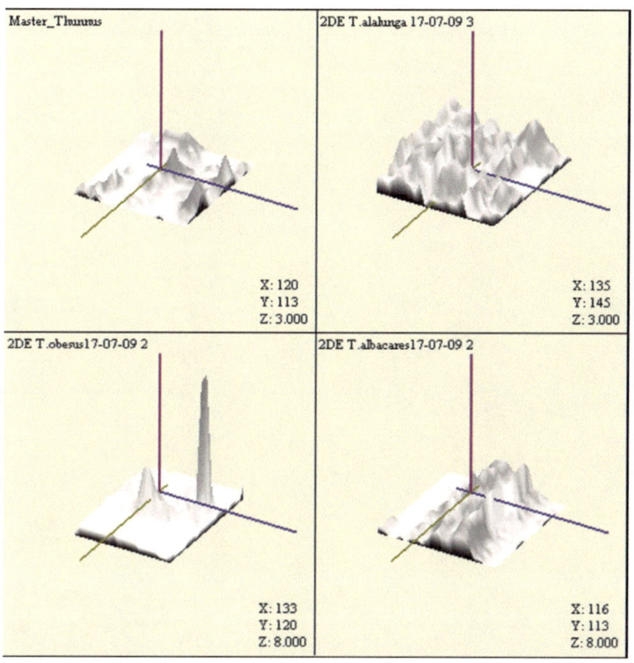

Fig. 11. 3-D view of the *T. obesus* species-specific spots and 3-D view of the same area from the other species gels. No spot/protein is present in the gel arrayed with samples from the other species.

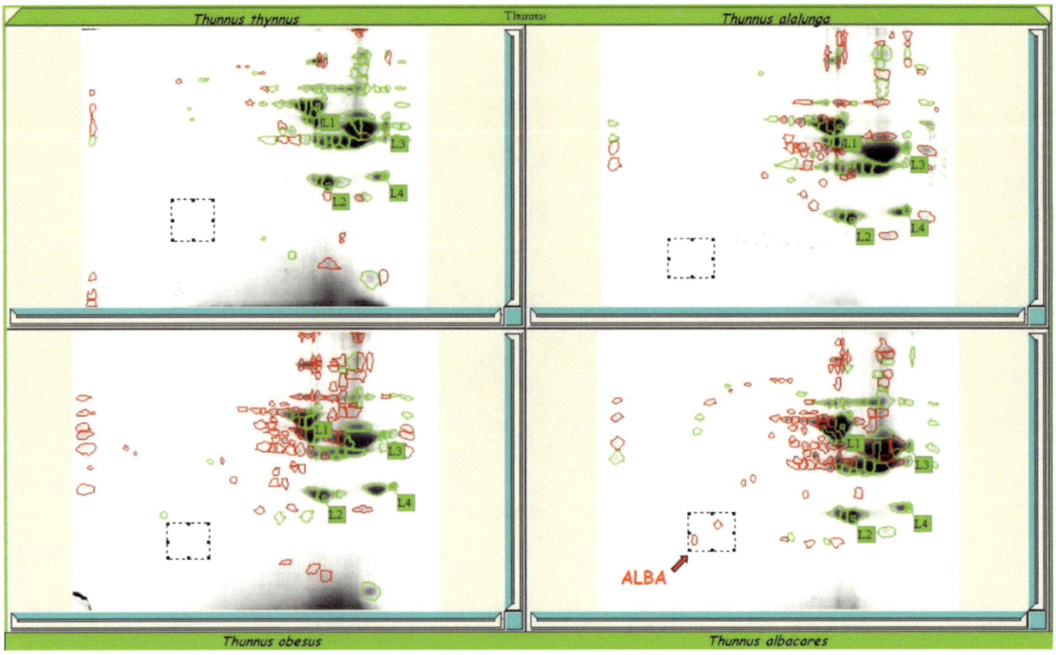

Fig. 12. An example of *T. albacares* spots that might be a specific marker (labeled ALBA). Equivalent areas on all gels highlighted with a box.

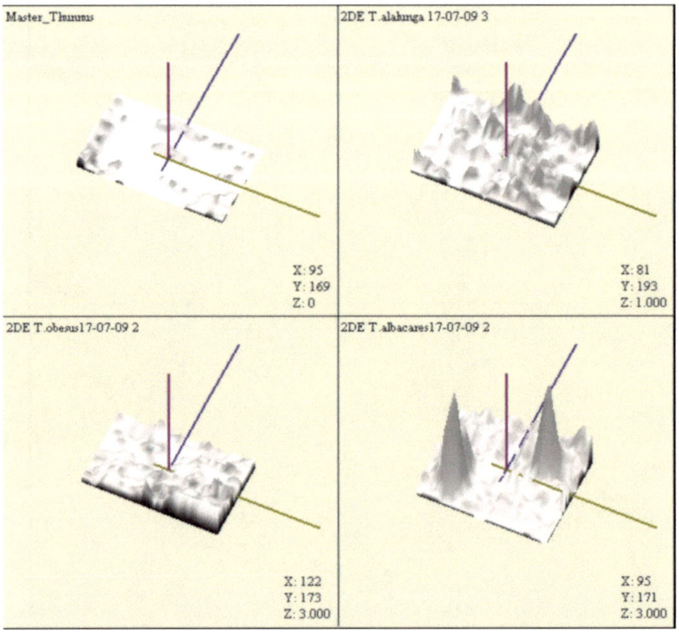

Fig. 13. 3-D view of the *T. albacares* species-specific spots and 3-D view of the same area from the other species gels. No spot/protein is present in the gel arrayed with samples from the other species.

BLAST alignment was done for: Triosephosphate isomerase (THY), Pyruvate kinase (ALA) and Fast skeletal muscle troponin T (ALBA). The results of mass spectrometric analysis of the species-specific markers are shown in Table 1.

Spot ID	Protein	Accession number	Species	MW	pI	Analysis Method	Score
THY	**Triosephosphate isomerase** [Priapulus caudatus]	gi46909469	*T. thynnus*	22.9 kDa	6.51	LC-MS/MS	MASCOT 83 BLAST 421
ALA	**Pyruvate kinase muscle** [Danio rerio]	gi40786398	*T. alalunga*	58.6 kDa	6.54	LC-MS/MS	MASCOT 93 BLAST 1052
ALBA	**Fast skeletal muscle troponin T Subunits** [Gadus morhua]	gi20386541	*T. albacares*	27.2 kDa	9.48	LC-MS/MS	MASCOT 92 BLAST 269
OBE	**Beta-enolase** [Epinephelus coioides]	gi295792264	*T. obesus*	47.5 kDa	6.29	MALDI	MASCOT 159

Table 1. Identification of potential species-marker proteins from 2-DE arrays of the four *Thunnus* species.

4. Discussion

The specific proteins have important metabolic functions. Pyruvate kinase identified in *T. alalunga* is an enzyme involved in glycolysis. This protein catalyzes the transfer of a phosphate group from phosphoenolpyruvate (PEP) to ADP, yielding pyruvate and ATP. The specific protein identified in *T. Thynnus* is triose phosphate isomerase (TPI), a glycolytic enzyme which catalyzes the interconversion of dihydroxyacetone phosphate (DHAP) with D-glyceraldehyde-3-phosphate. TPI plays an important role in glycolysis and is essential for efficient energy production. Beta-enolase was identified in *T. obesus* and is a muscle-specific enolase (MSE) and is an enzyme of the lyase class that catalyzes the dehydration of 2-phosphoglycerate to form phosphoenolpyruvate. It appears to have an important function in striated muscle development and regeneration. The species-specific *T. albacares* protein is troponin T, fast skeletal muscle subtype. Troponin T (also symbolized TNTF) is the tropomyosin-binding subunit of troponin, the thin filament regulatory complex which confers calcium-sensitivity to striated muscle actomyosin ATPase activity.

Therefore, all the identified species-specific proteins have an important metabolic function. For this reason it is not reasonable to think that these proteins do not exist in the other *Thunnus* species. But the ImageMaster™ 2D Platinum software did not find these proteins in the same localization of the other species 2D gels, which means that these proteins have a different isoelectric point and molecular weight in the other analysed species. The image analysis was correct for species identification, and it was confirmed by the 3-D view; the different spots are proteins and not artifacts caused by aberrant staining of the gel. So, the presence of the species-specific spot in a different area of the gel could indicate (e. g. pyruvate kinase) a higher rate for glycolysis in *T. alalunga.* It is important to continue the studies to enhance the knowledge of the identified species-specific spots and to identify other spots that could have species-specific function.

5. Conclusion

Proteomics has been demonstrated to be a useful method to increase scientific knowledge on animals and plants (Pandey & Mann, 2000). The progress in proteomic analytical techniques has enabled more accurate and reliable information for determining species differences (Tyers & Mann, 2003). The realization of a unique fingerprint for a given species is possible through the separation and subsequent identification of specific proteins.

In this study, a proteomic assay for the identification of species-specific markers of commercially important species of the genus *Thunnus* was undertaken. The proteomic fingerprinting of four species of the genus *Thunnus* was obtained using two dimensional electrophoresis followed by protein identification using mass spectrometry. The analysis of the 2-DE images revealed significant differences between the four tuna species investigated.

The gel matching (Figure 5) shows that there are several different spots between the species, circled in red. The number of species-specific spots identified by the software is substantial for each *Thunnus* species (28 out of 107 total spots for *T. thhynnus;* 48 out of 93 total spots for *T. alalunga;* 65 out of No 115 total spots for *T. albacares* and 60 out of No 123 total spots for *T. obesus*).

The 3-D view of the gels revealed the presence of some red circled spots absent in the same areas from the other species gels. These spots were chosen as species-specific.

The occurrence of species-specific protein spots may be due to differentially expressed proteins only present at low levels or absent in the other species. Thus, 2-DE analysis helped us to identify species-specific proteins, which could be used as specific markers to delineate each species.

The value of a proteomics approach to differentiate tuna species relies on both the ability to obtain the visualization of different protein spots in a 2D map but also the unique identification of the protein candidate by using mass spectral and bioinformatics procedures. Analyses were further enhanced through morphological visualization by 3-D reconstruction of differential spots from the four tuna species. In this way, it was possible to enhance differences and identify highly unique proteins from the *Thunnus* species. This second phase of study further validated the proteomic analysis technique as it confirmed that spots found in different locations and morphology on the 2D gels also corresponded to different proteins.

We have demonstrated that proteomics could be employed to differentiate species when they show contain high degrees of genetic homology (e.g. *Thunnus*). The DNA sequences normally used as species molecular markers are not discriminating for *Thunnus* species (Bottero et al., 2007). Moreover, without the option of a proteomic investigation, it would be necessary to further investigate the genome of each species, to identify genes that may differ between the species. This study shows how the use of proteomics tools is important for species identification.

The future developments of this study should be the identification of other species-specific proteins with metabolic functions characteristic of each species, to then identify species-specific genes. Primers can be subsequently designed for routine molecular biology methods to identify raw and processed fish products by PCR. In fact, PCR is currently routinely used for species identification and maintains this role due to practical attributes such as speed and cost. However, proteomics can provide an immediate and unambiguous identification of protein biomarkers, and in cases where the genomes are similar between species, the analysis of the proteome has a decisive advantage.

6. Acknowledgment

The authors are grateful to the "Pozzuoli fish market" , and Salerno P.I.F. (Posto di Ispezione Frontaliera) for providing fish samples. The assistance of Alberto Nunez who helped us in performing the MALDI-MS is acknowledged. We also appreciate the suggestions made by Pina Fratamico to improve this manuscript.

7. References

Acuña, G.; Ortiz-Riaño, E.; Vinagre, J.; García, L.; Kettlun, A.M.; Puente, J.; Collados, L. & Valenzuela, M.A. (2008). Application of capillary electrophoresis for the identification of Atlantic salmon and rainbow trout under raw and heat treatment. J Capill Electrophor Microchip Technol., Vol. 10, No.5-6, pp. 93-9, ISSN 10795383

Addis, MF.; Cappuccinelli, R.; Tedde, V.; Pagnozzi, D.; Viale, I.; Meloni, M.; Salati, F.; Roggio, T & Uzzau, S. (2010). Influence of Moraxella sp. colonization on the kidney proteome of farmed gilthead sea breams (Sparus aurata, L.). Proteome Sci., Vol. 8, No 50, (12 Oct 2010), pp. 1-8, ISSN 1477-5956

Agusa, T.; Kunito, T.; Yasunaga, G.; Iwata, H.; Subramanian, A.; Ismail, A. & Tanabe, S. (2005). Concentrations of trace elements in marine fish and its risk assessment in Malaysia. *Mar Pollut Bull.* Vol. 51, No. 8-12, pp. 896-911, ISSN 0025-326X

Besada, V.; Gonzalez, J.J.; Schultze & F. (2006). Mercury, cadmium, lead, arsenic, copper and zinc concentrations in albacore, yellowfin tuna and bigeye tuna from the Atlantic Ocean. *Cienc. Mar.* Vol. 32, pp. 439–445, ISSN 0185-3880

Bottero, M.T.; Dalmasso, A.; Cappelletti, M.; Secchi, C. & Civera, T. (2007). Differentiation of five tuna species by a multiplex primer-extension assay. J Biotechnol. May 1; Vol. 129, No. 3, (May 2007), pp. 575-80, ISSN 0168-1656

Bradford, M. (1976). A rapid and sensitive method for the quantification of mg quantities of protein. *Anal Chem,* Vol 72, pp. 248-254, ISSN 0003-2700

Carrera, M.; Cañas, B.; Piñeiro, C.; Vázquez, J. & Gallardo, J. M. (2006). Identification of commercial hake and grenadier species by proteomic analysis of the parvalbumin fraction. *Proteomics,* Vol. 6, pp. 5278–5287, ISSN 1615-9853

Carrera, M.; Cañas, B.; Piñeiro, C.; Vázquez, J. & Gallardo, J. M. (2007). De Novo Mass Spectrometry Sequencing and Characterization of Species-Specific Peptides from Nucleoside Diphosphate Kinase B for the Classification of Commercial Fish Species Belonging to the Family Merlucciidae. *J. Proteome Res.* Vol. 6, No. 8, pp. 3070-3080, ISSN 1535-3893

Chen, T.Y.; Shiau, C.Y.; Wei, C.I. & Hwang, D.F. (2004) Preliminary Study on Puffer Fish Proteome Species Identification of Puffer Fish by Two-Dimensional Electrophoresis. *J. Agric. Food Chem.* Vol. 52, No 8, pp. 2236-2241, ISSN 0021-8561

Chow, S. & Kishino, H. (1995). Phylogenetic relationships between tuna species of the genus Thunnus (Scombridae: Teleostei): inconsistent implications from morphology, nuclear and mitochondrial genomes. J Mol Evol. Vol. 41, No. 6 (December 1995), pp. 741-8, ISSN 0022-2844

Commission Regulation (EC) No 2065/2001 of 22 October 2001 laying down detailed rules for the application of Council Regulation (EC) No 104/2000 as regards informing consumers about fishery and aquaculture products. Off. Jour. of the European Communities. L 278/6.

D.M. MIPAAF, 31 Gennaio 2008
Espiñeira, M.; González-Lavín, N.; Vieites, JM. & Santaclara, FJ. (2008). Authentication of Anglerfish Species (Lophius spp) by Means of Polymerase Chain Reaction-Restriction Fragment Length Polymorphism (PCR-RFLP) and Forensically Informative Nucleotide Sequencing (FINS) Methodologies. J Agric Food Chem. Vol. 56, No 22, (26Nov 2008), pp. 10594-9, ISSN 0021-8561
. Jérôme, M.; Fleurence, J.; Rehbein, H.; Kündiger, R.; Mendes, R.; Costa, H.; Pérez-Martín, R.; Piñeiro-González, C.; Craig, A.; Mackie, I.; Malmheden Yman, I.; Ferm, M.; Martínez, I.; Jessen, F.; Smelt, A. & Luten, J. Identification of fish species after cooking by SDS-PAGE and urea IEF: a collaborative study. (2000). J Agric Food Chem. Vol. 48, No 7 (Jul 2000), pp. 2653-8, ISSN 0021-8561
Hochstrasser, D.; Augsburger,V.; Pun, T.; Weber, D.; Pellegrini, C. & Muller, AF. (1988). "High-resolution" mini-two-dimensional gel electrophoresis automatically run and stained in less than 6 h with small, ready-to-use slab gels. Clin Chem. Vol. 34, No 1 (Jan 1988), pp. 166-70, ISSN 0009-9147
Hubalkova, Z.; Kralik, P.; Kasalova, J. & Rencova, E. (2008). Identification of gadoid species in fish meat by polymerase chain reaction (PCR) on genomic DNA. J Agric Food Chem., Vol. 56, No 10, (28 May 2008), pp. 3454-9, ISSN 0021-8561
Kochzius, M.; Seidel, C.; Antoniou, A.; Botla, SK.; Campo, D.; Cariani, A.; Vazquez, EG.; Hauschild, J.; Hervet, C.; Hjörleifsdottir, S.; Hreggvidsson, G.; Kappel, K.; Landi, M.; Magoulas, A.; Marteinsson, V.; Nölte, M.; Planes, S.; Tinti, F.; Turan, C.; Venugopal, MN.; Weber, H. & Blohm, D. (2010). Identifying Fishes through DNA Barcodes and Microarrays. PLoS One, Vol. 5, No. 7, (7 Sep 2010), pp. 12620, ISSN 1932-6203
Kojadinovic, J.; Potier, M.; Le Corre, M.; Cosson, R.P. & Bustamante, P. (2007). Bioaccumulation of trace elements in pelagic fish from the Western Indian Ocean. Environ Pollut. Vol. 146, No. 2 (Mar 2007), pp. 548-66, ISSN 0269-7491
López, JL.; Marina, A.; Álvarez, G. & Vázquez, Jesús. (2002). Application of proteomics for fast identification of species-specific peptides from marine species. *Proteomics*, Vol. 2, No. 12, (Dec 2002), pp. 1658-65, ISSN 1615-9853
Lopez, I. & Pardo, MA. (2005). Application of relative quantification TaqMan real-time polymerase chain reaction technology for the identification and quantification of Thunnus alalunga and Thunnus albacares. J Agric Food Chem., Vol. 53, No 11, (1 Jun 2005), pp. 4554-60, ISSN 0021-8561
Mackie, I. M.; Pryde, S. E.; Gonzales-Sotelo,C.; Medina,I.; Peréz-Martín, R.; Quinteiro,J. ; Rey-Mendez,M. & Rehbein, H. (2000). Possibilità di identificazione delle specie di pesce in scatola. *Ind. Conserve*, Vol. 75, pp. 59–66, ISSN 00197483
Marko, P. B.; Lee, S. C.; Rice, A. M.; Gramling, J. M.; Fitzhenry, T. M.; Mc Aalister, J. S.; Harper, G. R. & Moran, A. L. (2004). Fisheries: mislabelling of a depleted reef fish. *Nature*, Vol. 430, No 6997, (Jul 15 2004), pp. 309-10, ISSN 0028-0836
Meyer, R.; Hofeleien, C.; Luphy, J. & Candrian, U. (1995). Polymerase chain reaction–restriction fragment length polymorphism analysis: a simple method for species identification in food. *J. AOAC Int.*, Vol. 78, pp. 1542–1545, ISSN 1060-3271
Michelini, E.; Cevenini, L.; Mezzanotte, L.; Simoni, P.; Baraldini, M.; De Laude, L. & Roda, A. (2007). One-step triplex-polymerase chain reaction assay for the authentication of yellowfin (Thunnus albacares), bigeye (Thunnus obesus), and skipjack

(Katsuwonus pelamis) tuna DNA from fresh, frozen, and canned tuna samples. J Agric Food Chem., Vol. 55, No 19, (19 Sep 2007), pp.7638-47, ISSN 0021-8561

Monti, G.; De Napoli, L.; Mainolfi, P.; Barone, R.; Guida, M.; Marino, G. & Amoresano, A. (2005). Monitoring Food Quality by Microfluidic Electrophoresis, Gas Chromatography, and Mass Spectrometry Techniques: Effects of Aquaculture on the Sea Bass (Dicentrarchus labrax). *Anal. Chem.,* Vol. 77, pp. 2587-2594, ISSN 0003-2700

Pandey, A. & Mann, M. (2000). Proteomics to study genes and genomes. *Nature.* Vol. 405, No 6788, pp. 837-846, ISSN 0028-0836

Pappin, DJ.; Hojrup, P. & Bleasby, AJ. (1993). Rapid identification of proteins by peptide-mass fingerprinting. Curr Biol. Vol. 3, No 6, (Jun 1993), pp. 327-32, ISSN 0960-9822

Pardo, MA. & Pérez-Villareal, B. (2004). Identification of commercial canned tuna species by restriction site analysis of mitochondrial DNA products obtained by nested primer PCR. *Food Chemistry.* Vol. 86, pp. 143–150, ISSN 0308-8146

Pepe, T.; Trotta, M.; di Marco, I.; Cennamo, P.; Anastasio, A. & Cortesi, ML. (2005). Mitochondrial cytochrome b DNA sequenze variations: An approach to fish species identification in processed fish products. *J. Food Prot.* Vol 68, pp. 421-425, ISSN 0362-028X

Pepe, T.; Trotta, M.; Di Marco, I.; Anastasio, A.; Bautista, J.M. & Cortesi, M.L. (2007). Fish species identification in surimi-based products. J Agric Food Chem. Vol. 55, No. 9 (May 2007), pp. 3681-5, ISSN 0021-8561

Pepe, T.; Ceruso, M.; Carpentieri, A.; Ventrone, I.; Amoresano, A. & Anastasio A. (2010). Proteomic analysis for the identification of *Thunnus* genus three species. *Vet Res Commun.* Vol. 34, Suppl. 1, pp. 153–S155, ISSN 0165-7380

Piñeiro, C.; Barros-Velázquez. J.; Stelo, CG. & Gallardo, JM. (1999). The use of two-dimensional electrophoresis in the characterization of the water-soluble protein fraction of commercial flat fish species. *Eur. Food Res. Technol.,* Vol. 208, Numbers 5-6, pp. 342-348, ISSN 1438-2377

Piñeiro, C.; Vázquez, Jesús.; Marina, AI.; Barros-Velázquez, J. & Gallardo, JM. (2001). Characterization andpartial sequencing of species-specific sarcoplasmic polypeptides from commercial hake species by mass spectrometry following two-dimensional electrophoresis. *Electrophoresis,* Vol. 22, No. 8, (May 2001), pp. 1545-52, ISSN 0173-0835

Rehbein, H.; Kündiger, R.; Pineiro, C. & Perez-Martin, RI. (2000). Fish muscle parvalbumins as marker proteins for native and urea isoelectric focusing. *Electrophoresis,* Vol. 21, No 8, (May 2000), pp.1458-63, ISSN 0173-0835

Renon, P.; Colombo, M.; Colombo F.; Malandra, R. & Biondi, P. A. (2001). Computer-assisted evaluation of isoelectric focusing patterns in electrophoretic gels: Identification of smoothhounds *(Mustelus mustelus, Mustelus asterias)* and comparison with lower value shark species. *Electrophoresis.* Vol. 22, pp. 1534–1538, ISSN 0173-0835

Russo, C.; Takezaki, N. & Nei, M. (1996). Efficiencies of different genes and tree-building methods in recovering a known vertebrate phylogeny. *Mol. Biol. EVol.,* Vol. 13, No. 3, (Mar 1996), pp. 525-536, ISSN 0737-4038

Storelli, M.M.; Barone, G.; Cuttone, G.; Giungato, D. & Garofalo, R. (2010) Occurrence of toxic metals (Hg, Cd and Pb) in fresh and canned tuna: public health implications. Food Chem Toxicol. Vol. 48, No. 11, (Nov 2010), pp. 3167-70, ISSN 0278-6915

Terio, V.; Di Pinto, P.; Decaro, N.; Parisi, A.; Desario, C.; Martella, V.; Buonavoglia, C. & Tantillo, M.G. (2010) Identification of tuna species in commercial cans by minor groove binder probe real-time polymerase chain reaction analysis of mitochondrial DNA sequences. Mol Cell Probes. Vol. 24, No.6, pp. 352-6, ISSN 0890-8508

Trotta, M.; Schönhuth, S.; Pepe, T.; Cortesi, ML.; Puyet, A. & Bautista, JM. (2005). Multiplex PCR method for use in real-time PCR for identification of fish fillets from grouper (*Epinephelus* and *Mycteroperca* species) and common substitute species; J Agric Food Chem., Vol. 53, No 6 (23 Mar 2005), pp. 2039-45, ISSN 0021-8561

Tyers, M. & Mann, M. (2003). From genomics to proteomics. Nature, Vol. 422, No 6928, (13 Mar 2003), pp. 193-7, ISSN 0028-0836

Viñas, J. & Tudela, S. (2009). A validated methodology for genetic identification of tuna species (genus Thunnus). PLoS One. Vol. 4, No. 10, (Oct 2009), pp. 7606, ISSN 1932-6203

Zehner, R.; Zimmermann, S. & Mebs, D. (1998). RFLP and sequence analysis of the cytochrome b gene of selected animals and man: Methodology and forensic application. *Int. J. Leg. Med.* Vol. *111*, pp. 323-327, ISSN 0937-9827

10

Identification of Proteins Involved in pH Adaptation in Extremophile Yeast *Yarrowia lipolytica*

Ekaterina Epova[1,2] et al.*
[1]*K.I. Skryabin Moscow state academy of veterinary medicine and biotechnology, Moscow*
[2]*Federal Center for Toxicological and Radiation Safety of Animals, Kazan*
Russia

1. Introduction

Extremophile yeast *Yarrowia* is now commonly acknowledged as a prospective industrial microorganism and a highly promising cell model. This is due to several unique properties of this organism. First, it is able to grow rapidly on a broad range of organic substrates including waste water, oil paraffin and non-natural substances. This property, due to the presence of peroxisomes, allows the application of *Y. lipolytica* for waste management, water and soil bioremediation and for conversion of fossil organic compounds and pollutants to feed ingredients. Second, due to a high metabolic activity and resistance to chemical stresses, *Y. lipolytica* is able to produce aggressive organic compounds (e.g. succinate) at high yields. Third, *Y. lipolytica* provides a well-established model of a dimorphic transition between a yeast-like state and a mycelium forming fungi. This property enables the application of *Y. lipolytica* as a model for drug discovery for therapeutic control of *Candida albicans* (the most common fungal pathogen in humans) and other pathogenic fungi. In contrast to *C.albicans*, *Y.lipolytica* is easily cultivated and has a complete sexual cycle. Thus, genetic mating analyses are readily applicable. Dimorphic transition in *Y. lipolytica* is also considered as the simplest model of cell differentiation in eukaryotes. Taken together, these factors provided the incentive for complete sequencing of the *Y. lipolytica* genome. This has been carried out by GenoLevures Consortium in France (Dujon et al, 2004). Availability of the complete genomic sequence opened access to proteomic assays to be combined with functional studies of *Y. lipolytica*. The proteomic approach has been successfully used by Morin et al (2007) for the identification of major proteins involved in the dimorphic transition of *Y. lipolytica*.

* Marina Guseva[2,3], Leonid Kovalyov[4], Elena Isakova[4], Yulia Deryabina[4], Alla Belyakova[1,2,3], Marina Zylkova[3] and Alexei Shevelev[1,2,3*]
[1]*K.I. Skryabin Moscow state academy of veterinary medicine and biotechnology, Moscow, Russia*
[2]*Federal Center for Toxicological and Radiation Safety of Animals, Kazan, Russia*
[3]*M.P. Chumakov Institute of poliomyelitis and viral encephalitides of RAMS, Moscow, Russia*
[4]*A.N. Bach Institute of Biochemistry RAS, Moscow, Russia*

Beyond the abovementioned properties, *Y. lipolytica* remains the only known ascomycetes yeast readily growing on alkaline media and in the presence of salts at near-saturating concentrations. These phenomena have been studied by both genetic and biophysical methods. Genetic and molecular biology data implicated the involvement of Rim101- and calcineurine-dependent signal pathways in the high pH adaptation (Lambert et al, 1997). Biophysical data by Zvyagilskaya et al (2000) demonstrated the exchange of proton-dependent machineries involved in metabolite symport (e.g. phosphate ion) through the plasma membrane as a mechanism for Na^+-dependent adaptability. Rim101- and calcineurine-dependent regulatory pathways as well as the proton/Na^+ symport switch are ubiquitous in all studied yeast including *Saccharomyces cerevisiae*. Under normal conditions, these mechanisms usually provide a launch of emergency responses to stress, allowing only a short-term survival of the cells under the alkaline / high salt conditions. In contrast, *Y. lipolytica* permanently grows on media with a pH up to 10. On the other hand, similar to other ascomycetes, *Y. lipolytica* is considered to prefer an acidic pH media. Many strains of this species demonstrate an exclusive resistance to low pH (Yuzbashev et al, 2010). Taken together, these data show that the ambivalent pH adaptation molecular mechanisms in *Y. lipolytica* coupled to an extreme halotolerance, remains obscure. Their discovery may significantly contribute to practical applicability of *Y. lipolytica.*

2. Research objectives

Taking into account the availability of a complete genomic sequence, we aimed to apply proteomics technique for the identification of *Y. lipolytica* proteins whose occurrence depends on pH medium and apparently contributes to global mechanisms of pH adaptation.

3. Methods

3.1 Yeast strain and culture conditions

Y. lipolytica strain PO1f (MatA, leu2-270, ura3-302, xpr2-322, axp-2) was purchased from CIRM-Levures collection (France) where it was deposited under accession number CLIB-724. The strain differs from the wild type *Y. lipolytica* by auxotrophy towards Leu and Ura and by an ability to grow on sucrose. Y. lipolytica basic strain was maintained on solid media of the following composition (g/l): yeast extract – 2.5; bactopeptone – 5.0; glycerol – 15.0; malt-extract– 3.0; agar - 20.0; pH 4.0-4.2 or 8.9-9.0. Liquid nutrient broths were prepared as follows (g/l): - $MgSO_4 \times 7H_2O$ - 0.5; NaCl - 0.1; $CaCl_2$ - 0.05; KH_2PO_4 - 2; $K_2HPO_4 \times 3H_2O$ – 0,5; $(NH_4)_2SO_4$ - 0.3; Ca pantotenate - 0.4; inositol - 2.0; nicotinic acid - 0.4; n-amino benzoic - 0.2; pyridoxine -0.4; riboflavin - 0.2; thiamine - 0.1; biotin - 0.002; folic acid - 0.002; H_3BO_4 -0.5; $CuSO_4 \times 5H_2O$ -0.04; KI - 0.1; $FeCl_3 \times 6H_2O$ - 0.2; $MnSO_4 \times H_2O$ - 0.4; $NaMoO_4 \times 2H_2O$ - 0.2; $ZnSO_4$ - 0.4; pH – 4.0-4.2 or 8.9-9.0, yeast extract "Difco" - 2.0. 1% glycerol was used as a principal carbon and energy supply. pH was controlled permanently during cultivation.

3.2 Cell extract preparation

Cell cultures (24 h) were used for proteomic studies (average A_{590} =7.5-8.0). The biomass was harvested by centrifugation at 4000g for 10 min. The cells were washed twice with ice-cold deionized water and eventually pelleted.

To prepare protein extracts, 100 mg of the cell pellet was transferred to a vial containing 2ml lysis buffer (9M urea, 5% β-mercaptoethanol, 2% Triton X-100, and 2% ampholytes, pH 3.5-10 (Sigma, USA)) and thoroughly suspended. The sample was either immediately heated in a boiling bath for 3-5 min or placed on ice and sonicated in an ultrasonic desintegrator (MSE-Pharmacia) for 2 min (4 cycles 30 sec each). In both cases the homogenate was clarified by centrifugation in a microfuge for 20 min at maximum speed. The pellet was discarded and 100 μl of the clear supernatant was used for isoelectrofocusing (IEF).

3.3 Two-dimensional gel electrophoresis (2DE)

The first dimension separation employed IEF in glass tubes (2.4 × 180mm) filled with 4% polyacrylamide gel prepared with 9M urea, 2% Triton X-100 and 2% ampholyte mixture. Ampholytes of 5-7 and 3.5-10 pH ranges mixed at 4:1 ratio were used in all experiments. The protein extracts (100μl) were applied at the acidic end of the gel, and IEF was carried out using a Model 175 electrophoretic cell (Bio-Rad, USA) until 2400 V/h was achieved. The polyacrylamide gel columns with protein samples separated by IEF were applied as a starting point for separation in the second dimension, for which slab electrophoresis in polyacrylamide gel (200 × 200 × 1 mm) was used with a linear 7.5-20% gradient of acrylamide in the presence of 0.1% SDS using a vertical electrophoretic cell (Helicon Company, Russia). A well was created for protein marker application at the edge of each gel slab. Further details of the modified 2DE approach are described earlier (Kovalyova et al, 1994; Laptev et al, 1995; Kovalyov et al, 1995).

For protein visualization, the polyacrylamide gel slabs were stained with Coomassie Blue R-250 and then with silver nitrate according to the well-described methods (Blum et al, 1987) and modified by the addition of 0.8% acetic acid to sodium thiosulfate. The stained gels were documented by scanning on an Epson Expression 1680 scanner, and densitometry was carried out using the Melanie software (GeneBio, Switzerland) according to the manufacturer's protocol.

Molecular masses (*M*) of the fractionated proteins were determined by their electrophoretic mobility in the second dimension as compared to protein markers from standard heart muscle lysates (Kovalyova et al, 1994). The results of the mass determinations were verified by a calibration curve plotted using a marker kit (MBI Fermentas, Lithuania) with *M* ranging 10-200kDa. Isoelectric points (p*I*) of fractionated proteins were determined from their electrophoretic location in the first dimension, as described earlier (Kovalyova et al, 1994; Laptev et al, 1995), taking into account the known localization of identified reference proteins. Theoretical values of *M* were also taken from the Swiss-Prot database taking into account evidence for posttranslational processing of signal sequences (when available).

3.4 Protein identification by mass spectrometry

Isolation of protein fractions from polyacrylamide gel slabs, hydrolysis with trypsin, and peptide extraction for protein identification by matrix assisted laser desorption/ionization time of flight mass-spectrometry (MALDI-TOF MS) were carried out according to published protocols (Shevchenko et al, 1996) with some modifications (Govorun et al, 2003). A sample (0.5 μl) was mixed on the target with equal volume of 20% acetonitrile containing 0.1% trifluoroacetic acid and 20 mg/ml of 2,5-dihydroxybenzoic acid (Sigma-Aldrich, USA) and air dried. Mass spectra were recorded on a Reflex III MALDI-TOF mass spectrometer (Bruker Daltonics, USA) equipped with a UV-laser (336nm) in the positive mode with masses ranging

from 500-8000Da. The mass spectra were internally calibrated using trypsin autolysis products. The proteins were identified with Mascot software (Matrix Science, USA) using databases of the US National Center of Biotechnological Information (ncbi.nhm.nih.gov).
The NCBI database was searched within a mass tolerance of ±70 ppm for the appropriate species proteins; with one missed cleavage allowed. Protein score > 84 are significant ($p<0.05$). Carbamidomethylation ion of a cysteine residue and the oxidation of methionine are considered modification. Proteins were evaluated by considering the number of matched tryptic peptides, the percentage coverage of the entire protein sequence, the apparent MW, and the pI of the protein.

4. Results

4.1 Equalizing culture growth conditions

Previously we reported data about pH adaptation of *Y. lipolytica* carried out in minimal synthetic medium with succinate as the single source of carbon and energy (Guseva et al, 2010). However, elucidation of principles enabling *Y. lipolytica* to survive under strong alkaline conditions requires discrimination of partial physiological reactions of certain media components. This is possible only if several media pairs (each with acidic and alkaline pH) are compared. Therefore, we aimed to reproduce the experiments in a complete liquid medium containing 2% yeast extract and 1% glycerol. It was prepared in three versions with pH 4.0, 5.5 and 9.0. Growth curves were plotted using A_{600} as a criterion (Fig. 1). The inoculums for each culture were produced on a solid medium using the same pH as the main experiment. Inoculation dosage was ≈10^4 cells per ml.
Surprisingly, retardation of *Y. lipolytica* growth at pH 4.0 and 5.5 versus pH 9.0 was found during periods of 1-20 h after inoculation. During periods of 20-24 h A_{600} as well as cfu contents, determined by microbiological method, were the same in all three cases. Consequently, only 24 h old cultures were subjected to further proteomic studies.

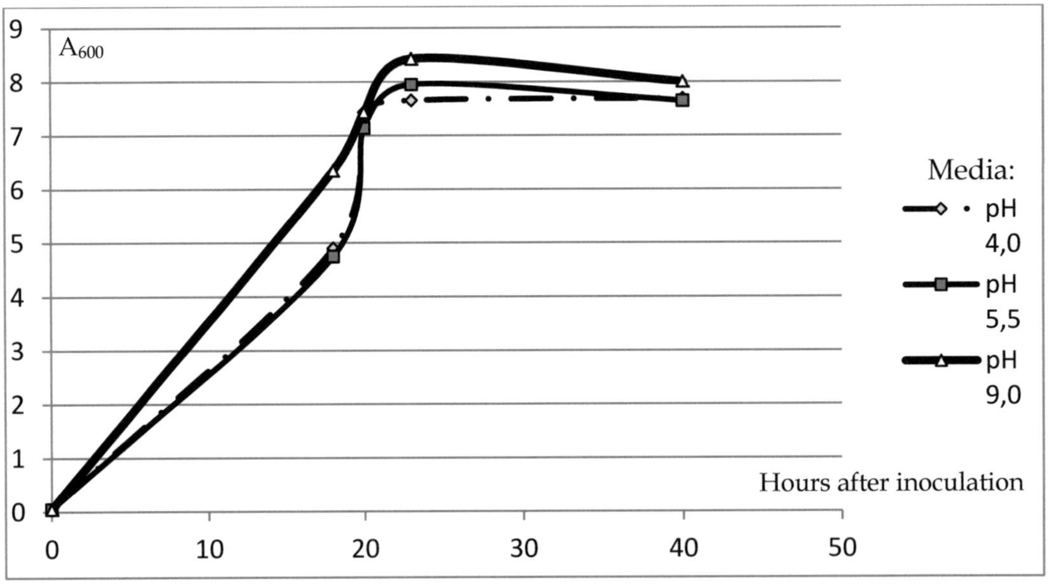

Fig. 1. Growth curves of *Y. lipolytica* at rich media with different pH's.

4.2 Analysing morphological differences of *Y. lipolytica* culture by microscopy

Measuring A_{600} of the culture is a precise and simple qualitative technique. However, it does not allow the visualization of putative morphological cell changes under different pH conditions. These changes may compromise the accuracy of A_{600} data conversion to cell number.

In order to track morphological changes in *Y. lipolytica* cells in liquid media at pH 4.0 and 9.0 cultures were subjected to visual phase-contrast microscopy ($100^{\times}$ magnification) with no fixation. The data (Fig. 2) demonstrate that average cell volume was 2-4 times larger in the culture at pH 4.0 when compared to pH 9.0. The cells grown in alkaline media contained massive vacuoles occupying most cell volume.

Taken together, these observations lead to conclusion that the volume of the cytoplasm relative to the total volume of the cells is much reduced when growing under alkaline conditions. One could also presume that the ratio between proteins in the cytoplasm and intracellular membrane compartments (vacuoles, mitochondria, Golgi apparatus) may also be altered (Brett & Merz, 2008).

4.3 Preparing *Y. lipolytica* protein extracts

Accurate pair-wise comparison of proteomes requires thorough equalizing and normalizing of source biological material. Massive and tightly cross-linked polysaccharide cell walls are a specific attribute of all yeast species including *Y. lipolytica*. It protects the cells from rapid changes in environmental conditions but also substantially hinders experimental processing of yeast samples (Dagley et al, 2011). This problem is commonly addressed in transcriptomic studies, but proteomic research also requires optimal extraction procedures. Fortunately, even mechanically durable cell walls are susceptible to mechanical crushing (ultrasonic treatment, French-press, glass beads) but such procedures take time. In the course of mechanical homogenization, intracellular lysosomes are broken, and thus incapsulated cathepsins come in contact with cytoplasmic proteins. Taken together these issues may result in the degradation of proteins that intend to be subjected to further analysis. On the other hand, many membrane and cell-wall associated proteins are poorly extracted by water or buffers. Moreover, detergent treatment does not always provide an exhaustive extraction technique. Heavily glycosylated proteins located in ER, Golgi apparatus and in the cell wall are often excluded by such processes (Morelle et al, 2009; Pascal et al, 2006).

These two problems substantially preclude complete characterization of the yeast proteome and may compromise validity of the obtained data. Thus far, only a single report has undertaken a proteomic study of *Y. lipolytica* (Morin et al, 2007). These authors analyzed proteins from water soluble cell fractions produced by mechanical disintegration and the subsequent removal of the insoluble fraction by centrifugation. Hence, the membrane, cell-wall and cytoskeleton associated proteins were excluded from consideration. Taking into account presumed contribution of membrane transport machinery to pH adaptation in *Y. lipolytica* (Zvyagilskaya et al, 2000) a complete proteome assay seemed to be more relevant to our research objectives.

To address this problem, we proposed two modifications of a chemical lysis method adapted from the preparation of human muscle tissue (Kovalyova et al, 2009). The first modification (Fig. 3, 4 and 5) included the instant resuspension of the yeast cells in a hot lysis buffer containing urea, reducing agent, Triton X-100 and ampholytes. The second included a preliminary ultrasonic treatment of the cells suspended in the same buffer on ice

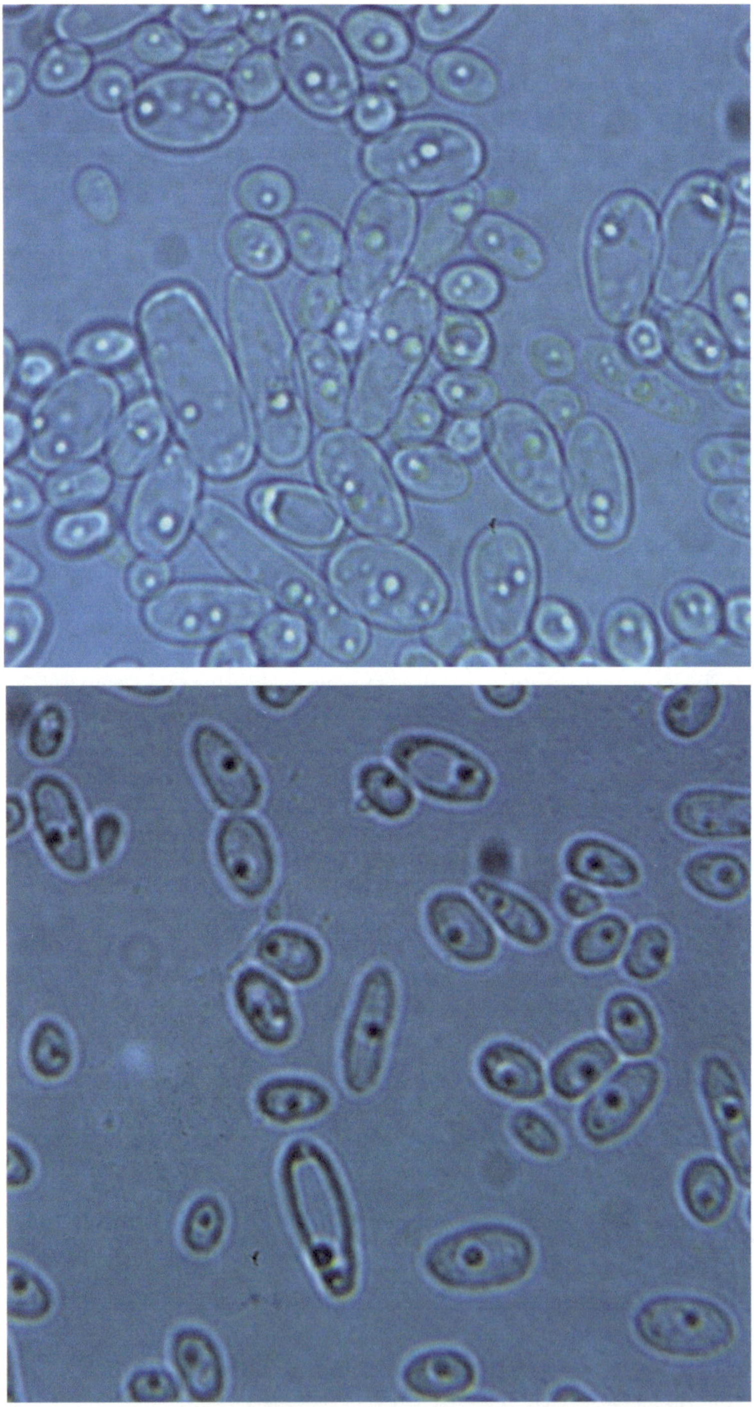

A

B

Fig. 2. Y. lipolytica cells cultured in growth media under acidic (A; pH 4.0) and alkaline (B; pH 9.0) conditions (growth time 24 h). Images from an optical microscope with 100x magnification.

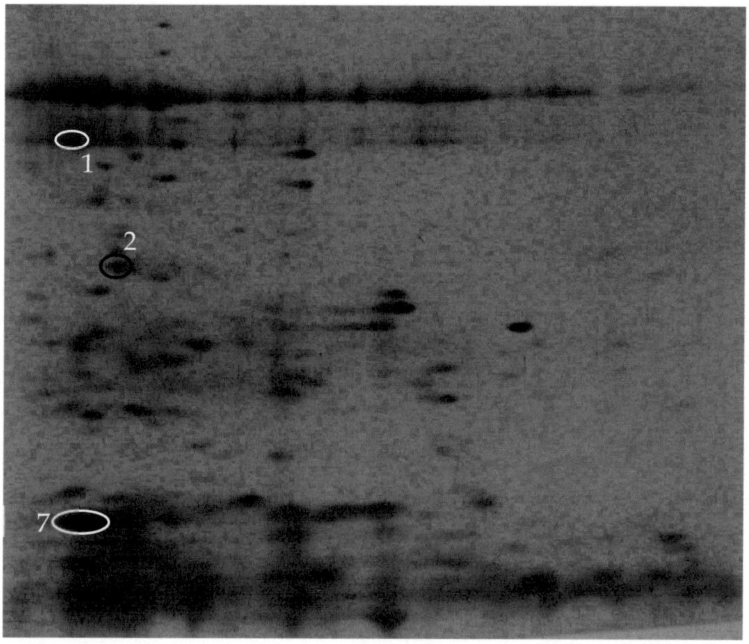

Fig. 3. 2D electophoregarm of *Y. lipolytica* proteome cultured on pH 4.0 medium (double silver/Coomassie R-250 staining). The cells were lysed in the denaturing buffer without mechanical disintegration. MALDI-TOF MS analysis of the spots specific for this specimen (not found in Fig. 4 or 5).

Fig. 4. 2D electophoregarm of *Y. lipolytica* proteome cultured on pH 5.5 medium (double silver/Coomassie R-250 staining). The cells were lysed in the denaturing buffer without mechanical disintegration. MALDI-TOF MS analysis of the spots specific for this specimen (not found on Fig. 3 and 5).

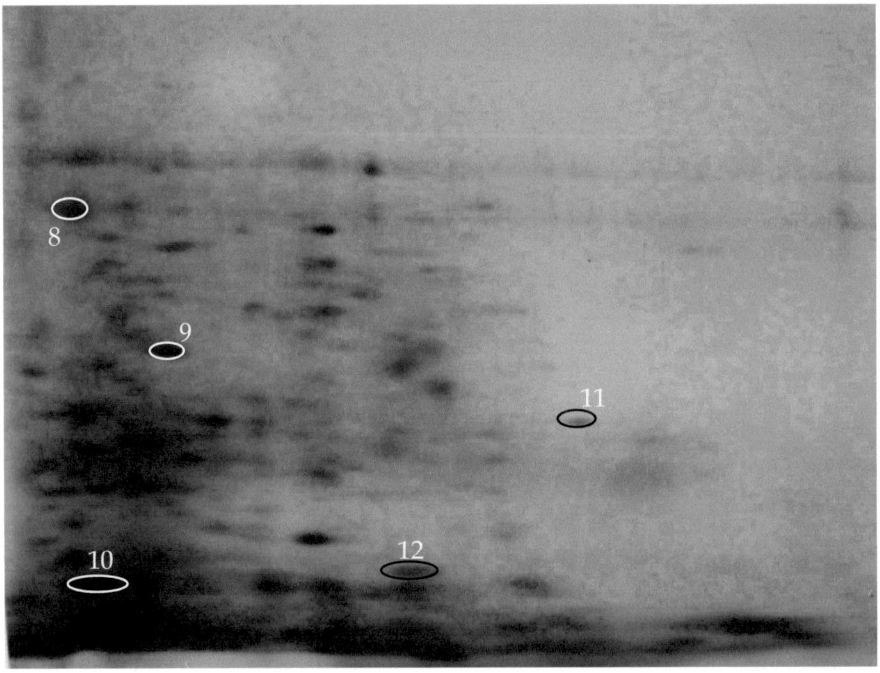

Fig. 5. 2D electophoregarm of *Y. lipolytica* proteome cultured on pH 9.0 medium (double silver/Coomassie R-250 staining). The cells were lysed in the denaturing buffer without mechanical disintegration. MALDI-TOF MS analysis of the spots specific for this specimen (not found on Fig. 3 and 4).

(Fig. 6 and 7). The volume ratio between cell pellet and the lysis buffer must be about 1:20. The cells must be placed into a vial containing the buffer to provide instant resuspension of the sample. After homogenization, the non-soluble pellet containing polysaccharides must be discarded by an intensive centrifugation step to avoid clogging of IEF tubes.

Both methods resulted in gels that produced ≈1000 individual spots, compared to other tested methods which rendered <100 spots (data not shown). However, the overall spot pattern obtained by two methods from the same biological material was significantly different (compare Fig. 3 to Fig. 6 and Fig. 5 to Fig. 7). Moreover, the quality of the protein extract produced under alkaline conditions was always less than in samples produced under acidic conditions. However, the results were highly reproducible for the same method even when applied to independently cultured material.

4.4 Studies of *Y. lipolytica* protein extracts by 2DE and MALDI-TOF MS

A total of 5 types of extracts were analyzed. Three samples were produced using hot buffer extraction from whole cells (the cultures were produced in media at pH 4.0, 5.5 and 9.0). Two samples were obtained from the cells subjected to ultrasonic disintegration directly in the ice-cold lysis buffer (the cultures were produced in media at pH 4.0 and 9.0). The unique spots specific for each sample were identified by comparison with the samples obtained by the same technique. Only intense spots corresponding to abundant cell proteins were analyzed by MALDI-TOF MS. Although cultures produced at pH 4.0 and 5.5 were analyzed separately, we suggest that differences between them must be considered as the "base-line

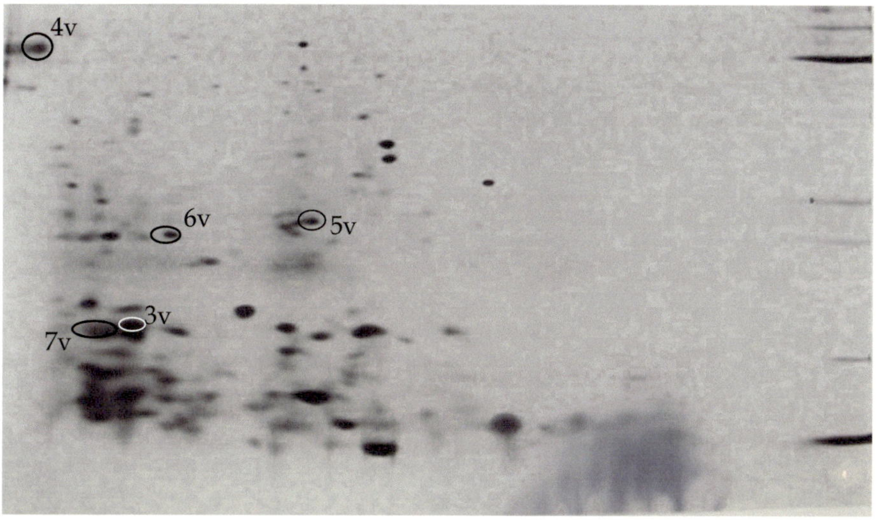

Fig. 6. 2D electophoregarm of *Y. lipolytica* proteome cultured on pH 4.0 medium (double silver/Coomassie R-250 staining). The cells were homogenized by ultrasonic treatment with subsequent denaturing buffer without mechanical disintegration. MALDI-TOF MS analysis of the spots specific for this specimen (not found on Fig. 7).

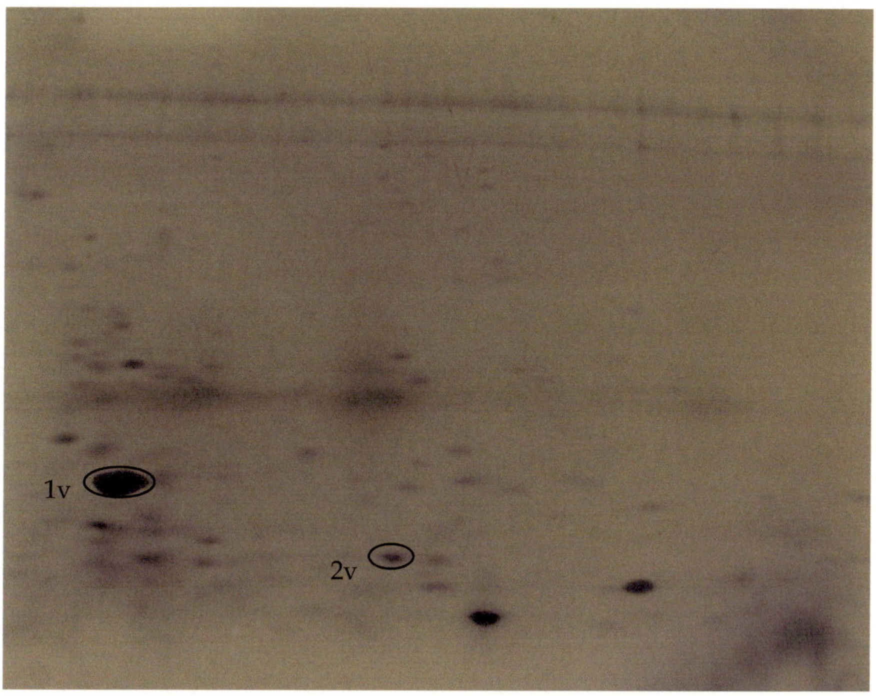

Fig. 7. 2D electophoregarm of *Y. lipolytica* proteome cultured on pH 9.0 medium (double silver/Coomassie R-250 staining). The cells were homogenized by ultrasonic treatment with subsequent denaturing buffer without mechanical disintegration. MALDI-TOF MS analysis of the spots specific for this specimen (not found on Fig. 6).

fluctuation" since both pH ranges are considerably below the pH of the cytoplasm. Comparison within this pair may allow an estimation of the reproducibility of the employed techniques e.g. as described by (Huang et al, 2011).
The data shown demonstrates that many selected spots from the 2D electophoregrams were not able to be identified by MALDI-TOF MS analysis (Table 1). Consequently, only two

Code	Exp. Mr kDa	YL protein identified by Mascot	Mascot Score	Calc. Mr Da	Homologue with known function
1	72	Invalid data			
2	48	Invalid data			
7	12	Invalid data			
3	39	YALI0B03564p	106	34031	P43070 C. albicans Glucan 1,3- □-glucosidase precursor (EC 3.2.1.58) (Exo-1-3-β- glucanase)
4	23	YALI0B15125p	247	21311	P34760 S. cerevisiae YML028w TSA1 thiol-specific antioxidant
5	25	Invalid data			
6	13	YALI0F09229p	99	17031	P36010 S. cerevisiae YKL067w YNK1 nucleoside diphosphate kinase
8	75	Invalid data			
9	38	Invalid data			
10	11	Invalid data			
11	29	YALI0F17314p	163	29514	P04840 S. cerevisiae YNL055c POR1 mitochondrial outer membrane porin
3v	13	YALI0E19723p	95	17290	P04037 S. cerevisiae YGL187c COX4 cytochrome-c oxidase chain IV
5v	24	YALI0F05214p	151	26679	P00942 S. cerevisiae YDR050c TPI1 triose-phosphate isomerase singleton
6v	21	YALI0B03366p	97	20957	P14306 S. cerevisiae YLR178C carboxypeptidase Y inhibitor (CPY inhibitor) (Ic)(DKA1/NSP1/TFS1)
7v	12	Invalid data			
1v	72	Invalid data			
2v	12	YALI0D20526p	106	13681	P22943 S. cerevisiae YFL014W 12 kDa heat shock protein (Glucose and lipid-regulated protein)

Table 1. 2DE protein spots subjected to identification by MALDI-TOF MS

clearly alkaline-inducible proteins were identified. The most prominent candidate proteins exhibiting great pH-inducibility and high overall expression levels (e.g. 1v, 8, 9 and 10) could not be identified. A higher proportion of spots were successfully identified from the samples originating from pH 5.5 medium compared to the samples from pH 4.0 medium. Furthermore, gel resolution and total number of resolved spots also increased under pH 5.5 conditions. This could be explained by the observation that the share of cytoplasm proteins in the total cell volume is proportionally higher under optimal conditions (pH 5.5) and decreases under acidic or alkaline stress in favor of the membrane compartments (vacuoles, mitochondria, ER, Golgi apparatus) (see Fig. 2). This idea is supported by observation that 6 out of 8 proteins represented in Table 1 are "pH-reactive" and are allocated to non-cytoplasm compartments. It is also in a good agreement with numerous communications about involvement of ER and mitochondria to anti-stress adaptation of organisms from all kingdoms (Hoepflinger et al 2011; Rodriguez-Colman et al, 2010). Reactive oxygen species (ROS) formation accompanies all responses to stresses and cross-talk between ER and mitochondria contributes to abatement of damage caused by uncontrolled oxidation (Bravo et al, 2011; Tikunov et al, 2010).

4.5 Functions and genomic organisation of the genes encoding potential "pH-reactive proteins" in *Y. lipolytica*

In order to systematically assess properties of the up-and down-regulated alkaline-sensitive proteins, we arranged the available functional data from Swiss-Prot records for each identified protein (Table 2).

Genomic localization of the pH-regulated proteins is not uniform. However, one can make an observation that no pH-reactive genes were found on chromosomes A or C.

The data demonstrate an important role of non-cytoplasmic cell compartments in the pH adaptation of *Y. lipolytica*. Only two proteins (4 and 5v) from the eight identified have annotated subcellular locations corresponding to the cytoplasm. While it is possible that adaptation to the acidic and alkaline pH depends on these polypeptide structures, one must take into account that many potentially important pH-reactive proteins failed to be identified. Therefore, we cannot conclude that all major pH-reactive proteins were found. It is worth noting that this and other studies (Guseva et al, 2010) have failed to identify plasma membrane components (ATPase subunits and pumps) responsible for direct ion exchange between the cytoplasm and the environment.

A comparison of this study with pH-reactive proteins identified previously (Guseva et al, 2010) in *Y. lipolytica* cultivated on a minimal medium with succinate was undertaken. Two proteins YALI0F17314p and YALI0B03366p were found in both cases. YALI0F17314p (outer membrane mitochondrial porin, VDAC) was the only alkaline-inducible protein found in both cases. In contrast, YALI0B03366p (carboxypeptidase Y inhibitor, a lysosomal component) was found to be an alkaline-inducible on minimal medium with succinate and alkaline-repressible in complete medium with glycerol (present study). This comparison leads to the conclusion that the outer membrane mitochondrial porin is possibly an essential part of *Y. lipolytica* pH-adaptation machinery, independent of the utilized nutrient source.

Another identified alkaline-inducible component of *Y. lipolytica*, Hsp12 is an intrinsically unstructured stress protein that folds upon membrane association and modulates membrane function (Welker et al, 2010). Hsp12 of *S. cerevisiae* is upregulated several 100-fold in response to stress. Our phenotypic analysis showed that this protein is important for survival under a variety of stress conditions, including high temperature. In the absence of

a. Alkaline-inducible proteins

Code	YL Swiss -Prot acc. #	Function	Protein cell localization	Gene (Gene Bank acc. Number)	Chromosomal localization
11	YALI 0F173 14p	POR1 mitochondrial outer membrane porin	Outer mebrane of mitochondria	gi|50556244	F (2311796-2313207)
2v	YALI 0D20 526p	12 kDa heat shock protein	Cytoplasm/internal membrans	gi|50551205	D (2604298-2604907)

b. Alkaline-repressible proteins

Code	YL Swiss -Prot acc. #	Function	Protein cell localization	Gene (Gene Bank acc. Number)	Chromosomal localization
3	YALI 0B035 64p	Glucan 1,3- beta - glucosidase precursor	Golgi appartus	gi|50545854	B (498811-499752)
4	YALI 0B151 25p	Peroxiredoxin (PRX) family, Typical 2-Cys PRX subfamily	Cytoplasm	gi|50546891	B (2015063-2015653)
6	YALI 0F092 29p	nucleoside diphosphate kinase	Mitochondria matrix	gi|50555578	F (1287080-1287730)
3v	YALI 0E197 23p	COX4 cytochrome-c oxidase chain IV	Inner membrane of mitochondria	gi|50553496	E (2354436-2354924)
5v	YALI 0F052 14p	TPI1 triose-phosphate isomerase singleton (glycolysis)	Cytoplasm	gi|50555229	F (783915-784734)
6v	YALI 0B033 66p	carboxypeptidase Y inhibitor	Vacuole	gi|50545840	B (481138-482256)

Table 2. Functions and genomic organisation of the genes encoding potential "pH-reactive proteins" in *Y. lipolytica*

Hsp12, we observed changes in cell morphology under stress conditions. Surprisingly, in the cell, Hsp12 exists both as a soluble cytosolic protein and associated with the plasma membrane. The *in vitro* analysis revealed that Hsp12, unlike all other Hsps studied so far, is completely unfolded; however, in the presence of certain lipids, it adopts a helical structure.

The presence of Hsp12 does not alter the overall lipid composition of the plasma membrane but increases membrane stability (Welker et al, 2010). This information allows us to hypothesize that the biological function of Hsp12 is in rearranging and repairing membrane compartments under the stress conditions. This point of view is in perfect agreement with observations about the key role of the inner membrane compartments in the alkaline adaptation in *Y. lipolytica*. Unfortunately the involvement of this protein in many types of stress responses may result in data concerning its expression pattern poorly reproducible.

5. Conclusion

A new yeast cell extraction procedure enabled the resolution of more than 1000 individual protein spots of *Y. lipolytica* samples for each gel. This is ~2-fold more than in outlined by previous studies (Morin et al, 2007) where water soluble cell fractions were analyzed. In total, two proteins were up-regulated at pH 9.0, the mitochondrial outer membrane porin (VDAC) and 12 kDa heat shock protein.

These data complement the conclusions by Morin et al (2007) who emphasized the occurrence of energy metabolism proteins within the proteome portion as up-regulated in *Y. lipolytica* hyphae during the dimorphic transition. Similar conclusions were reported for stress adaptation in *S. cerevisiae* (Martínez-Pastor et al, 2010; Rodriguez-Colman et al, 2010) and *Candida albicans* (Dagley et al, 2011). VDAC is not only responsible for protein import into mitochondria but essentially contributes to antioxidant resistance of mitochondria (Tikunov et al, 2010).

To the best of our knowledge, we provide the first report about the application of proteomic techniques to address the problem of *Y. lipolytica* adaptation to growth in alkaline conditions. In contrast to the previously hypothesized involvement of plasma membrane transporters and global transcription regulators (e.g. Rim101) in high pH adaptation, our study elucidated a key role for mitochondrial proteins and represents a new result for *Y. lipolytica*. On the other hand, this observation is in a good agreement with reports concerning the pivotal role of non-cytoplasmic compartments in stress adaptation in other biological systems e.g. yeast and plants (Bravo et al, 2011; Hoepflinger et al 2011; Rodriguez-Colman et al, 2010).

In our opinion, this work exemplifies a prompt and inexpensive study which could be easily undertaken for any physiological experiment with the organisms whose genome has been recently sequenced. Finally, it should be noted that a total cell proteome assay is strongly recommended for this kind of study, although some effort to determine an appropriate lysis buffer for protein extraction must be undertaken.

6. References

Blum, H.; Beir, H. & Cross, H.G. (1987) Improved silver staining of plant proteins, RNA and DNA in polyacrylamide gels. *Electrophoresis*, Vol. 8, No. 2, (February 1987), pp. 93-99, ISSN 0173-0835.

Bravo, R.; Vicencio, JM.; Parra, V.; Troncoso, R.; Munoz, JP.; Bui, M.; Quiroga, C.; Rodriguez, AE.; Verdejo, H.E.; Ferreira, J.; Iglewski, M.; Chiong, M.; Simmen, T.; Zorzano, A.; Hill, J.A.; Rothermel, B.A.; Szabadkai, G.; Lavandero, S. (2011) Increased ER-mitochondrial coupling promotes mitochondrial respiration and bioenergetics during early phases of ER stress. *Journal of Cell Science,* Vol. 124, No. 13, (July 2011), pp. 2143-2152, ISSN 2157-7013.

Brett, C.L. & Merz, A.J. (2008) Osmotic regulation of Rab-mediated organelle docking. *Current Biology,* Vol. 18, No. 14, pp. 1072-1077, (July 2008), ISSN 0960-9822.

Dagley, M.J.; Gentle, I.E.; Beilharz, T.H.; Pettolino, F.A.; Djordjevic, J.T.; Lo, T.L.; Uwamahoro, N.; Rupasinghe, T.; Tull, D.L.; McConville, M.; Beaurepaire, C.; Nantel, A.; Lithgow, T.; Mitchell, A.P. & Traven, A. (2011) Cell wall integrity is linked to mitochondria and phospholipid homeostasis in *Candida albicans* through the activity of the post-transcriptional regulator Ccr4-Pop2. *Molecular Microbiolgy,* Vol. 79, No. 4, (February 2011), pp. 968-989, ISSN 0950-382X.

Dujon, B., Sherman, D., Fischer, G., Durrens, P., Casaregola, S., Lafontaine, I., De Montigny, J., Marck, C., Neuvéglise, C., Talla, E., Goffard, N., Frangeul, L., Aigle, M., Anthouard, V., Babour, A.; Barbe, V.; Barnay, S.; Blanchin, S.; Beckerich, J.M.; Beyne, E.; Bleykasten, C.; Boisramé, A.; Boyer, J.; Cattolico, L.; Confanioleri, F.; De Daruvar, A.; Despons, L.; Fabre, E.; Fairhead, C.; Ferry-Dumazet, H.; Groppi, A.; Hantraye, F.; Hennequin, C.; Jauniaux, N.; Joyet, P.; Kachouri, R.; Kerrest, A.; Koszul, R.; Lemaire, M.; Lesur, I.; Ma, L.; Muller, H.; Nicaud, JM.; Nikolski, M.; Oztas, S.; Ozier-Kalogeropoulos, O.; Pellenz, S.; Potier, S.; Richard, G.F.; Straub, M.L.; Suleau, A.; Swennen, D.; Tekaia, F.; Wésolowski-Louvel, M.; Westhof, E.; Wirth, B.; Zeniou-Meyer, M.; Zivanovic, I.; Bolotin-Fukuhara, M.; Thierry, A.; Bouchier, C.; Caudron, B.; Scarpelli, C.; Gaillardin, C.; Weissenbach, J.; Wincker, P. & Souciet, J.L. (2004) Genome evolution in yeasts. *Nature,* Vol. 430, No. 6995, (July 2004) pp. 35-44, ISSN 0028-0836.

Govorun, V.M.; Moshkovskii, S.A.; Tikhonova, O.V.; Goufman, E.I.; Serebryakova, M.V.; Momynaliev, K.T.; Lokhov, P.G.; Khryapova, E.V.; Kudryavtseva, L.V.; Smirnova, O.V.; Toropygin, I.Yu.; Maksimov, B.I. & Archakov, A.I. (2003) Comparative analysis of proteome maps of Helicobacter pylori clinical isolates. *Biochemistry (Moscow),* Vol. 68, No. 1, (January 2003), pp. 42-49, ISSN 0006-2979.

Guseva, M.A.; Epova, E.Iu.; Kovalev, L.I. & Shevelev, A.B. (2010). The study of adaptation mechanisms of *Yarrowia lipolytica* yeast to alkaline conditions by means of proteomics. *Prikladnaia Biokhimiia i Mikrobiologiia (Moscow),* Vol. 46, No. 3, (May-June 2010), pp. 336-341, ISSN 1521-6543.

Hoepflinger, M.C.; Pieslinger, A.M. & Tenhaken, R. (2011) Investigations on N-rich protein (NRP) of Arabidopsis thaliana under different stress conditions. Plant Physiology and Biochemistry (March 2011) Vol. 49, No. 3, pp. 293-302, ISSN 0981-9428

Huang C.J.; Damasceno L.M.; Anderson K.A.; Zhang S.; Old L.J. & Batt C.A. (2011) A proteomic analysis of the *Pichia pastoris* secretome in methanol-induced cultures. *Appl Microbiol Biotechnol,* Vol. 90, No. 1, (April 2011), pp. 235-247, ISSN 0175-7598.

Kovalyov, L.I.; Shishkin, S.S.; Efimochkin, A.S.; Kovalyova, M.A.; Ershova, E.S.; Egorov, T.A., & Musalyamov, A.K. (1995) The major protein expression profile and two-dimensional protein database of human heart. *Electrophoresis,* Vol. 16, No. 7, (July 1995), pp. 1160-1169, ISSN 0173-0835.

Kovalyova, M.A.; Kovalyov, L.I.; Khudaidatov, A.I.; Efimochkin, A.S. & Shishkin, S.S. (1994) Comparative analysis of protein composition of human skeleton and cardiac muscle by 2D electrophoresis. *Biochemistry (Moscow),* Vol. 59, No. 5, (May 1994), pp. 493-498, ISSN 0006-2979.

Kovalyova M.A.; Kovalyov L.I.; Toropygin I.Y.; Shigeev S.V.; Ivanov A.V. & Shishkin S.S. (2009) Proteomic analysis of human skeletal muscle (m. vastus lateralis) proteins: identification of 89 gene expression products. *Biochemistry (Moscow)*, Vol. 74, No. 11, (November 2009), pp. 1239-1252, ISSN 0006-2979.

Lambert, M.; Blanchin-Roland, S.; Le Louedec F.; Lepingle, A. & Gaillardin, C. (1997) Genetic analysis of regulatory mutants affecting synthesis of extracellular proteinases in the yeast *Yarrowia lipolytica*: identification of a RIM101/pacC homolog. *Molecular and Cellular Biology*, Vol. 17, No. 7, (July 1997), pp. 3966-3976, ISSN 0270-7306.

Laptev, A.V.; Shishkin, S.S.; Egorov, Ts.A.; Kovalyov, L.I.; Tsvetkova, M.N.; Galyuk, M.A.; Musalyamov, A.Kh. & Efimochkin, A.S. (1994) Searching new gene products in human cardiac muscle. Microsequencing proteins after 2D-electrophoresis. *Molecular Biology (Moscow)*, Vol. 28, (Jan-Feb 1994), pp. 52-58, ISSN□0026-8933.

Martínez-Pastor, M.; Proft, M. & Pascual-Ahuir, A. (2010) Adaptive changes of the yeast mitochondrial proteome in response to salt stress. *OMICS*, Vol. 14, No. 5, (October 2010), pp. 541-552, ISSN 1557-8100.

Morelle, W.; Faid, V.; Chirat, F. & Michalski, J.C. (2009) Analysis of N- and O-linked glycans from glycoproteins using MALDI-TOF mass spectrometry. *Methods in Molecular Biology*, Vol. 5, No. 34, (Mach 2009), pp. 5-21, ISSN:1064-3745

Morín, M.; Monteoliva, L.; Insenser, M.; Gil, C. & Domínguez, A. (2007) Proteomic analysis reveals metabolic changes during yeast to hypha transition in *Yarrowia lipolytica*. *Journal of Mass Spectrometry*, Vol. 42, No. 11, (November 2007), pp. 1453-1462, ISSN 1076-5174.

Pascal, C.; Bigey, F.; Ratomahenina, R.; Boze, H.; Moulin, G. & Sarni-Manchado P. (2006) Overexpression and characterization of two human salivary proline rich proteins. *Protein expression and purification*, Vol. 47, No. 2, (June 2006), pp. 524-532, ISSN 1046-5928.

Rodriguez-Colman M.J.; Reverter-Branchat, G.; Sorolla, M.A.; Tamarit, J.; Ros, J. & Cabiscol, E. (2010) The forkhead transcription factor Hcm1 promotes mitochondrial biogenesis and stress resistance in yeast. *Journal of Biological Chemistry*, Vol. 285, No 47, (November 2010), pp. 37092-37101, ISSN 0021-9258.

Shevchenko, A.; Wilm, M.; Vorm, O. & Mann, M. (1996) Mass spectrometric sequencing of proteins from silver-stained polyacrylamide gels. *Analytical Chemistry*, Vol. 68, No. 5, (March 1996), pp. 850-858, ISSN 0003-2700.

Tikunov, A.; Johnson, C.B.; Pediaditakis, P.; Markevich, N.; Macdonald, J.M.; Lemasters, J.J. & Holmuhamedov E. (2010) Closure of VDAC causes oxidative stress and accelerates the Ca(2+)-induced mitochondrial permeability transition in rat liver mitochondria. *Archives of Biochemistry and Biophysics*, Vol. 495, No. 2, (March 2010), pp. 174-181, ISSN 0003-9861

Welker S.; Rudolph B.; Frenzel E.; Hagn F.; Liebisch G.; Schmitz G.; Scheuring J.; Kerth A.; Blume A.; Weinkauf S.; Haslbeck M.; Kessler H. & Buchner J. (2010) Hsp12 is an intrinsically unstructured stress protein that folds upon membrane association and modulates membrane function. *Molecular Cell*, Vol. 39, No. 4, (August 2010), pp. 507-520, ISSN 1097-2765.

Yuzbashev, T.V.; Yuzbasheva, E.Y.; Sobolevskaya, T.I.; Laptev, I.A.; Vybornaya, T.V.; Larina, A.S.; Matsui, K.; Fukui, K. & Sineoky, S.P. (2010). Production of succinic acid at low

pH by a recombinant strain of the aerobic yeast *Yarrowia lipolytica*. *Biotechnology and Bioengineering*, Vol. 107, No. 4, (November 2010), pp. 673-682, ISSN 1097-0290.
Zvyagilskaya, R.; Parchomenko, O. & Persson, B.L. (2000) Phosphate-uptake systems in *Yarrowia lipolytica* cells grown under alkaline conditions. *IUBMB Life*, Vol. 50, No. 2, (August 2000), pp. 151-155, ISSN 1521-6543.

11

The Role of Conventional Two-Dimensional Electrophoresis (2DE) and Its Newer Applications in the Study of Snake Venoms

Jaya Vejayan[1]*, Mei San Tang[1] and Ibrahim Halijah[2]
[1]*School of Medicine and Health Sciences, Monash University Sunway Campus, Jalan Lagoon Selatan, Selangor Darul Ehsan*
[2]*Institute of Biological Sciences, University of Malaya, Kuala Lumpur Malaysia*

1. Introduction

The objective of this chapter is to provide an overview of the different approaches that have been undertaken in our laboratory and by other researchers to investigate the different aspects of snake venoms using two-dimensional electrophoresis (2DE). It will also highlight the few novel modifications that we have employed to improve the protocol of 2DE, in order to further increase its versatility as a research tool in the study of snake venoms.

2. The utilization of proteomics to characterize snake venoms

The biological and pathological activities of snake venoms are associated with proteins and peptides in the venoms. These venom constituents are often conveniently classified as either neurotoxic or hemotoxic (Calvete *et al.*, 2009). The venoms of the Elapidae and Viperidae families are among the most thoroughly investigated. The main constituents of the Elapidae venoms are the neurotoxic proteins with lower molecular weights. On the other hand, the main constituents of the Viperidae venoms are the hemotoxic proteins with higher molecular weights. Nevertheless, this classification is not mutually exclusive, since in certain venoms, such as the Elapidae *Ophiophagus hannah*, the main constituents are the higher molecular weight enzymes, which are typically more characteristic of Viperidae venoms (Tan & Saifuddin, 1989). Apart from this widely accepted classification of neurotoxins and hemotoxins, the other aspect in the diversity of venom proteins includes the relative abundances of each protein family. High abundance proteins are important in generic killing and are generally the primary targets of immunotherapy while low abundance proteins are considered to be more important in evolutionary studies (Calvete *et al.*, 2009). Understanding the differences in venom proteins abundances is important as it also has an influence on the method that is required to study these proteins with different abundances in different venoms.

In the early studies of snake venoms, in order to dissect and to analyze the complexity of snake venom constituents, the typical workflow employed has been to isolate and subsequently characterize the biochemical characteristics of individual venom proteins

(Bougis *et al.*, 1986; Graham *et al.*, 2008; Ownby & Colberg, 1987; Tan & Saifuddin, 1989). For example, the crude venom of *O. hannah* was fractionated by Sephadex G75 gel filtration chromatography and DEAE-Sephacel ion-exchange chromatography and the biological properties of the individual chromatography fractions were subsequently determined by utilizing various biochemical assays (Tan & Saifuddin, 1989). The objectives of the study were to investigate the presence of toxic components in the *O. hannah* venom and to provide information for further investigations of the biochemistry and toxicology of *O. hannah* venom. Graham *et al* (2008) analyzed 30 venoms from the Elapidae and Viperidae families by G50 gel filtration chromatography and following comparison of the chromatography profiles, definitive patterns that could be used for preliminary analyses of venom components were established.

However, the comparison of elution profiles was limited by the less-than-optimum resolution of peaks, especially those containing venom components that were present at higher amount in the venom, resulting in broader peaks within the chromatography profiles, masking the presence of other components (Chippaux *et al.*, 1991). Biochemical analysis and characterization also did not allow for the differentiation and comparison of venom constituents in terms of protein structure (Chippaux *et al.*, 1991). Nevertheless, with the development and refinement of chromatographic techniques that allow for further detailed analyses of fraction components, such strategy of isolation and characterization of venom constituents remains the mainstay of toxinology (Graham *et al.*, 2008).

Notwithstanding the few limitations of 2DE, its recent revitalization and its utilization as part of the workflow to analyze venom complexity has encouraged a new direction in venom studies that uses a more global approach in visualizing venom complexity (Fox & Serrano, 2008). Separating proteins based on two independent parameters - pI value by isoelectric focusing (IEF) in the first dimension and molecular weight by SDS-PAGE in the second dimension - 2DE is able to resolve venom proteins into a few thousand individual spots, producing a specific profile for each venom analyzed via 2DE (Carrette *et al.*, 2006). The different 2DE profiles of venoms will then be used for comparison and this concept of between-gel comparison, or comparative proteomics, has largely been put into a few different practical applications of snake venom study.

2.1 Venom variation

Venom variation is one of the very important aspects in the study of snake venom. Snake venom variation is essential to both basic venom research and the management of snake envenomation (Fox & Serrano, 2008). During the selection of a snake donor for crude venom that is to be used for research purposes, it is essential that the chosen venoms are rich in the components of study interests (Chippaux *et al.*, 1991). Therapeutically, the knowledge of venom variations at all levels, including inter-species and intra-species variations, would aid in the decision of an appropriate antivenom and allow for more effective treatment of envenomation victims (Chippaux *et al.*, 1991). Subsequently, the production of antivenom is also reliant on the knowledge of venom variations.

Within our laboratory, we have attempted to develop a 2DE-based approach to investigate the variations among the venoms of eight Malaysian snakes (Vejayan *et al.*, 2010). Even though there were venom proteins distributed throughout the entire 2DE profiles, as expected with such a complex sample, a closer examination revealed that each venom profile had its own distinguishing features. For instance, each of the three Crotalinae venoms (*Trimeresurus sumatranus*, *Tropidolaemus wagleri*, *Calloselasma rhodostoma*) had profiles

with heavy spotting of proteins at the molecular weight range 15-37kDa. The Elapidae venoms, on the other hand, had profiles with protein spots at the molecular weight range of 15-20kDa (Fig.1). These results clearly elicited the differences of the patterns between the Elapidae and Viperidae venoms that have been so well documented in other literatures discussed above, thus proving the feasibility of 2DE as an ancillary taxonomic tool (Calvete *et al.*, 2009; Nawarak *et al.*, 2003). Apart from inter-family difference in venom compositions, the 2DE analysis of the eight Malaysian snake venoms also demonstrated the obvious pattern of train of spots due to post-translational modifications in venom proteins (Fig. 1).
Similarly, Guercio et al (2006) performed 2DE analysis on crude Bothrops atrox venoms obtained from three different stages of maturation - juveniles, sub-adults and adults . The 2DE profiles obtained demonstrated the alteration that occurred in the proteome composition of snake venoms following progression in developmental stages. Subsequently, the group identified new groups of ontogenetic molecular markers - for instance, P-III class metalloproteinases and serine proteinases were more abundant in juveniles while P-I class metalloproteinases were more abundant in adults.

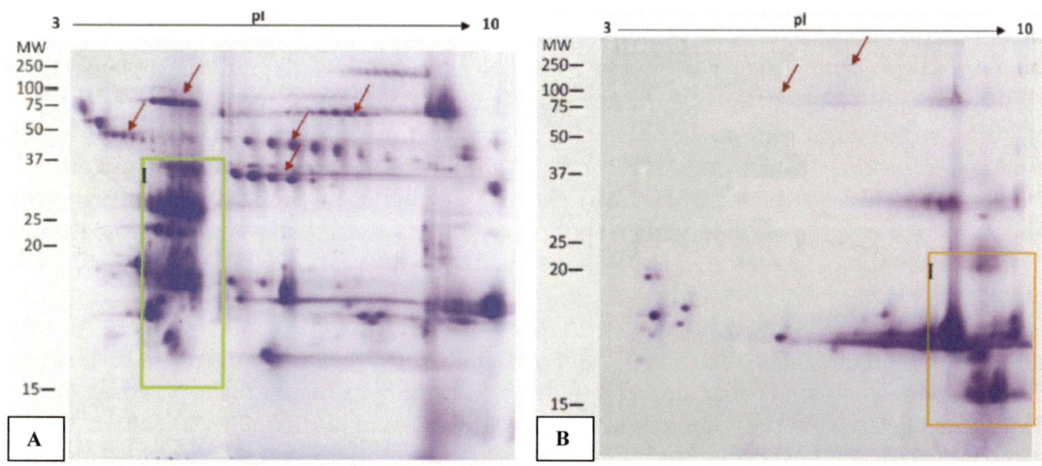

Fig. 1. 2DE images of venoms, (A) *Trimeresurus sumatranus* (Crotalinae) with scattered spots in region of 15 -37kDa while (B) *Bungarus fasciatus* (Elapidae) with spots predominantly at range of 15-20kDa. The boxed I zone show heavy spotting of acidic proteins for (A) compared to protein clustering in the lower right region, that is, the basic and lower molecular mass proteins (<20kDa) as shown for (B). Arrows indicate trains of spots due to post-translational modifications. Only two out of the eight 2DE venom profiles are shown here (Vejayan *et al.*, 2010).

2.2 Envenomation pathology

2DE also plays a role in the study of envenomation pathology and antivenom mechanisms. 2DE profiles can clearly demonstrate the venom components that are most immunogenic. In a study conducted by Correa-Netto *et al* (2010), crude *Bothrops jararacussu* venom fractionated by non-reducing 2DE was submitted to immunoblot analysis using anti-jararaca, anti-jararacussu and anti-crotalid sera. The results showed that the anti-jararaca and anti-jararacussu sera showed immune reactivity for venom proteins between 30 and

97kDa. The study also showed cross-reactivity of *B. jararacussu* venom with anti-crotalid serum. The importance of these results was that they allowed for the identification of the groups of proteins responsible for the horses' immune response in the process of antivenom production.

To investigate the effects of *Echis carinatus* envenomation on the human plasma proteome, an in vitro model utilizing 2DE as one of its core techniques has been established in two studies (Cortelazzo *et al.*, 2010; Guerranti *et al.*, 2010). The results from these two studies showed that 2DE was capable of demonstrating global proteomic changes when the human plasma was incubated with the *E. carinatus* venom. Upon comparison with the 2DE profile of the untreated control plasma, the 2DE profile of the human plasma treated with *E. carinatus* venom showed that some of the protein spots entirely disappeared or had a decreased level. Some of the protein spots in the 2DE profile of the venom-treated plasma also showed that the appearance of some new venom-dependent fragments. These proteins that were affected by the *E. carinatus* venom were identified by mass spectrometry analysis and have important functions in the blood coagulation process, which explained venom-induced thrombophilia in *E. carinatus* envenomation. The researchers, therefore, concluded that the 2DE proteomic approach was a valid method to study the molecular mechanism of envenomation on human blood proteins.

We tried to investigate the possibility of using the *Mimosa pudica* tannin (MPT) isolate as an antivenom of plant origin against the *Naja naja kaouthia* venom and have utilized 2DE as one of our core techniques in these studies. In an initial study, 2DE analysis was done on the crude *N. n. kaouthia* venom and MPT-treated *N. n. kaouthia* venom. In comparison, the 2DE results of the MPT-treated gel (Fig. 2A) showed a number of protein spots missing and within them we were able to identify 6 spots using mass spectrometry analysis to be isomers of the phospholipase A2 family of enzymes (Fig. 2) (Vejayan *et al.*, 2007). The disappearance of spots as detected via 2DE analysis indicated the binding mechanism of tannin from MPT that could potentially function as an antidote to the *N. n. kaouthia* venom. The results from this study, therefore, served as the preliminary findings before we progressed further into designing an *in vivo* study that looked into the efficacy of MPT to neutralize *N. n. kaouthia*.

In a following *in vivo* study, the crude *N. n. kaouthia* venom and *N. n. kaouthia* venom pre-incubated with MPT were injected into two different groups of mice (Ambikabothy *et al.*, 2011). There was a third group of mice in the study that served as the control group. The blood from all three groups of mice were collected with the cardiac puncture technique and centrifuged at 3000rpm for 10 min at 4°C to separate sera from cells. The sera from all three groups were then subjected to 2DE analysis in order to look into the different protein expression between the different groups of mice that had been given three different treatments respectively. Comparative analysis of the three 2DE profiles showed that a total of 5 protein spots were differentially expressed (>2 folds) (Fig. 3). It could be seen from the results that the serum of the venom-treated mice showed substantial proteomic changes that were absent from the serum of the mice treated with venom that was pre-incubated with MPT. Four of the five protein spots (Spots 2505, 4606, 3513, 3303) were identified as serine protease inhibitors, gelsolin, hemopexin and α2-macroglobulin respectively. These four proteins were found to be upregulated in the serum of the venom-treated mice and could be the most possible candidates that were causative of mortality in *N. n. kaouthia* envenomation. The fifth spot, Spot 4407, could not be identified.

Based on the few examples elaborated above, it can be seen that 2DE is a versatile research tool that can be used in various aspects of snake venom study. This is underlined by its

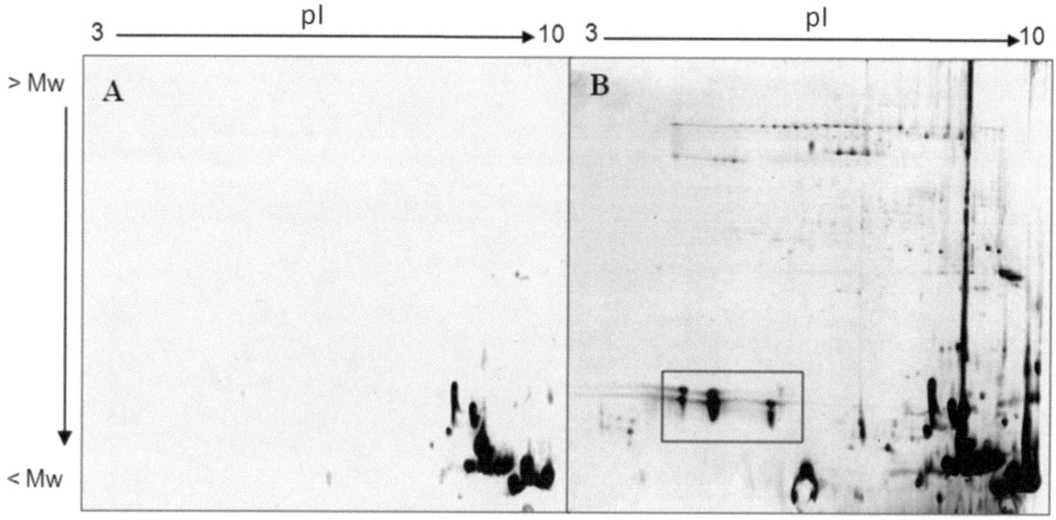

Fig. 2. The 2DE profiles of (A) MPT-treated *N. n. kaouthia* venom (B) crude *N. n. kaothia* venom alone. Both treatments, upon incubation, were centrifuged and the supernatants were collected for 2DE analysis. The rectangular box outlined the six protein spots that had disappeared after the venom was treated with MPT (Vejayan *et al.*, 2007).

powerful resolving capability of separating venom proteins into hundreds of spots to provide a realistic overview of venom complexity. Such 2DE profiles can then be further subjected to comparative proteomic analysis for venom variation investigations. 2DE technique can also be easily and efficiently interfaced with other biochemical techniques (Rabilloud *et al.*, 2010), such as immunoblot analysis, to elucidate the mechanism of antivenom and pathology of envenomation.

In the next section, we will discuss how the resolution power of 2DE enables it to be used as a useful sample fractionation tool before mass spectrometry analysis, thus establishing the concept of integrative proteomics in venom study that is especially essential in novel protein identification.

3. 2DE-MS venom proteome mapping: Its importance and challenges

While 2DE has high resolving capability, detailed analysis of venom constituents and identification of novel proteins would not have been made possible without advances in mass spectrometry techniques. Subsequent to these combined efforts of 2DE mapping and protein identification by mass spectrometry, the concept of integrated proteomics has also been established, allowing for the systematic characterization of venom proteins.

The principles of integrated proteomics can largely be described in two stages - the first stage involves optimum sample fractionation by either electrophorectic methods (1D-PAGE or 2DE) or trypsin digest followed by liquid chromatography, while the second stage involves mass spectrometry analysis and database searches for protein identification (Brewis & Brennan, 2010). The concept of integrated proteomics in snake venom study was first pioneered with the study of venom glands from the sea snake *Laticauda colubrina* and the terrestrial *Vipera russelli* (Rioux et al., 1998). The workflow employed was sample fractionation by 2DE followed by a combination of Edman sequencing and amino acid

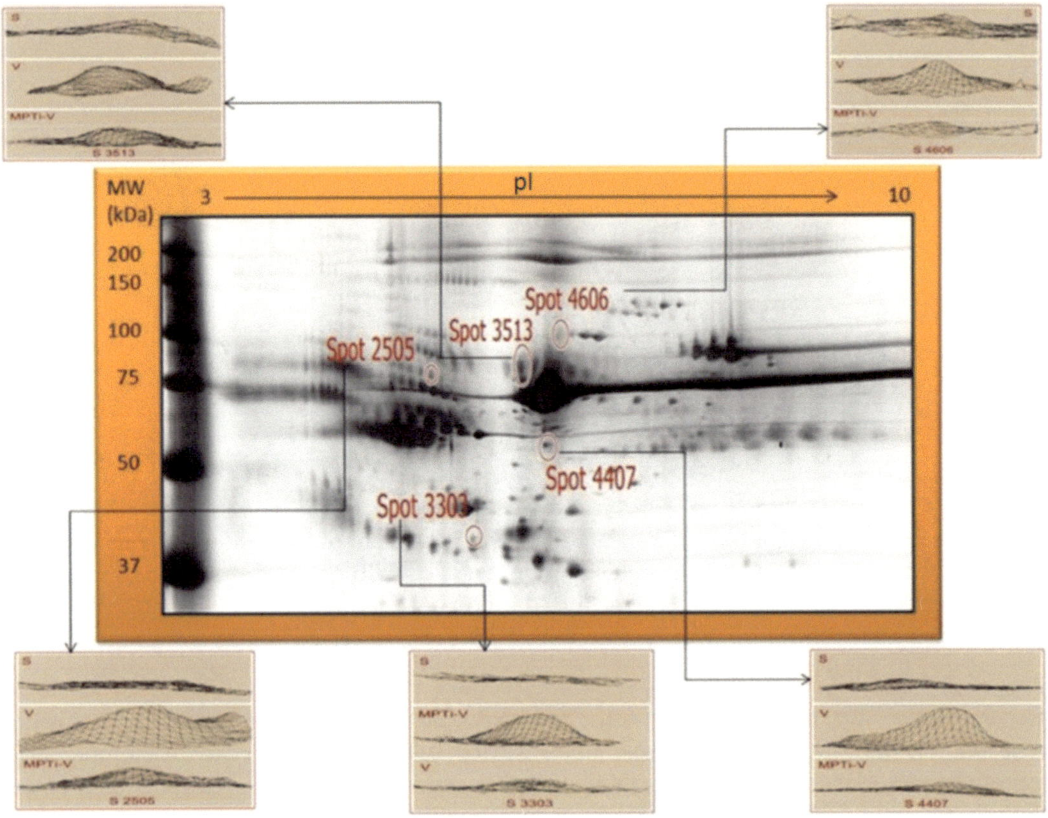

Fig. 3. Proteomic pattern of venom injected mice serum. 20µg mice serum protein after 2DE analysis at non-linear pH 3-10, stained by silver nitrate and 3D view of Spots 3513, 4606, 2505, 3303 and 4407 respectively in three different groups - control group (S), venom group (V) and MPT-treated group (MPT-V) (Ambikabothy *et al.*, 2011).

analysis. Subsequent to that, beginning in 2003, there was a rapid increase in literatures reporting on proteomic analysis of snake venoms. Using a variety of approaches, venom proteomes of 55 snake venoms have been analyzed (Fox & Serrano, 2008).

3.1 2DE-MS mapping of four venom proteomes

Following our initial 2DE analysis of the eight Malaysian snake venoms described earlier, we selected four of the venoms for further study on their constituents via mass spectrometry techniques. The four venoms selected were *N. n. kaouthia* (NK), *O. hannah* (KC), *Bungarus fasciatus* (BF) and *C. rhodostoma* (CR) (Fig. 4). We successfully identified a total of 64 proteins from the four venoms - 16 from *N. n. kaouthia*, 15 from *O. hannah*, 6 from *B. fasciatus* and 27 from *C. rhodostoma* respectively. All these proteins have biochemically and pharmacologically important properties. Evidently, based on the results from this study, the typical proteomic routine of 2DE and mass spectrometry, which was also the prototype workflow introduced when the concept of integrated proteomics was first pioneered, has been useful in investigating important venom constituents. Some of the major groups of proteins are discussed in the following.

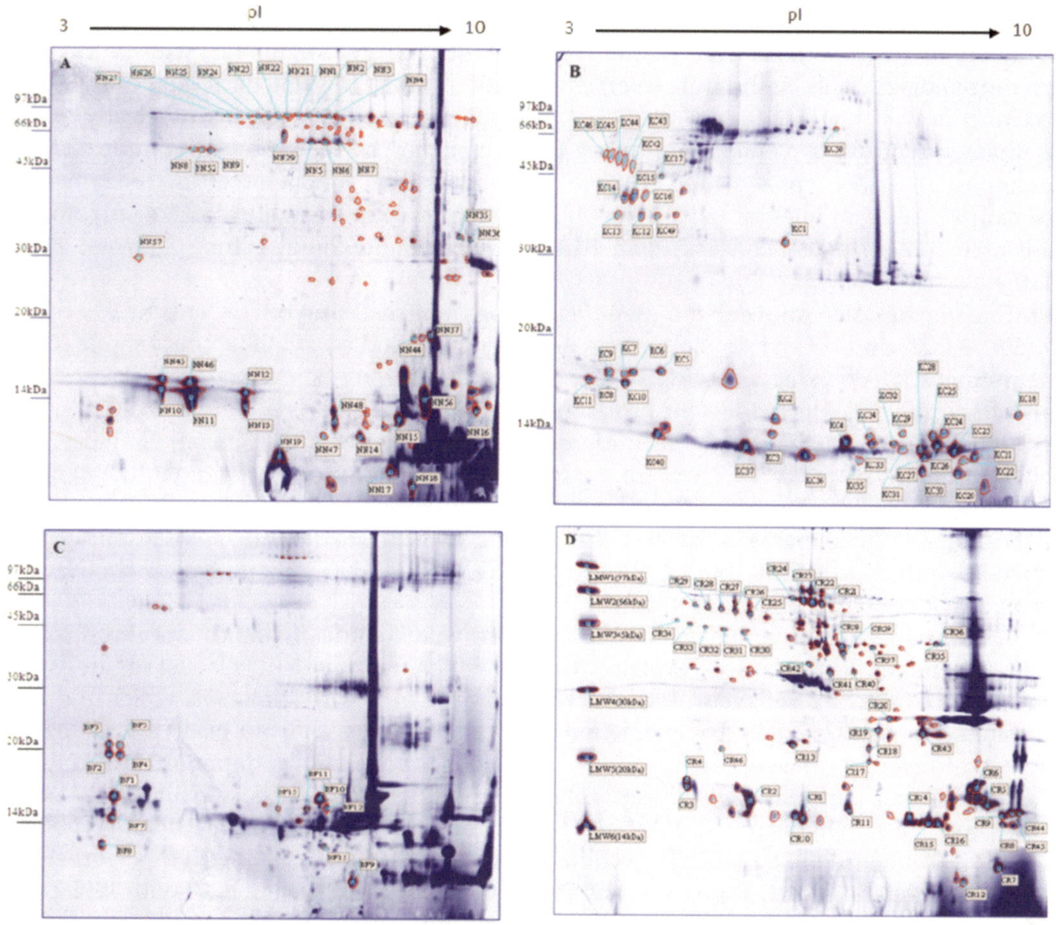

Fig. 4. Proteome maps of the snake venoms of *N. kaouthia* (A), *O. hannah* (B), *B. fasciatus* (C), and *C. rhodostoma* (D) with IPG 3-10 (300µg protein, 18cm), 15%T, Coomassie blue staining showing the annotated spots of the 4 venoms. The spots were detected and annotated by using the Image Master 2D Platinum software. These protein spots picked from the 2-DE gel stained by Coomassie blue, tryptic-digested, extracted and analyzed in duplicates per spot with MALDI-TOF MS in reflectron mode. The figure displays in (D) 6 annotated spots of the low molecular weight markers (Amersham Biosciences - Uppsala, Sweden).

The phospholipase A_2 (PLA_2) group of enzymes was identified in all 4 species (annotated as NN10, NN11, NN12, NN13, NN45, NN46, KC5, KC6, KC9, KC10, BF1, BF2, BF4, BF7, BF9, BF15, CR1, CR5 and CR6). At least one type of these enzymes was identified in the 4 species studied. It is not surprising as extensive studies on venom have demonstrated the accelerated natural selection force that drives the evolution of a multitude of extremely potent snake toxins from an ancestral PLA_2 with digestive function (Montecucco *et al.*, 2008; Ogawa *et al.*, 1996). Hundreds of species of venomous snakes of the families Elapidae and Viperidae were shown to have evolved a wide variety of venoms which contain varying proportions of toxins endowed with PLA_2 activity, characterized by their neurotoxicity, myotoxicity, as well as anticoagulant and edema-inducing properties (Boffa *et al.*, 1976;

Montecucco *et al.*, 2008; Vishwanath *et al.*, 1987). Hence, in snake venom, PLA_2 enzymes, in addition to their possible role in the digestion of the prey, exhibit a wide variety of pharmacological effects through interfering with normal physiological processes (Kini, 2003). It is well known that some of the most toxic and potent pharmacologically active components of snake venoms are either PLA_2 enzymes or their protein complexes. For example, all known presynaptic neurotoxins from snake venom are PLA_2 enzymes or contain PLA_2 as an integral part (Bon, 1997; Fletcher & Rosenberg, 1997). PLA_2 myotoxins are more potent and fast-acting than their non-enzymatic counterparts (Gubenek *et al.*, 1997).

PMF using MS also allowed the identification of long neurotoxins (NN14, KC23, KC25, KC26, KC27 and KC33) in both the cobra species studied. Unlike presynaptic beta-neurotoxins which exhibit varying PLA_2 activities, these long neurotoxins are classified as alpha-neurotoxins which affect the post-synaptic membrane (Lewis & Gutmann, 2004). Post-synaptic neurotoxins have only been identified in venoms from the families Elapidae and Hydrophiidae (sea snakes). They are antagonists of the nicotinic receptor on the skeletal muscle and display different binding kinetics and affinity for subtypes of nicotinic receptors (Hodgson & Wickramaratna, 2002). Other significant neurotoxins, such as the muscarinic toxin-like proteins (MTLP) (NN15, NN17 and KC32), ohanin (KC18) and thaicobrin (NN37), were also identified in the cobra venoms.

In *C. rhodostoma* venom another important protein was identified - the rhodocetin (CR2), a Ca^{2+}-dependent lectin-related protein (CLPs), which is a potent platelet aggregation inhibitor induced by collagen. It is a prime example of a CLP dimer, in which the two subunits are held together by interactions and act synergistically to elicit the biological activity, affecting platelet aggregation and blood coagulation, important in cellular thrombosis and non-cellular processes in homeostasis (Kornalik, 1991; R Wang *et al.*, 1999). This protein also demonstrated the ability to antagonize stromal tumor invasion *in vitro* and other $\alpha 2\beta 1$ integrin-mediated cell functions (Eble *et al.*, 2002). According to this worker, its ability to inhibit tumor cell invasion through a collagen matrix, combined with its lack of cytotoxicity, high solubility, diffusibility and biochemical stability, may qualify rhodocetin to be one of the first snake venom disintegrins of potential practical importance in tumor therapy, *e.g.* in attempts to interfere with stromal invasion and metastasis.

The proteomic approach applied in this study also successfully identified a significant number of enzymes, namely zinc metalloproteinase or disintegrin (CR17, CR18 and CR43), Ancrod or venombin A (CR25, CR27, CR30, CR31, CR32, CR33, CR34, CR40 and CR41), and L-amino-acid oxidases (CR21, CR22, CR23, CR39), that display the well-documented hemotoxic properties of the viper's venom. Disintegrin exhibits hemorrhagic activities by binding to the glycoprotein IIb-IIIa receptor on the platelet surface, thus inhibiting fibrinogen interaction with platelet receptors while L-amino-acid oxidases exhibit hemorrhagic activities by catalyzing oxidative deamination of hydrophobic and aromatic L-amino acids (Au, 1993; Dennis *et al.*, 1990; Gould *et al.*; 1990, Macheroux *et al.*, 2001). L-amino-acid oxidases have also recently been shown to display antibacterial properties (Tonismagi *et al.*, 2006). On the other hand, ancrod is a thrombin-like serine protease that selectively cleaves the fibrinopeptides, resulting in aberrant fibrinogen that is unable to form dispersible blood clots (Au *et al.*, 1993).

As discussed earlier, venom proteins have different abundances and, thus, the identification of each of these proteins may require different methods. The high abundance proteins, for example, causes incomplete resolubilization during equilibration, resulting in vertical streaking and tailing of the most intense protein spots (Berkelman *et al.*, 2004). During the mapping of the four selected venoms, we noticed that these vertical streaks so commonly observed on 2DE profiles of snake venoms were particularly prominent in the 2DE profile of *B. fasciatus*. The presence of these vertical streaks led to the lack of complete visualization of the separated protein spots on the 2DE profiles, limiting the identification of these proteins by mass spectrometry analysis.

Therefore, in order to eliminate these streaks for protein identification, various measures were taken, for instance, prolonging the equilibration time to facilitate sufficient equilibration, scavenging any excess or residual thiol reducing agent with iodoacetamide before loading the IPG strips onto the 2nd dimension gel (as this reducing agent known to exacerbate this effect) or by loading lesser content of protein. However, apart from loading far lesser amount of venom, none of the other measures produced the desired results. As shown in Fig. 5a and Fig. 5b, loading of only 0.8µg and 1.06µg protein, respectively, resulted in elimination of vertical streaks. As the load of protein was increased to 1.6µg (Fig. 5c), an apparent vertical streak begins to show up on the 2DE gel. Apart from highlighting the presence of certain highly abundant proteins in snake venoms, these results also demonstrated another important aspect of venom proteome mapping by 2DE-MS - the complete visualization of a venom proteome, additional steps must be taken, including loading different amounts of venom, use of different staining techniques and the use of a variety of pI ranges (Fox *et al.*, 2002). The end result will probably show that venom constituents are much more complex than originally shown.

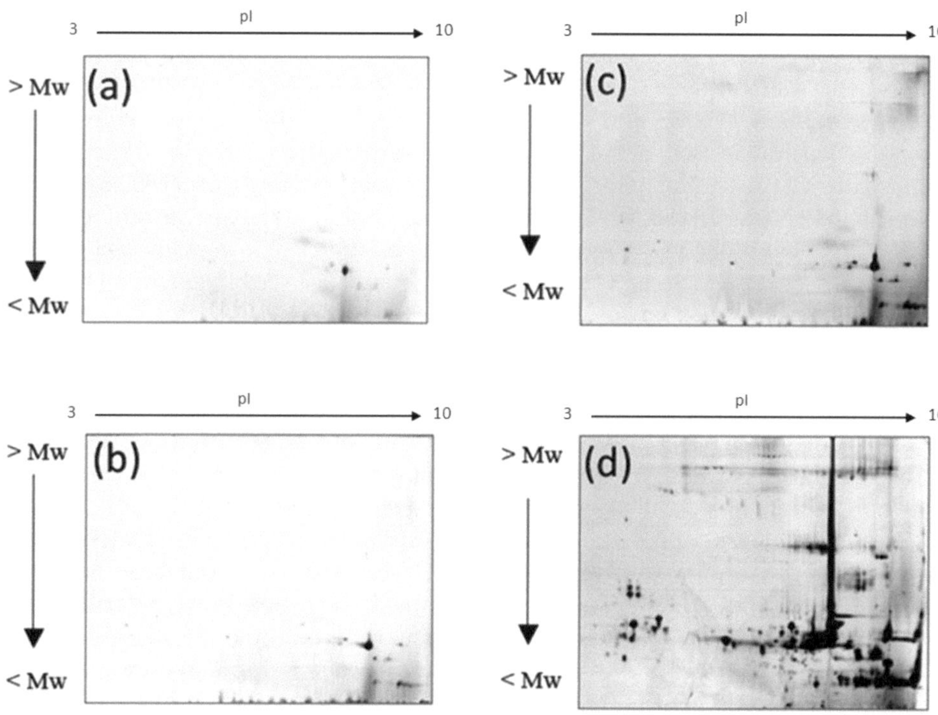

Fig. 5. 2DE profiles of *B. fasciatus* venom at different loading protein content during first dimension IEF separation, IPG 3-10, 15%T, Coomassie blue staining. (a) 0.8µg protein (b) 1.06µg protein (c) 1.6µg protein (d) 35µg protein.

3.2 Challenges in 2DE-MS venom proteome mapping

The main objective of integrated proteomics in venom study is to achieve full proteome coverage of snake venoms. Nevertheless, despite our successful identification of major proteins in the four venoms, a large number of other spots could not be identified, despite displaying high-quality mass spectra. As such, researchers are constantly establishing new, additional workflows to meet this objective and it is no longer limited to only 2DE followed by mass spectrometry analysis in the process of proteome coverage. For example, Li *et al* (2004) has described the novel identification of 124 and 74 proteins from the *Naja naja atra* and *Agkistrodon halys* venoms respectively through the utilization of four combined proteomic approaches, namely – (1) shotgun digestion plus HPLC with ion-trap tandem mass spectrometry, (2) 1D-PAGE plus HPLC with MS/MS, (3) gel filtration plus HPLC with MS/MS, (4) 2DE plus MALDI-TOF-MS. By using four different workflows to characterize the constituents of the same venoms, each approach could compensate the detection coverage of the venomous proteins, since it was found that a few proteins could only be identified by one specific approach (Li *et al.*, 2004).

While conducting database searches for protein identification during the profiling of the four venoms, we found that the limited database for snake venom proteins could pose as a major limitation to this approach of 2DE and mass spectrometry for protein identification. The Taxonomy Browser contained in the Entrez database (URL: http://www.ncbi.nlm.nih.gov) provided the total number of proteins available for matching for each of the 4 snake venoms. The number of known proteins available, as of 16 June 2011, was 122 for *N. n. kaouthia,* 176 for *O. hannah,* 119 for *B. fasciatus* and 76 for *C. rhodostoma.* The protein database was constructed based on the sequence data from the translated coding regions from DNA sequences in GenBank, EMBL, and DDBJ as well as protein sequences submitted to Protein Information Resource (PIR), SWISS-PROT, Protein Research Foundation (PRF), and Protein Data Bank (PDB) (sequences from solved structures). It was obvious that the database was incomplete for snake venom matching, compared with 534,370 available for *Homo sapiens* (human) proteins, or 88,882 for *Bos Taurus* (cattle) proteins. Nevertheless, the snake venom protein database is still of value in comparison to some other venomous species such as *Chironexyamaguchii* (sea wasp) with only 2 proteins, *Hadrurusaztecus* (scorpion) with 1, and *Dolomedesplantarius* (spider) with none, to name a few, for matching.

A closer examination of the snake venom protein database revealed the following information. Some of the proteins were not found in the venom but were evident elsewhere in the snake's body, for example, the nerve growth factor (located in the nervous system), NADH dehydrogenase and cytochrome b (located in mitochondria) and oocyte maturation factor (located in the reproductive system). Also, most of the venom proteins were precursor forms or polypeptide chains of the same protein, example: phospholipase PLA_2 precursor and Chain A, crystal structure of L-Amino Acid Oxidase. Based on these factors, therefore, in reality, the snake venom proteins available for matching in the database were, in fact, much more limited. This limitation may be overcome in the future as the database for snake venoms is developing. This is evident by comparing the number of known proteins dated back to 3rd October 2005 (67 for *N. kaouthia;* 78 for *O. hannah;* 34 for *B. fasciatus;* and 60 for *C. rhodostoma*) to the most recent Entrez protein database. The protein database of *B. fasciatus* has shown its most promising development, displaying increment of up 250% while that of *O. hannah* increased by 126%, *N. kaouthia* by 82%, and *C. rhodostoma* by 27%. If not for the reason of incomplete database another common possibility is due to post-translational modification (PTM) undergone by some of the proteins. PTM of venomous protein are common phenomenon in snake venoms. Of a number of mechanisms inducing chemical modification of protein, glycosylation is one of the most frequently found (Li *et al.*, 2004). It is expected the advances in mass spectrometry combined with posttranslational modification database specific for venoms may solve this shortcoming. A PTM database comprising of phosphorylation, N-glycosylation or acetylation sites is already existing and growing steadily for nine different species (none yet for snake)(Gnad *et al.*, 2011).

3.3 Relevance of 2DE-MS venom proteome mapping in the present proteomic landscape

With the advances in mass spectrometry techniques, non gel-based proteomic techniques such as LC/MS are now generally being considered as state-of-the-art, thus leaving the gel-based 2DE technique in a questionable state as to whether it still has any relevance as a proteomic method. Nevertheless, while the issue at hand is one that will be of continuous debate, we should keep in mind that the limitations of 2DE are well known and are probably most thoroughly investigated than any other proteomic techniques. Hence, researchers who choose to utilize the technique are generally aware of the limitations and are usually able to adapt the complexity of the sample to the resolution power of the 2DE method by narrowing the study subject to one focus in order to reduce the sample complexity (Rabilloud *et al.*, 2010). For instance, in a study done by Nawarak *et al* (Nawarak *et al.*, 2003) to investigate the proteomes of a number of selected Elapidae and Viperidae venoms, the group first fractionated the crude venoms on RP-HPLC before subjecting only the major eluted peaks to 2DE analysis.

In addition, the robustness of the 2DE technique has been tested thoroughly and the influence of the various parameters on the intra-laboratory reproducibility has been investigated (Choe & Lee, 2003), thus making the 2DE process a strong technique for proteome profile building and for subsequent deposition in databases to be accessed by researchers worldwide to be used for reference.

Specifically in the field of venom proteome study, the 2DE technique remains important when it comes to the study of (1) subpopulation of venom proteins and (2) post-translational modifications (PTM) in venom proteins. In two studies that were done in similar manner, Serrano *et al* (2005) and Birrell *et al* (2006) investigated the diversity of venom proteins in the viperid venoms (*Bothrops jararaca* and *Crotalid atrox*) and the Australian Brown Snake venom (*Pseudonaja textilis*) respectively. Taking advantage of the efficient interface of 2DE with other biochemical techniques, the two groups of researchers subjected the obtained 2DE profiles to immunoblot analysis with antisera raised against specific venom protein groups that were of study interests. Serrano *et al* (2005) did immunoblot analysis with antisera raised against metalloproteinase's, serine protease and phospholipase A2. Birrell *et al* (2006), on the other hand, did immunoblot analysis with antisera raised against prothrombinase complex, heavy chain of the Factor Xa-like protease, Gla residues, textilinins and textilotoxins. The results were specific 2DE profiles of these protein groups, giving rise to a more thorough understanding of venom complexity and providing insights for investigators who want to focus on these subpopulations of proteins in future studies.

Large portions of venom proteins undergo PTM (Nawarak *et al.*, 2004). Protein spots that have undergone PTM appeared as train of spots on the 2DE profiles, owing to their differences in pI values that gave rise to their non-identical migration profiles (Birrell *et al.*, 2006). Both groups employed specific fluorescent dyes, Pro-Q Emerald and Pro-Q Diamond, for the study of post-translational modifications (PTM) in venom proteins (Birrell *et al.*, 2006; Serrano *et al.*, 2005). Apart from specific staining methods, other modifications can also be applied on the conventional 2DE protocol for the study of PTM. Nawarak *et al* (2004) performed lectin-affinity purification using Sepharose-bound Con A to fractionate venom glycoproteins. The bound glycoproteins were then eluted and studied using 2DE followed by mass spectrometry.

Furthermore, through efforts of venom proteome mapping by 2DE-MS, major protein groups have been identified, as evident from our results elaborated in Section 3.1. These proteins have potential values when it comes to pharmaceutical and diagnostic uses. The purification of these proteins is thus slowly emerging as another new aspect of investigation when it comes to snake venom research. Therefore, the profiling of these proteins on 2DE maps that can be stored in a reference library for the use of researchers worldwide will be very useful when a researcher wants to use these profiled protein spots for the purpose of 2DE-guided purification, a technique that will be described in greater details in a following section.

4. Modifications to the 2DE protocol

4.1 Spiking

In performing high-performance liquid chromatography (HPLC), a concept known as co-injection can be used to help identify unknown compounds. To perform co-injection, a process known as spiking has to be done. Spiking is done by first mixing a synthesized or isolated standard (some are available commercially) with the sample containing the compound to be identified and subsequently, if the co-injected standard and unknown compounds co-elute, then the relative peak intensity of the unknown compound on the chromatogram of the spiked sample will be higher than that for the unspiked sample (Peters *et al.*, 2005). Co-elution supports, but does not prove, the idea that the compounds are identical.

The spiking concept from HPLC was then adapted to proteomics and 2DE. It can be demonstrated by using alpha-bungarotoxin from the *Bungarus multicinctus* (Many banded krait) venom (Vejayan *et al.*, 2008) using the following steps:

1. Crude *B. multicinctus* venom was first subjected to 2DE analysis on an 18cm format gel, using the conventional 2DE protocol without any modification (Fig.6). Crude venom, containing 0.8µg protein, was loaded onto the IPG strip via a sample-loading cup on the anodic end.
2. A second 2DE analysis is done on the same venom, but with one modification – instead of using a single sample loading cup in IEF, two sample loading cups were used. One of the cups was placed at the anodic end (designated as Cup A) and loaded with crude *B. multicinctus* venom (0.8µg protein) alone while the other cup was placed at the cathodic end (designated as Cup B) and loaded with commercially purchased purified *B. multicinctus* alpha-bungarotoxin (0.1µg protein) (Fig. 7). The remaining of the first and second dimension separations was done as per the usual conventional 2DE protocol.
3. Both the 2DE profiles obtained from Steps 1 and 2 were then subjected to comparative analysis using the Image Master 2D Platinum software. A spot of increased intensity was identified on the 2DE profile obtained from Step 2. The increase of intensity was quantified as a 2.5 fold increase in the % volume (Fig. 8).
4. The spot with increased intensity was cleaved and subjected to MALDI-TOF-MS peptide mass fingerprinting. The protein was identified and confirmed as alpha-bungarotoxin.

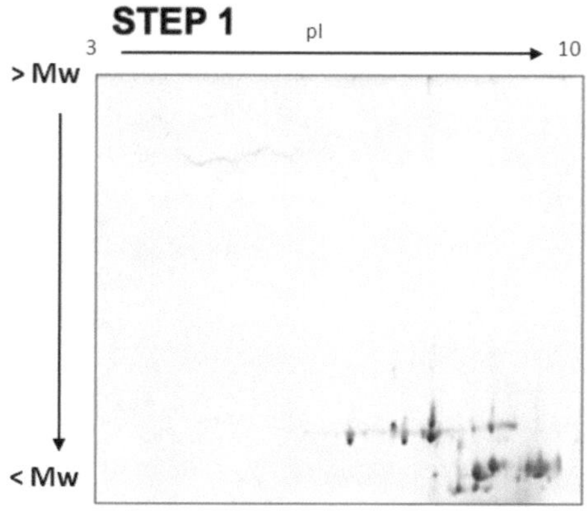

Fig. 6. Step 1 of spiking technique: 2DE analysis of the crude *B. multicinctus* alone (0.8μg protein) was done on an 18cm format gel using the conventional 2DE protocol.

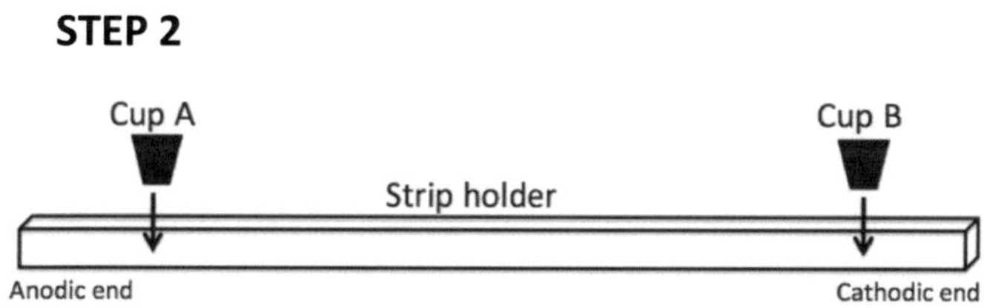

Fig. 7. Step 2 in spiking technique: A second 2DE analysis was done using two sample loading cups - Cup A at the anodic end is loaded with crude *B. multicinctus* venom while Cup B at the cathodic end is loaded with pure alpha-bungarotoxin.

This method was initially conceptualized with the intention to locate the spot of alpha-bungarotoxin on the 2DE profile of *B. multicinctus* by simply identifying the "spiked spot", which was the spot of increased intensity, without using any mass spectrometry technique. However, we have also found the technique useful when we needed to affirm the identity of a compound purified from snake venom. After we have successfully purified rhodocetin from the venom of *C. rhodostoma*, we spiked the purified protein on the crude *C. rhodostoma* venom and found a spot of increased intensity that matches to the rhodocetin spot on the 2DE profile of *C. rhodostoma*. The usefulness of spiking for this purpose can be better demonstrated in the following section.

4.2 2DE-guided purification

In recent years, natural products drug discoveries have received a renewed interest and snake venoms are also being investigated for pharmacologically important components. Once a crude venom has been identified as an active source, bioassay-guided purification is typically used to isolate the active component. While the general paradigm of this process can be relatively straightforward in the academic laboratory setting, the design of a suitable bioassay can present as a practical challenge to the whole purification process. Notwithstanding the fact that the design of a bioassay has to take into consideration some

STEP 3

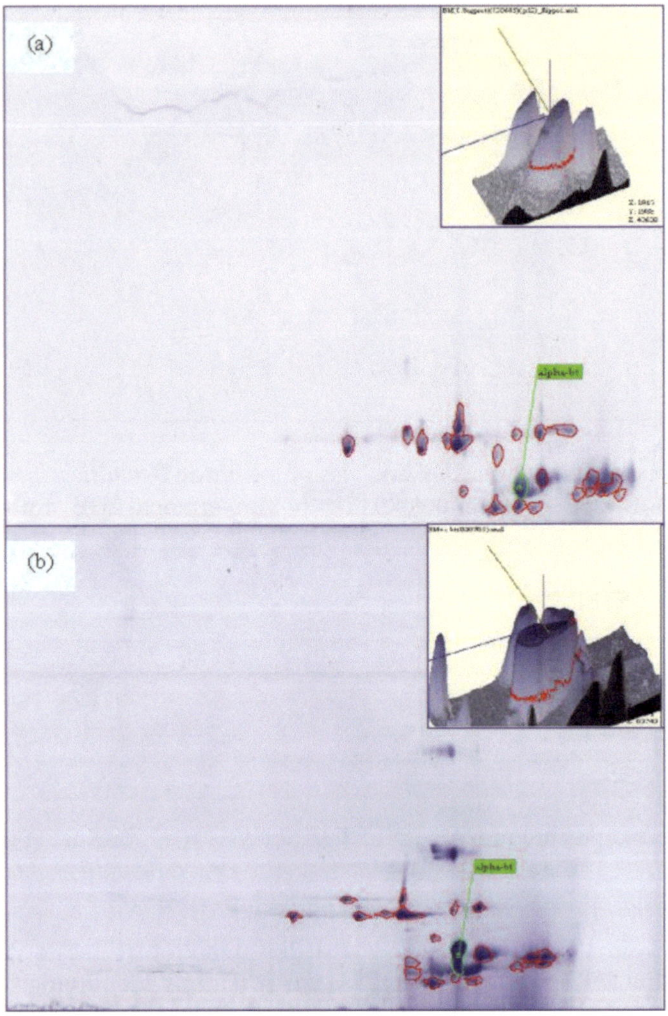

Fig. 8. Step 3 in spiking technique: Comparative analysis between the unspiked (a) and alpha bungarotoxin-spiked (b) 2DE profiles showed a spot with increased intensity (the spiked spot). The inset 3D figures showed a % volume increase quantified at 2.5 fold. The spiked spot was later confirmed as alpha-bungarotoxin by mass spectrometry.

important criteria - sensitivity, specificity, lack of ambiguity, accuracy, reproducibility and a reasonable cost - some bioassays also have lengthy turnaround time and require high amount of the active component, thus posing as challenges to the progression of the fractionation process.

All the above elaboration on the challenges of purifying therapeutically important proteins from snake venoms can perhaps be best illustrated using rhodocetin as an example. Rhodocetin is a CLP from the venom of *C. rhodostoma* and has been investigated for important therapeutic properties including platelet aggregation inhibition and stromal invasion inhibition. Two previous groups of researchers have employed bioassay-guided purification to purify this important protein. Wang *et al* (R Wang *et al.*, 1999) has employed the use of a platelet aggregation bioassay that required the use of human and rabbit blood collection and the inhibition of platelet aggregation would prove the presence of rhodocetin. Eble *et al* (2002) on the other hand, utilized the ELISA-like procedure to detect the inhibition of the binding between α1β2 integrin with type I collagen by rhodocetin. Given the importance of rhodocetin as a potentially therapeutic component, we would like to have purified rhodocetin in our laboratory for the purpose of further investigations. However, we were confronted by a few challenges in wanting to do so. Firstly, rhodocetin has yet to be made available as a commercially sold purified protein. Secondly, the purification of rhodocetin, as described above, required the design of complicated bioassays.

In view of all these mentioned challenges in rhodocetin purification, we attempted to find an alternative to bypass the requirement of a bioassay to purify rhodocetin. As such, we hypothesized that a concept known as "2DE-guided purification" would be able to achieve this purpose (Tang *et al.*, 2011). The concept of 2DE-guided purification is essentially using the presence of a protein spot on the 2DE gel as an indication of the presence of a protein within a particular sample. As discussed in Section 3, following our efforts of venom proteome profiling, rhodocetin has been successfully identified on the 2DE profile of *C. rhodostoma*. We could, therefore, take advantage of this result and use it for the 2DE-guided purification of rhodocetin, of which will be described in a detailed step-by-step description as follows:

1. The crude venom of *C. rhodostoma* was subjected to 2DE analysis on a 7cm format minigel. The 2DE profile obtained was then compared with our previous work of profiling the crude *C. rhodostoma* venom on a larger 18cm format 2DE gel. The rhodocetin spot was identified on the minigel, cleaved and sent for MALDI-TOF-MS peptide mass fingerprinting (Fig. 9). It was successfully identified as the alpha subunit of rhodocetin.
2. The crude *C. rhodostoma* venom was then subjected to fractionation by liquid chromatography. Out of the various chromatography techniques for protein purification, we have selected anion-exchange using the column Mono Q 5/50 GL (1ml). The chromatography profile showed six eluted peaks and we designated the peaks as U, P1, P2, P3, P4 and P5 (Fig. 10). The fractions of these peaks were collected for the subsequent desalting process.
3. The desalting process involves size exclusion chromatography using G25 HiTrap Desalting column. Interestingly, while each of the peaks typically gave a single protein peak upon desalting, P2 was an exception as it produced two distinct peaks on its desalting profile. As such, we assumed that the G25 gel filtration has further fractionated the contents of P2 and we designated the two peaks as DP1 and DP2 (Fig. 11).
4. After the two-step protein fractionation process, we collected seven peaks based on the chromatography profiles. All the seven peaks were then subjected to 2DE analysis respectively on minigels. The resulting profiles were then compared to the *C. rhodostoma* profile done in Step 1. Out of the seven profiles, only the profile of DP2 showed the presence of the rhodocetin spot. Therefore, we could conclude that the peak DP2 contained rhodocetin (Fig. 12).
5. To conclude the purification process, it was essential for us to determine the homogeneity of DP2, in order to decide if any further cycle of fractionation was necessary. The SDS-PAGE of DP2 showed two distinct bands at the low molecular weight region (Fig. 13) that was characteristic of purified rhodocetin, as shown by previous groups of researchers (Eble *et al.*, 2002; R Wang *et al.*, 1999).

6. Finally, to further confirm the identity of the purified compound as rhodocetin, we employed the spiking technique, as described in Section 4.1, by loading the crude *C. rhodostoma* venom into a sample loading cup located at the anodic end and the purified compound into another sample loading cup at the cathodic end. A spiked spot was identified and upon comparison with the 2DE profile of the unspiked crude *C. rhodostoma* venom, the location of the spiked spot correspond to the location of the rhodocetin (alpha subunit) and the increase in intensity was quantified at 1.6 fold (Fig. 14). Together, these confirmed the identity of the purified compound as rhodocetin.

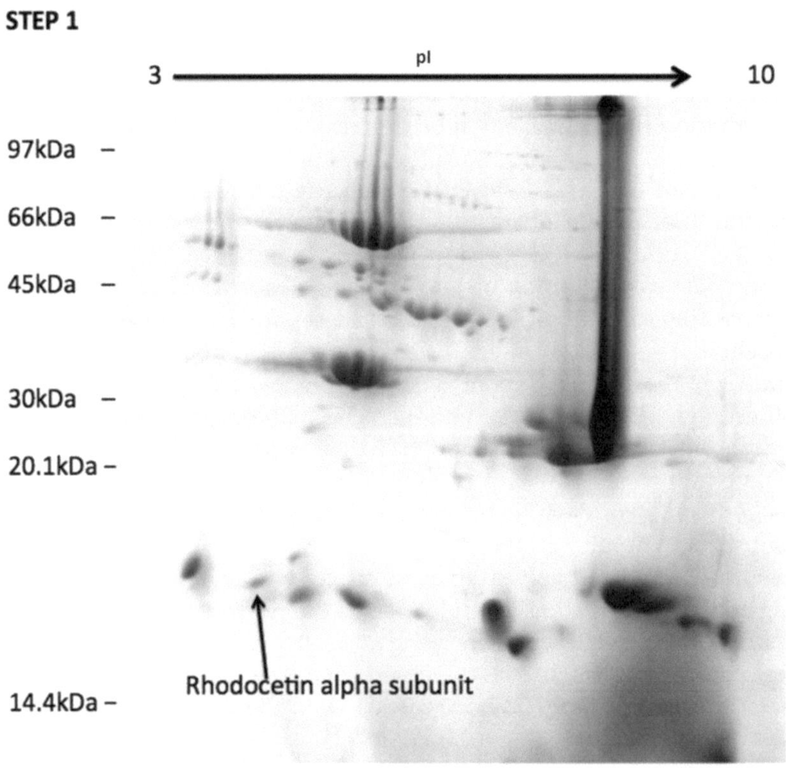

Fig. 9. Step 1 of 2DE-guided purification: 2DE profile of *C. rhodostoma* (60µg protein) with the rhodocetin (alpha subunit) spot annotated. The profile was obtained by IEF on a 7cm IPG strip (pH 3-10) and the proteins subsequently separated in the second dimension by 15% SDS-PAGE. The separated proteins were visualized by Coomassie Brilliant Blue staining.

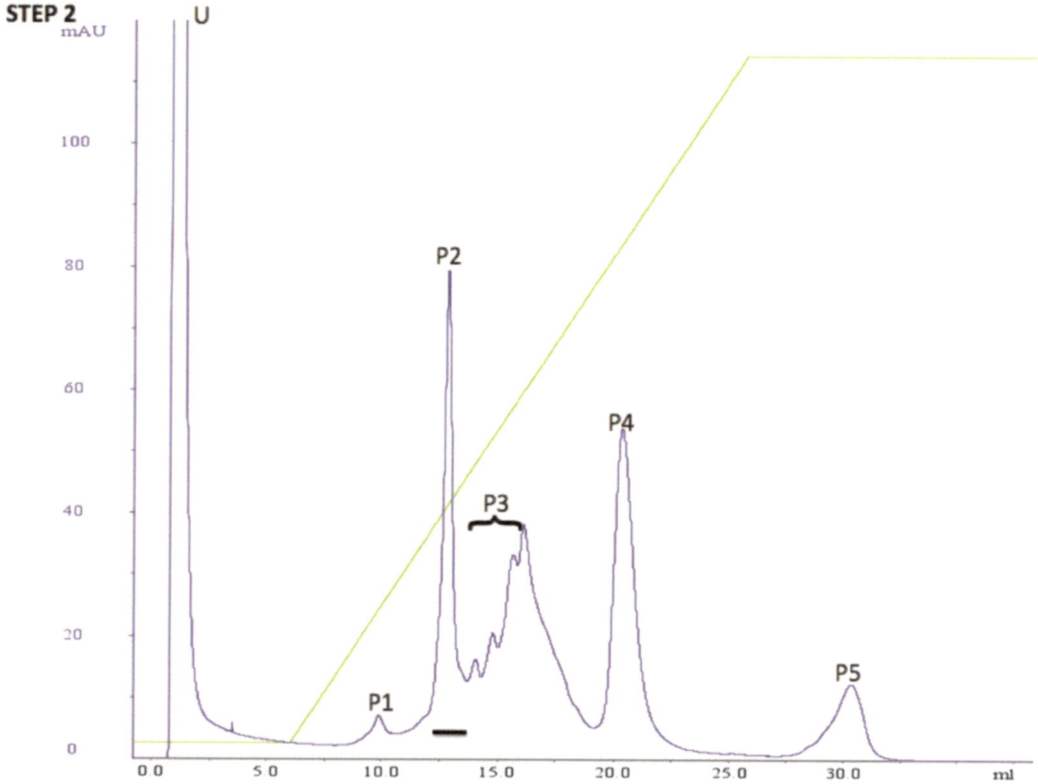

Fig. 10. Step 2 of 2DE-guided purification: Anion-exchange to fractionate crude *C. rhodostoma* venom – 5mg of crude *C. rhodostoma* venom dissolved in 250µl of 20mM Tris-HCl, pH 8.5 and loaded into a Mono Q 5/50 GL (1ml) column, equilibrated with 20mM Tris-HCl, pH 8.5. Six peaks (U, P1, P2, P3, P4 and P5) were obtained.

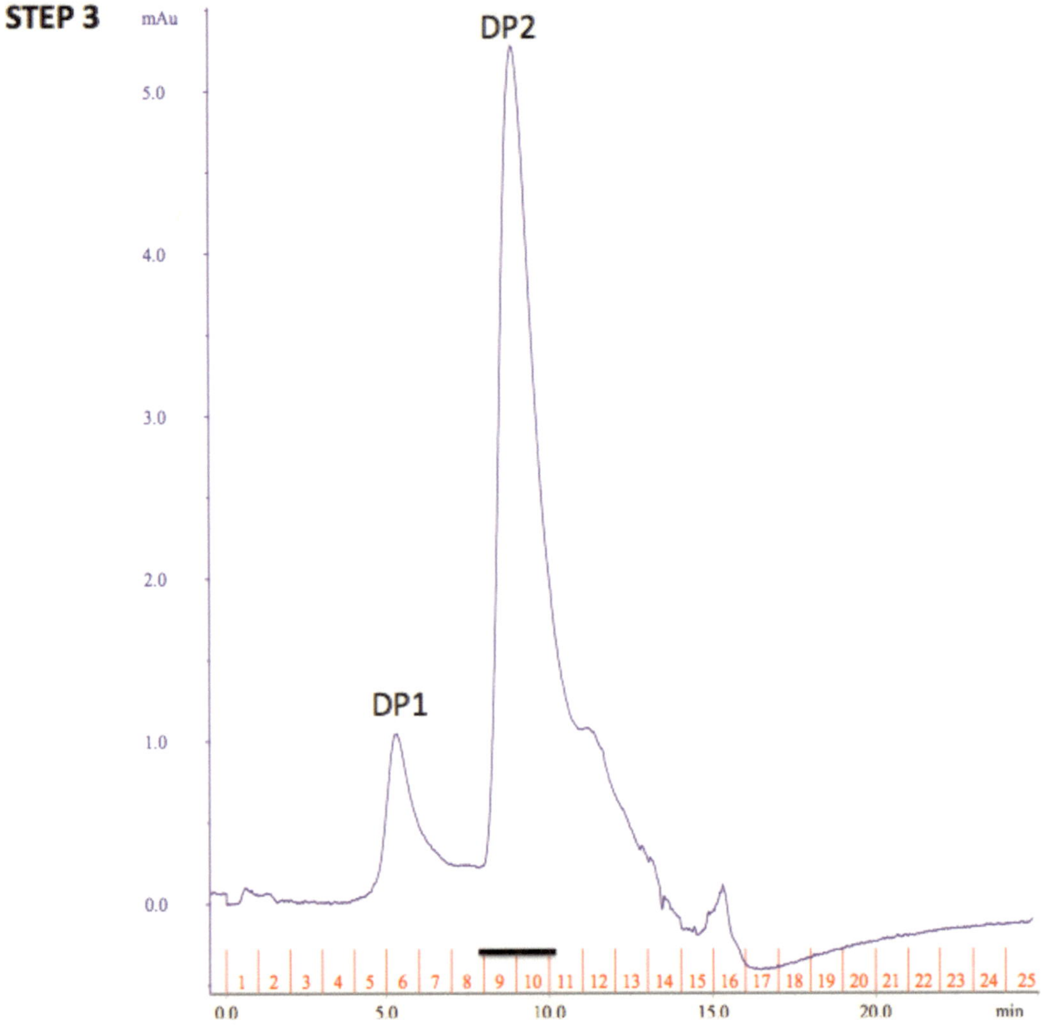

Fig. 11. Step 3 of 2DE-guided purification: The chromatography profile obtained when P2 fraction collected from the Mono Q column was directly injected into a G25 HiTrap Desalting column, equilibrated with distilled water. Two peaks of DP1 and DP2 were obtained.

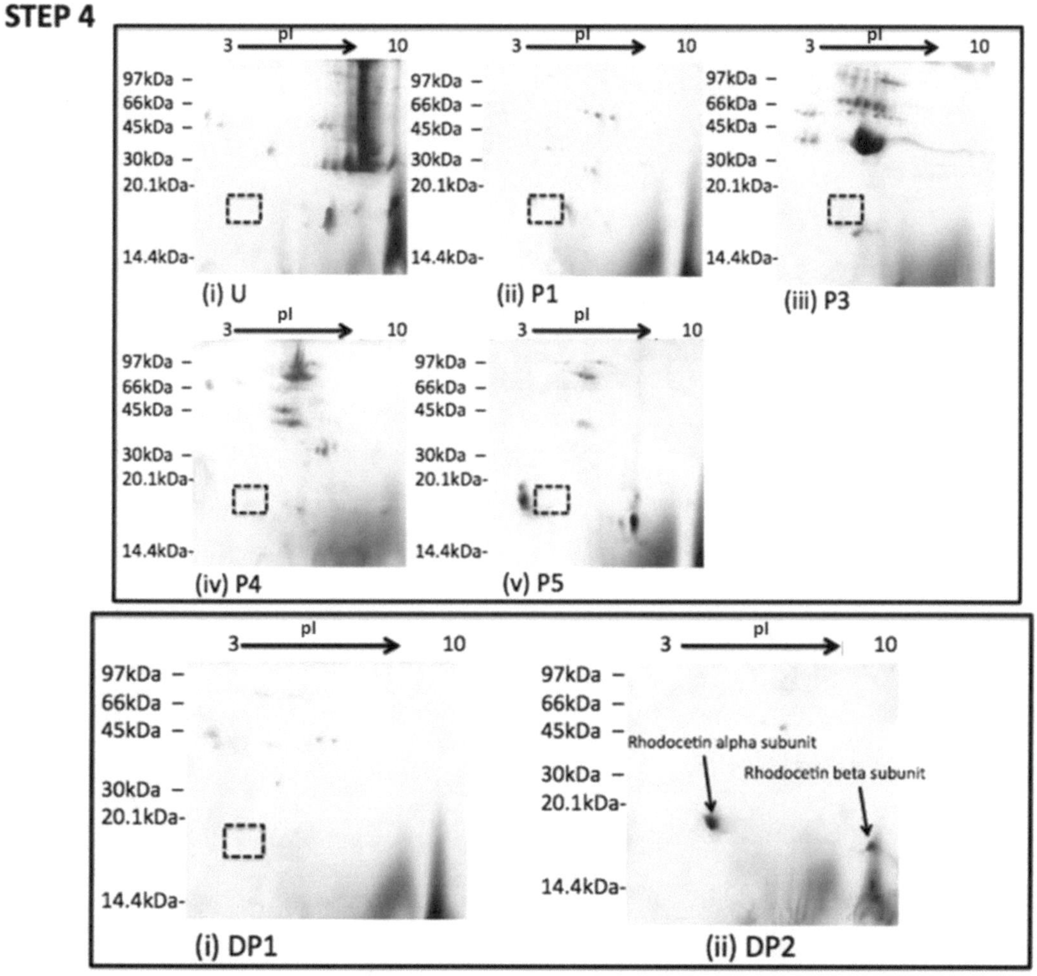

Fig. 12. Step 4 of 2DE-guided purification: 2DE assay was done on the seven collected peaks after the fractions were desalted and lyophilized. The 2DE profile of each peak is shown here and the small area outlined by the black grids on each 2DE gel represents our area of interest in which rhodocetin (alpha subunit) spot should have been present. The 2DE profile of DP2 clearly showed the presence of the alpha and beta subunits of rhodocetin.

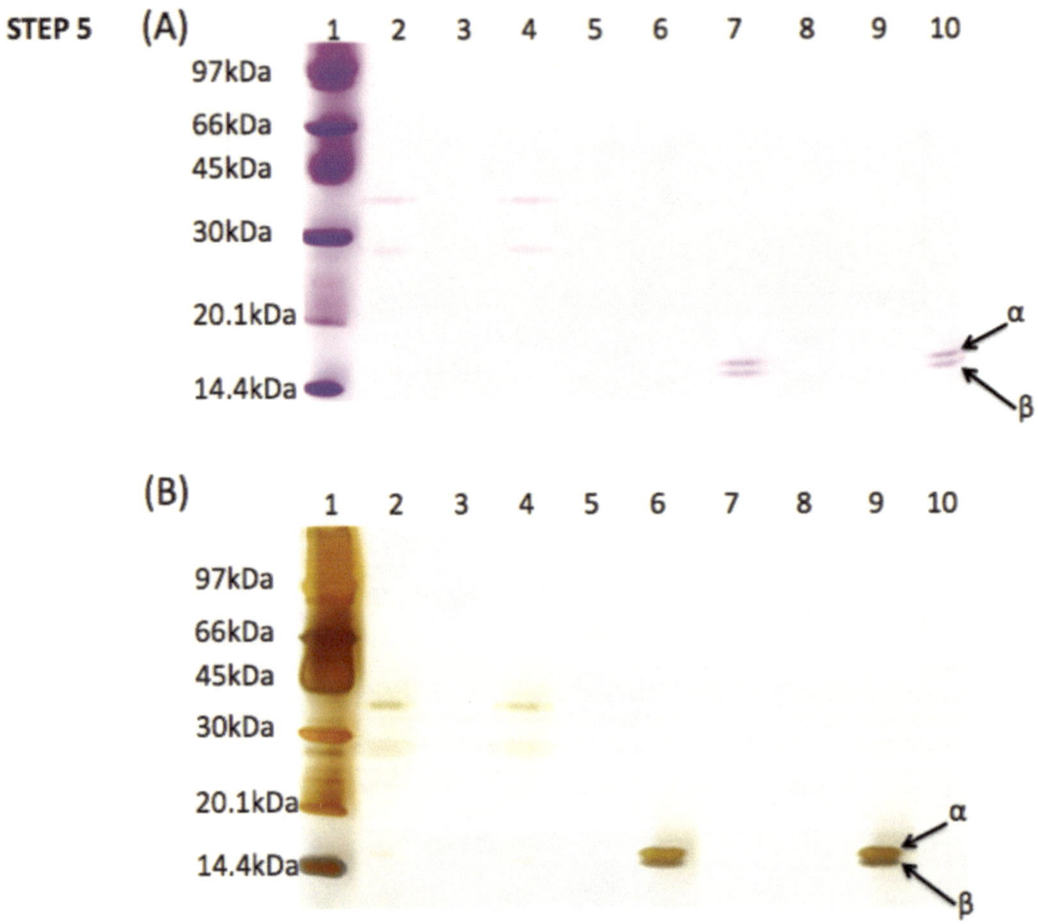

Fig. 13. Step 5 of 2DE-guided purification (SDS-PAGE): Homogeneity of purified rhodocetin from DP2 assessed using 15% SDS-PAGE. The purified rhodocetin showed two distinct bands due to the separation of the heterodimer into its alpha and beta subunits by SDS denaturation. The separated bands were visualized with both (A) Coomassie Brilliant Blue and (B) silver staining. (A) Lane 1: GE Healthcare Low Molecular Weight (LMW) markers; Lane 2: DP1; Lane 3: *blank*; Lane 4: DP1; Lane 5 and 6: *blank*; Lane 7: DP2; Lane 8 and 9: *blank*; Lane 10: DP2. (B) Lane 1: GE Healthcare LMW markers; Lane 2: DP1; Lane 3: *blank*; Lane 4: DP1; Lane 5: *blank*; Lane 6: DP2; Lane 7 and 8: *blank*; Lane 9: DP2; Lane 10: *blank*. The blank wells were intentionally skipped to prevent any effect of inter-well spillage.

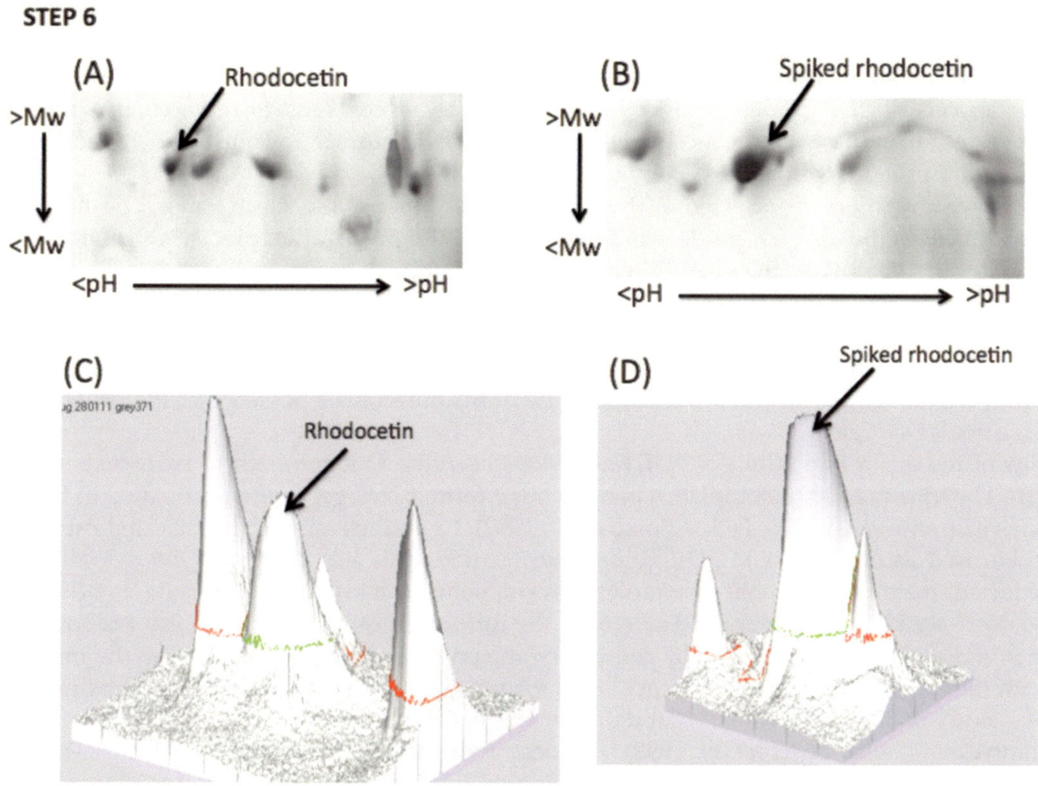

Fig. 14. Step 6 of 2DE-guided purification (spiking): (A) Area of interest on the 2DE profile of crude *C. rhodostoma* venom with the rhodocetin (alpha subunit) spot labelled. (B) The same area showing the spot of spiked rhodocetin with an observed increased intensity. (C) 3D representation views of the rhodocetin (alpha subunit) spot on the crude venom alone and (D) the spiked rhodocetin (alpha subunit) spot, with the latter spot having a quantified 1.6 fold increase in intensity.

Based on our results, we have successfully proved that rhodocetin could be purified using 2DE-guided purification. 2DE profile, in place of an assay, is sufficiently selective and specific to determine which peak contained rhodocetin, therefore allowing us to decide which peak should be selected for further fractionation. While we have only described the use of this method using rhodocetin and *C. rhodostoma*, 2DE is a versatile technique that can be applied to any sample, as long as it is protein containing (Carrette *et al.*, 2006; O' Farrell, 1975). Therefore, we see that this concept is probably one of the most important innovations that we have developed for our laboratory; especially given the fact that 2DE has undergone much development and effort of standardization since its initiation. These efforts have helped to improve 2DE to become a method with a standardized protocol that requires little optimization and is often reproducible. Hence, the following few paragraphs will discuss a few aspects of 2DE-guided purification that may be of concerns to researchers who are interested to utilize this concept in their own laboratories to purify therapeutically important proteins from snake venoms.

We have intentionally selected 2DE over the one-dimensional electrophorectic method SDS-PAGE as the assay to guide our progression in the purification process of rhodocetin, despite the fact that SDS-PAGE could be done much more easily. Given its one-dimensional separation capability, SDS-PAGE has only limited differentiation efficiency of crude venom proteins, owing to the overlapping of protein bands with similar molecular weights (Soares *et al.*, 1998). The protein spots on the 2DE profile, on the other hand, are more specific and are more definite indications of the presence of the proteins in a particular sample.

One of the major limitations of 2DE has always been the time required to perform a single run. The time needed to complete a general large format 2DE gel is often estimated to be 3-5 days (Carrette *et al.*, 2006; Felley-Bosco *et al.*, 1999). Nevertheless, we have selected minigels to be used as our assays in 2DE-guided purification. This has decreased the overall time required, making it possible to complete several simultaneous runs in a single day (Felley-Bosco *et al.*, 1999). In our context of study, the utilization of minigels was also adequate in identifying the rhodocetin spot by comparing the crude *C. rhodostoma* profile on the minigel with that previously done on a larger 18cm format 2DE gel. This is in line with the findings of a study that has also shown that data transfer between large format gel and minigel was compatible (Felley-Bosco *et al.*, 1999). Besides, with the recent advent of 2DE innovations such as the bench top proteomics system ZOOM® IPGRunner™ System (Invitrogen) that allows for rapid first and second dimension protein separation in 2DE, any laboratory can achieve high-resolution 2DE faster, simpler and easier (Pisano *et al.*, 2002).

The detection of spots in 2DE relies critically on the staining method and our utilization of Coomassie Brilliant Blue has been sufficiently sensitive for our progression. The two common staining methods, silver staining and Coomassie Brilliant Blue, stain between 0.04-2ng/mm^2 and 10-200ng/mm^2 respectively (Wittman-Liebold *et al.*, 2006). Several recent modifications to the Coomassie Brilliant Blue staining protocol has also greatly increased its sensitivity (Pink *et al.*, 2010; X Wang *et al.*, 2007). As such, the 2DE assay is a sensitive one requiring relatively low amount of sample, as compared to certain bioassays. In addition, the sensitivity of this technique is expected to improve with the development of fluorescent staining (Yan *et al.*, 2000). This is especially important, since progression into further cycle of fractionation only results in reduction of the available sample while bioassay-guided purification of venom's neurotoxins utilizing animal assays require fairly large amount of the sample material (Escoubas *et al.*, 1995). Although a microinjection technique has been described to address this issue, this technique can be labour intensive and time consuming (Escoubas *et al.*, 1995).

Since liquid chromatography frequently employs salt gradient and utilizes non-volatile buffer (such as Tris-HCl), salt can still be present even after desalting and lyophilisation of the peaks. This was evident by our inability to increase the voltage during IEF resulting in underfocusing of the protein spots. Subsequently, whenever this problem appeared, we prolonged the IEF protocol to an overnight running by introducing an additional first step of 50V at step and hold for 12h. This was found to improve IEF and voltage could be increased up to 5000V. This is in line with the concept of electrophoretic desalting described by Gorg *et al* (1995) in which samples with high salt concentration were directly desalted in the IPG strip using a low voltage during the first few hours of IEF. Davidsson *et al* (2002) also previously reported that such prolonging of IEF run could improve the problem of incomplete focusing due to the presence of ampholytes in cerebrospinal fluid samples.

The biggest limitation of 2DE-guided purification is its dependence on protein profiling efforts and publications of 2DE reference maps. In our study, without prior profiling of rhodocetin into the 2DE reference map of CR, the rhodocetin spot will not be located and consequently, it will be impossible to determine the presence of rhodocetin in the chromatography peaks by 2DE testing. However, this challenge show prospects of improvisation as protein profiling efforts continue to be on the rise in recent years.

5. Conclusion

We hope that the role of 2DE in snake venom study has been effectively underlined in this chapter. While the present setting in the field of proteomic methods is one that tends to incline towards the rapidly advancing non-gel based proteomic methods, it is obvious that 2DE still has the advantages of being a robust technique with high resolution power. In terms of investigating the complexity of snake venoms, it is evident that the application of 2DE is not limited to only whole proteome analysis for taxonomic and envenomation pathology investigations, but is also feasible as an assay in the multistep protein purification process for pharmacologically important venom proteins. There is no standardized workflow as to how 2DE should be used in the investigation of snake venoms. Depending on the objective of the study, 2DE should be innovatively used along with other proteomic methods and its protocol should be appropriately modified in order to meet the study objectives.

6. Acknowledgement

The authors are very grateful to Mr Zainuddin from Bukit Bintang Enterprise Sdn Bhd for enabling the milking and purchasing of all venoms used in this study. The work was conducted utilizing chemicals and consumables supplemented from grants: Malaysian Ministry of Science and Technology (project number 36-02-03-6005 & 02-02-10-SF0033) and Monash University Sunway Campus Internal Grant (514004400000).

7. References

Ambikabothy, J., Srikumar, Khalijah, A. & Vejayan, J. (2011). Efficacy evaluation of *Mimosa pudica* tannin isolate (MPT) for its anti-ophidian properties. *Journal of Ethnopharmacology*, In Press.

Au, L.-C. (1993). Nucleotide sequence of a full-length cDNA encoding a common precursor of platelet aggregation inhibitor and hemorrhagic protein from Calloselasma rhodostoma venom. *Biochimica et biophysica acta,* Vol.1173, No.2, pp.243-245.

Au, L.-C., Lin, S.-B., Chou, J.-S., The, G.-W., Chang, K.-J. & Shih, C.-M. (1993). Molecular cloning and sequence analysis of the cDNA for ancrod, a thrombin-like enzyme from the venom of Calloselasma rhodostoma. *Biochemical Journal,* Vol.294, No.2, pp.387-390.

Berkelman, T., Brubacher, M. C. & Chang, H. (2004). Prevention of Vertical Streaking. *In: BioRadiations: A Resources for Life Science Research.* pp.23, Bio-Rad Laboratories, Inc.: U.S.A.

Birrell, G. W., Earl, S., Masci, P. P., de Jersey, J., Wallis, T. P., Gorman, J. J. & Lavin, M. F. (2006). Molecular diversity in venom from the Australian Brown Snake, *Pseudonaja textilis. Molecular and Cellular Proteomics,* Vol.5, No.2, pp.379-389.

Boffa, G. A., Boffa, M. C. & Winchenne, J. J. (1976). A phospholipase A2 with anticoagulant activity. I. Isolation from *Vipera berus* venom and properties *Biochimica et biophysica acta,* Vol.429, No.3, pp.839-852.

Bon, C. (1997). Multicomponent neurotoxic phospholipase A2. *In: Venom Phospholipase A2 Enzymes: Structure, Function and Mechanism,* Kini, R. M. (ed.). pp.269-285, Wiley: Chichester, England.

Bougis, P. E., Marchot, P. & Rochat, H. (1986). Characterization of Elapidae snake venom components using optimized reverse-phase high-performance liquid chromatographic conditions and screening assays for. alpha.-neurotoxin and phospholipase A2 activities. *Biochemistry,* Vol.25, No.22, pp.7235-7243.

Brewis, I. A. & Brennan, P. (2010). Proteomics Technologies for the Global Identification and Quantification of Proteins. *Advances in Protein Chemistry and Structural Biology,* Vol.80, pp.1-44.

Calvete, J. J., Sanz, L., Angulo, Y., Lomonte, B. & GutiÈrrez, J. M. (2009). Venoms, venomics, antivenomics. *FEBS letters,* Vol.583, No.11, pp.1736-1743.

Carrette, O., Burkhard, P. R., Sanchez, J. C. & Hochstrasser, D. S. F. (2006). State-of-the-art two-dimensional gel electrophoresis: a key tool of proteomics research. *Nature Protocols,* Vol.1, No.2, pp.812-823.

Chippaux, J., Williams, V. & White, J. (1991). Snake venom variability: methods of study, results and interpretation. *Toxicon,* Vol.29, No.11, pp.1279-1303.

Choe, L. H. & Lee, K. H. (2003). Quantitative and qualitative measure of intralaboratory two-dimensional protein gel reproducibility and the effects of sample preparation, sample load and image analysis. *Electrophoresis,* Vol.24, No.19-20, pp.3500-3507.

Correa-Netto, C., Teixeira- Araujo, R., Aguiar, A. S., Melgarejo, A. R., De-Simone, S. G., Soares, M. R., Foguel, D. & Zingali, R. B. (2010). Immunome and venome of Bothrops jararacussu: A proteomic approach to study the molecular immunology of snake toxins. *Toxicon,* Vol.55, 1222-1235.

Cortelazzo, A., Guerranti, R., Bini, L., Hope-Onyekwere, N., Muzzi, C., Leoncini, R. & Pagani, R. (2010). Effects of snake venom proteases on human fibrinogen chains. *Blood Transfusion,* Vol.8, No.Suppl 3, pp.s120-s125.

Davidsson, P., Folkensson, S., Christiansson, M., Lindbjer, M., Dellheden, B., Blennow, K. & Westman-Brinkman, A. (2002). Identification of proteins in human cerebrospinal fluid using liquid-phase isoelectric focusing as a prefractionation step followed by two-dimensional gel electrophoresis and matrix-assisted laser desorption/ionisation mass spectrometry. *Rapid Communication in Mass Spectrometry,* Vol.16, No.22, pp.2083-2088.

Dennis, M. S., Henzel, W. J., Pitti, R. M., Lipari, M. T., Napier, M. A., Deisher, T. A., Bunting, S. & Lazarus, R. A. (1990). Platelet glycoprotein IIb-IIIa protein antagonists from snake venoms: evidence for a family of platelet aggregation inhibitors. *Proceedings of the National Academy of Sciences,* Vol.87, No.7, pp.2471-2475.

Eble, J. A., Niland, S., Dennes, A., Schmidt-Hederich, A., Bruckner, P. & Brunner, G. (2002). Rhodocetin antagonizes stromal tumor invasion in vitro and other alpha2beta1 integrin-mediated cell functions. *Matrix Biology,* Vol.21, No.7, pp.547-558.

Escoubas, P., Palma, M. F. & Nakajima, T. (1995). A microinjection technique using *Drasophila melanogaster* for bioassay-guided purification of neurotoxins in arthropod venoms. *Toxicon,* Vol.33, No.12, pp.1549-1555.

Felley-Bosco, E., Demalte, I., Barcelo, S., Sanchez, J. C., Hochstrasser, D. F. & Schlegel, W. (1999). Information transfer between large and small two-dimensional polyacrylamide gel electrophoresis. *Electrophoresis,* Vol.20, No.18, pp.3508-3513.

Fletcher, J. E. & Rosenberg, P. (1997). The cellular effects and mechanisms of action of presynaptically acting phospholipase A2. *In: Venom Phospholipase A2 Enzymes: Structure, Function and Mechanism,* Kini, R. M. (ed.). pp.413-454, Wiley: Chichester, England.

Fox, J. W. & Serrano, S. M. T. (2008). Exploring snake venom proteomes: multifaceted analyses for complex toxin mixtures. *Proteomics,* Vol.8, No.4, pp.909-920.

Fox, J. W., Shannon, J. D., Steffansson, B., Kamiguti, A. S., Theakston, R. D. G., Serrano, S. M. T., Camargo, A. C. M. & Sherman, N. (2002). Role of Discovery Science in Toxicology: Examples in Venom Proteomics. *In: Perspective in Molecular Toxinology,* Menez, A. (ed.). pp.97-105, John Wiley & Sons: West Sussex.

Gnad, F., Gunawardena, J., Mann, M. (2011). "PHOSIDA 2011: the posttranslational modification database" (in eng). *Nucleic Acids Res.,* Vol. 39, pp.253-260.

Gorg, A., Boguth, G., Obermaier, C., Posch, A. & Weiss, W. (1995). Two-dimensional polyacrylamide gel electrophoresis with immobilized pH gradients in the first dimension (IPG-Dalt): the state of the art and the controversy of vertical versus horizontal systems. *Electrophoresis,* Vol.16, No.7, pp.1079-1086.

Gould, R. J., Polokoff, M. A., Friedman, P. A., Huang, T.-F., Holt, J. C., Cook, J. J. & Niecviarowski, S. (1990). Disintegrins: a family of integrin inhibitory proteins. *Proceedings of the Society for Experimental Biology and Medicine,* Vol.195, No.2, pp.168-171.

Graham, R. L. J., Graham, C., Theakston, D., McMullan, G. & Shaw, C. (2008). Elucidation of trends within venom components from the snake families Elapidae and Viperidae using gel filtration chromatography. *Toxicon,* Vol.51, No.1, pp.121-129.

Gubenek, F., Kriaj, I. & Pungerar, J. (1997). Monomeric phospholipase A2 neurotoxins. *In: Venom Phospholipase A2: Structure, Function and Mechanism,* Kini, R. M. (ed.). pp.245-268, Wiley: Chichester, England.

Guercio, R. A. P., Shevchenki, A., Shevchenko, A., Lopez-Lozano, J. L., Paba, J., Sousa, M. V. & Ricart, C. A. O. (2006). Ontogenetic variations in the venom proteome of the Amazonian snake *Bothrops atrox*. *Proteome Science,* Vol.4, No.11, doi:10.1186/1477-5956-4-11.

Guerranti, R., Cortelazzo, A., Hope Onyekwere, N. S., Furlani, E., Cerutti, H., Puglia, M., Bini, L. & Leoncini, R. (2010). In vitro effects of *Echis carinatus* venom on the human plasma proteome. *Proteomics,* Vol.10, No.20, pp.3712-3722.

Hodgson, W. C. & Wickramaratna, J. C. (2002). In vitro neuromuscular activity of snake venoms. *Clinical and Experimental Pharmacology and Physiology,* Vol.29, No.9, pp.807-814.

Kini, R. M. (2003). Excitement ahead: structure, function and mechanism of snake venom Phospholipase A2 enzymes. *Toxicon,* Vol.42, No.8, pp.827-840.

Kornalik, F. (1991). The influence of snake venom proteins on blood coagulation. *In: Snake Toxin,* Harvery, A. L. (ed.). pp.323-383, Pergamon Press: New York.

Lewis, R. L. & Gutmann, L. (2004). Snake venoms and the neuromuscular junction. *Seminars in Neurology,* Vol.24, No.2, pp.175-179.

Li, S., Wang, J., Zhang, X., Ren, Y., Wang, N., Zhao, K., Chen, X., Zhao, C., Li, X., Shao, J., Yin, J., West, M. B., Xu, N. & Liu, S. (2004). Proteomic characterization of two snake venoms: *Naja naja atra* and *Agkistrodon halys. Biochemical Journal,* Vol.15, No.384 (Pt 1), pp.119-127.

Macheroux, P., Seth, O., Bollschweiler, C., Schwartz, M., Kurfuerst, M., Au, L.-C. & Ghisla, S. (2001). L-amino-acid oxidase from Malayan pit viper *Calloselasma rhodostoma*: comparative sequence analysis and characterization of active and inactive forms of the enzyme. *European Journal of Biochemistry,* Vol.268, No.6, pp.1679-1686.

Montecucco, C., Gutierrez, J. M. & Lomonte, B. (2008). Cellular pathology induced by snake venom phospholipase A2 myotoxins and neurotoxins: common aspects of their mechanisms of action. *Cellular and Molecular Life Sciences,* Vol.65, No.8, pp.2897-2912.

Nawarak, J., Phutrakul, S. & Chen, S.-T. (2004). Analysis of Lectin-bound Glycoproteins in Snake Venom from the Elapidae and Viperidae families. *Journal of Proteome Research,* Vol.3, No.3, pp.383-392.

Nawarak, J., Sinchaikul, S., Wu, C. Y., Liau, M. Y., Phutrakul, S. & Chen, S. T. (2003). Proteomics of snake venoms from Elapidae and Viperidae families by multidimensional chromatographic methods. *Electrophoresis,* Vol.24, No.16, pp.2838-2854.

O' Farrell, P. H. (1975). Two-dimensional electrophoresis of proteins. *Journal of Biological Chemistry,* Vol.250, No.10, pp.4007-4021.

Ogawa, T., Nakashima, K.-I., Nobuhisa, I., Deshimaru, M., Shimohigashi, Y., Fukumaki, Y., Sakaki, Y., Hattori, S. & Ohno, M. (1996). Accelerated evolution of snake venom Phospholipase A2 isozymes for acquisition of diverse physiological functions. *Toxicon,* Vol.34, No.11-12, pp.1229-1236.

Ownby, C. L. & Colberg, T. R. (1987). Characterization of the biological and immunological properties of fractions of prairie rattlesnake (Crotalus viridis viridis) venom. *Toxicon,* Vol.25, No.12, pp.1329-1342.

Peters, K. E., Walters, C. C. & Moldowan, J. M. (2005). *The Biomarker Guide: Biomarkers and isotopes in the environment and human industry,* Cambridge, Cambridge University Press.

Pink, M., Verma, N., Rettenmeier, A. W. & Schmitz-Spanke, S. (2010). CBB staining protocol with higher sensitivity and mass spectrometric compatibility *Electrophoresis,* Vol.31, No.4, pp.593-598.

Pisano, M., Allen, B. & Nunez, R. 2002. ZOOM® Proteomics: Rapid Methodology for 2D Protein Profiling. *Human Proteome Orgnization (HUPO) 10th World Congress.* Versailles, France: Molecular and Cellular Proteomics.

Rabilloud, T., Chevallet, M., Luche, S. & Lelong, C. (2010). Two-dimensional gel electrophoresis in proteomics: Past, present and future. *Journal of Proteomics,* Vol.73, No.11, pp.2364-2377.

Rioux, V., Gerbod, M.-C., Bouet, F., Menez, A. & Galat, A. (1998). Divergent and common groups of proteins in glands of venomous snakes. *Electrophoresis,* Vol.19, No.5, pp.788-796.

Serrano, S. M., Shannon, J. D., Wang, D., Camargo, A. C. & Fox, J. W. (2005). A multifaceted analysis of viperid snake venoms by two-dimensional gel electrophoresis: an approach to understanding venom proteomics. *Proteomics,* Vol.5, No.2, pp.501-510.

Soares, A. M., Anzaloni Perosa, L. H., Fontes, M. R. M., Da Silva, R. J. & Giglio, J. R. (1998). Polyacrylamide gel electrophroresis as a tool for taxonomic identification of snakes from Elapidae and Viperidae. *Journal of Venomous Animal Toxins including Tropical Disease,* Vol.4, No.2, pp.137-141.

Tan, N. H. & Saifuddin, M. N. H. (1989). Enzymatic and toxic properties of Ophiophagus hannah (King Cobra) and venom fractions. *Toxicon,* Vol.27, No.6, pp.689-695.

Tang, M. S., Vejayan, J. & Ibrahim, H. (2011). The concept of two-dimensional electrophoresis-guided purification proven by isolation of rhodocetin from Calloselasma rhodostoma (Malayan pit viper). *Journal of Venomous Animal Toxins including Tropical Disease,* In Press.

Tonismagi, K., Samel, M., Trummal, K., Ronnholm, G., Siigur, J., Kalkkinen, N. & Siigur, E. (2006). L-amino acid oxidase from *Vipera lebetina* venom: isolation, characterization, effects on platelets and bacteria. *Toxicon,* Vol.48, No.2, pp.227-237.

Vejayan, J., Ibrahim, H. & Othman, I. (2007). The Potential of Mimosa pudica (Mimosaceae) against snake envenomation. *Journal of Tropical Forest Science,* Vol.19, No.4, pp.189-197.

Vejayan, J., Ibrahim, H. & Othman, I. (2008). Locating Alpha-Bungarotoxin in 2-DE Gel of *Bungarus multicinctus* (Many Banded Krait) Venom. *Malaysian Journal of Science,* Vol.27, No.1, pp.27-34.

Vejayan, J., Shin Yee, L., Ponnudurai, G., Ambu, S. & Ibrahim, I. (2010). Protein profile analysis of Malaysian snake venoms by two-dimensional gel electrophoresis. *Journal of Venomous Animal Toxins including Tropical Disease,* Vol.16, No.4, pp.623-630.

Vishwanath, B. S., Kini, R. M. & Gowda, T. V. (1987). Characterization of three edema-inducing phospholipase A_2 enzymes from habu (*Trimeresurus flatroviridis*) venom and their interaction with alkaloid aristolochic acid. *Toxicon,* Vol.25, No.5, pp.501-515.

Wang, R., Kini, R. M. & Chung, M. C. M. (1999). Rhodocetin, a novel platelet aggregation inhibitor from the venom of *Calloselasma rhodostoma* (Malayan pit viper): synergistic and non-covalent interaction between its subunit. *Biochemistry,* Vol.38, No.23, pp.7584-7593.

Wang, X., Li, X. & Li, Y. (2007). A modified Coomasie Brillian Blue staining method at nanogram sensitivity compatible with proteomic analysis. *Biotechnology Letter,* Vol.29, No.10, pp.1599-1603.

Wittman-Liebold, B., Graack, H. R. & Pohl, T. (2006). Two-dimensional gel electrophoresis as tool for proteomics studies in combination with protein identification by mass spectrometry. *Proteomics,* Vol.6, No.17, pp.4688-4703.

Yan, X. J., Harry, R. A., Spibey, C. & Dunn, M. J. (2000). Postelectrophorectic staining of proteins separated by the two-dimensional gel electrophoresis using SYPRO dyes. *Electrophoresis,* Vol.21, No.17, pp.3657-3665.

12

Protein Homologous to Human CHD1, Which Interacts with Active Chromatin (HMTase) from Onion Plants

DongYun Hyun[1] and Hong-Yul Seo[2]
[1]*National Institute of Horticultural and Herbal Science, RDA*
[2]*National Institute of Biological Resources, Ministry of Environment, Republic of Korea*

1. Introduction

Onions are grown as an annual plant for commercial purposes; although since they are biennial it takes two seasons to grow from seed to seed. Bolting (flowering) of onion plants is determined by two factors, the size of the plant and cold temperatures. The critical size for bolting occurs when the onion reaches the five-leaf stage of growth. If onions are seeded in early fall, warm temperatures will result in sufficient size for bolting in the subsequent winter. Early transplants and some onion varieties are especially susceptible to bolting during cold temperatures. However, cold temperatures are not the sole prerequisite for bolting. If onions are not at the critical size in their development, they do not recognize cold as a signal to initiate bolting. Thus, sowing and transplanting at the correct time of year is the most important factor to avoid premature bolting.

Genetic and molecular studies of *Arabidopsis* have revealed a complicated network of signaling pathways involved in flowering time (Boss et al., 2004; Macknight et al., 2002; Putterill et al., 2004). Four genetic pathways, which are known as the photoperiod, autonomous, vernalization, and gibberellin (GA) pathway, have been identified based on the phenotypes of flowering time mutants (Koornneef et al., 1998). The photoperiod pathway includes genes whose mutants show a late flowering phenotype under long day (LD) conditions that is not responsive to vernalization treatments. This pathway contains genes encoding photoreceptors such as *PHYTOCHROME* (*PHY*), components of the circadian clock, clock associated genes such as *GIGANTEA* (*GI*) (Fowler et al., 1999; Park et al., 1999), and the transcriptional regulator *CONSTANS* (*CO*) (Putterill et al., 1995). *FLOWERING LOCUS T* (*FT*) (Kardailsky et al., 1999; Kobayashi et al., 1999) and *SUPPRESSOR OF OVEREXPRESSION OF CO 1* (*SOC1*) (Lee et al., 2000) are targets of *CO* (Samach et al., 2000). The autonomous pathway includes genes whose mutants show a late flowering independently of day length that can be rescued by vernalization. Genes included in this pathway are *FCA, FY, FVE, FLOWERING LOCUS D* (*FLD*), *FPA, FLOWERING LOCUS K* (*FLK*), and *LUMINIDEPENDENS* (*LD*) (Ausin et al., 2004; He et al., 2003; Kim et al., 2004; Lee et al., 1994; Lim et al., 2004; Macknight et al., 1997; Schomburg et al., 2001; Simpson et al., 2003). They regulate *FLOWERING LOCUS C* (*FLC*) (Michaels and Amasino, 1999), a floral repressor, through several different mechanisms

such as histone modification and RNA binding (Simpson, 2004). Some genes of this pathway are also involved in ambient temperature signaling (Blazquez et al., 2003; Lee et al., 2007). The vernalization pathway includes genes whose mutations inhibit the promotion of flowering by vernalization. Genes included in this pathway are *VERNALIZATION INSENSITIVE3* (*VIN3*), *VERNALIZATION1* (*VRN1*), and *VERNALIZATION2* (*VRN2*) (Gendall et al., 2001; Levy et al., 2002; Sung and Amasino, 2004). The GA pathway includes genes whose mutations show a late flowering especially under short day (SD) conditions. This pathway has GA biosynthesis genes, *FLOWERING PROMOTIVE FACTOR1* (*FPF1*), and genes involved in GA signal transduction (Huang et al., 1998; Kania et al., 1997). GAs have been known to positively regulate the expression of floral integrator genes such as *SOC1* and *LEAFY* (*LFY*) (Blazquez et al., 1998; Moon et al., 2003).
We report here, genetic and molecular evidences for regulation of bolting time in onion plants using a late bolting-type cultivar (MOS8) and an very early bolting-type cultivar (Guikum). We screened the proteins extracted from onion plants with different bolting times by using a proteomic approach and identified a protein with significant similarities to chromodomains of mammalian chromo-ATPase/helicase-DNA-binding 1 (CHD1) or heterochromatin protein 1 (HP1). Furthermore, we examined *in vitro* histone methyltransferase (HMTase) activity using purified protein isolated from onion plants. Our results suggest that a floral genetic pathway in controlling bolting time may be involved in onion plant.

2. Methodology

2.1 Plant growth and cutivars

Two onion cultivars, MOS8 (Eul-Tai Lee et al. 2009) with a late bolting phenotype and Guikum (provided by Kaneko seed Co., Japan) with a very early bolting phenotype, were used in this study. F_1 plants produced from crosses between MOS8 and Guikum were self-pollinated to produce F_2 populations. Based on the segregation ratio of bolting, inheritances of F_2 generations were evaluated. Bolting was assayed from the time of transplantation into the field to the first open flower.

2.2 Northern

Total RNA was extracted from leaves using an RNeasy plant Mini Kit (Qiagen, USA) according to the manufacturer's instructions. About 15 μg of total RNA was separated via electrophoresis on a 1.2% formaldehyde-agarose gel and then transferred onto a Hybond-N^+ membrane (Amersham, USA) by capillary action (Sambrook et al. 1989). The full-length open reading frame (ORF) regions of *Arabidopsis FRIGIDA* (*FRI*) and *FLC* were amplified from cDNAs prepared from *Arabidopsis* seedlings. These fragments were labeled with [α-^{32}P] and used as Northern blot probes. Hybridization was performed for 20 h at 68°C, and the filters were washed with 2 ×SSC, 0.1% SDS at 68°C for 20 min and 1×SSC, 0.1% SDS at 37°C for 30 min. The filters were exposed to X-ray film at -70°C for 3-7 days.

2.3 2-DE

The meristematically active parts (200 mg) isolated from onion plants were homogenized with lysis buffer containing 8 M urea, 2% NP-40, 5% β-mercaptoethanol, and 5% polyvinyl pyrrolidene, and then assayed by 2-DE (Yang et al. 2005). Extracted protein samples (100 μg) were separated in the first dimension by isoelectric focusing (IEF) tube gel and in the second dimension by SDS-PAGE. Electrophoresis was carried out at 500 V for 30 min, followed by 1000 V for 30 min and 5000 V for 1 h 40 min. The focusing strips were immediately used for SDS-PAGE or stored at −80°C. After electrophoresis of the first dimension, the focusing strips were incubated for 15 min in equilibration buffer I (6M urea, 2% SDS, 50mM Tris-HCl [pH 8.8], 30% glycerol, 1% DTT, and 0.002% bromophenol blue) and equilibration buffer II (6 M urea, 2% SDS, 50mM Tris–HCl [pH 8.8], 30% glycerol, 2.5% iodoacetamide, and bromophenol blue). Equilibrated strips were then run on an SDS-PAGE gel as the second dimension. The gels were stained with Silver Stain Plus and the image analysis was performed with a FluorS MAX multimager (Bio-Rad, Hercules, CA).

2.4 N-terminal sequencing analysis

Proteins were electroblotted onto a polyvinylidene difluoride (PVDF) membrane (Pall, Port Washington, NY) using a semidry transfer blotter (Nippon Eido) and visualized by Coomassie brilliant blue (CBB) staining. The stained protein spots were excised from the PVDF membrane and applied to the reaction chamber of a Procise protein sequencer (Applied Biosystems, Foster city, CA). Edman degradation was performed in accordance with the standard program supplied by Applied Biosystems. The amino acid sequences were compared to known proteins deposited in NCBIBLAST databases.

2.5 Mass spectrometry

Protein spots were excised, destained from 2-DE gels, dehydrated, reduced with DTT, alkylated with iodoacetamide, and digested with trypsin in accordance with the recommended procedures. Samples were then analyzed by Matrix-assisted laser desorption-ionization time-of-flight mass spectrometry (MALDI-TOF) MS on a Voyager-DE STR machine (Applied Biosystems, Framingham MA). Parent ion masses were measured in the reflectron/delayed extraction mode with an accelerating voltage of 20 kV, a grid voltage of 76.000%, a guide wire voltage of 0.01%, and a delay time of 150 ns. A two-point internal standard for calibration was used with des-Arg1-Bradykinin (m/z 904.4681) and angiotensin 1 (m/z 1296.6853). Peptides were selected in the mass range of 700 - 3000 Da. For data processing, the MoverZ software program was used. Peak annotations were checked manually to prevent non-monoisotopic peak labeling. Monoisotopic peptide masses were used to search the databases, allowing a peptide mass accuracy of 100 ppm and one partial cleavage. To determine the confidence of the identification results, the following criteria were used: minimum of four must be matched, and the sequence coverage must be greater than 15%. Database searches were performed using Protein Prospector (http:// prospector.ucsf.edu), ProFound (http:// www.unb.br/cbsp/paginiciais/profound.htm), and MASCOT (www.matrixscience.com).

2.6 Enzyme assays

HMTase assays were carried out at 30°C for 1 h in 20 μl volumes containing 50mM Tris-HCl (pH 8.5), 20mM KCl, 10mM $MgCl_2$, 10mM β-mercaptoethanol, 250mM sucrose, 8 μg/μl histone from calf thymus (Roche, USA), 220 nCi of *S*-adenosyl-L-[methyl-^{14}C]methionine ([^{14}C]SAM), and protein extracts prepared from onion plants. Methylation reactions were stopped by the addition of SDS-PAGE sample buffer, separated on a 16% polyacrylamide gel, and analyzed by autofluorography.

3. Results

3.1 Genetic inheritance of bolting in onion plants

In order to understand to the genetic control of bolting in onion plants, we crossed late bolting-type cultivar (MOS8, days to bolting=165-170 days) with very early bolting-type cultivar (Guikum, days to bolting=130-135 days). The bolting phenotypes of F_1 generations were similar to those of late bolting-type cultivars (data not shown). This suggests that genetic loci affecting bolting may be present in onion plants. Subsequent analysis of the inheritance distribution in F_2 generations is shown in Figure 1. Table 1 shows the distribution pattern and segregation ratio (late bolting:early bolting = about 3:1) indicating that bolting time depends on the segregation of any gene where the dominant allele confers lateness. Furthermore, bolting phenotypes of onion cultivars were reduced by long exposure to cold (E.T. Lee, personal communication). Given the crosses between late and very early bolting onion varieties, and effects of low temperature in onion plants, it appears likely that the genetic basis involved in the regulation of bolting time in onion is similar to that of vernalization requirement in plant species (Sung and Amasino, 2005). Genetic and molecular studies in various winter-annual and summer-annual accessions of *Arabidopsis* as a model plant have shown that *FRIGIDA* (*FRI*) and *FLC* have important functions in distinguishing winter-annual habits and summer-annual habits in *Arabidopsis* accessions (Clarke and Dean, 1994; Gazzani et al., 2003; Shindo et al., 2005). We assessed the expression patterns of these two genes in MOS8, Guikum, and F_1 plants (derived from crosses between MOS8 and Guikum) by northern hybridization (data not shown). The ORF regions of *FRI* and *FLC* amplified from *Arabidopsis* seedlings were used as probes. The mRNA levels of the *FRI* and *FLC* were strongly increased in the late-bolting-type cultivar, MOS8 ; however, the levels of *FRI* and *FLC* expression were significantly decreased in the very early-bolting-type cultivar, Guikum. These results suggest that the bolting time observed in onion plants may be affected by changes in *FRI* and *FLC* expression. However, we cannot dismiss the possibility that loci other than *FRI* and *FLC* may affect the bolting time of onion plants. Consistent with this idea, flowering in cereals is principally controlled by *VERNALIZATION 1* (*VRN1*) and *VERNALIZATION 2* (*VRN2*), which encode *APETALA1* (*AP1*)-like MADS box transcription factor and *CONSTANS* (*CO*)-like transcription factor, respectively (Trevaskis et al. 2003; Yan et al. 2004). Bolting time in other plant species are also determined by a relatively small number of loci, either dominant or recessive locus. With *Hyocyamus niger* (henbane), the biennial habit is governed by a single dominant locus, whereas this habit is governed by a single recessive locus in *Beta vulagris* (sugar beet) (Abegg, 1936; Lang, 1986).

3.2 2-DE analysis in onion plants

In order to examine the components involved in the control of bolting time in onion, we checked protein profiles of MOS8 and Guikum by using a 2-DE proteomics approach. The inner basal tissues of onion bulbs grown for 96 days after transplanting were used for proteomics analysis, because bolting is initiated in this region after cold treatment (Fig. 2a). Initial 2-DE analysis of soluble proteins from onion plants was performed using an IEF range of pH 3 to 6 (data not shown). Because the use of appropriate pH gradients is an effective way to reduce overlapping spots, additional analysis with pH 4 to 6 immobiline pH gradient (IPG) strips was performed (Fig. 2b). After CBB staining, several differences in protein accumulation profiles were detected in onion plants with different bolting times. Although many spots were differentially accumulated in onion plants, we failed to obtain

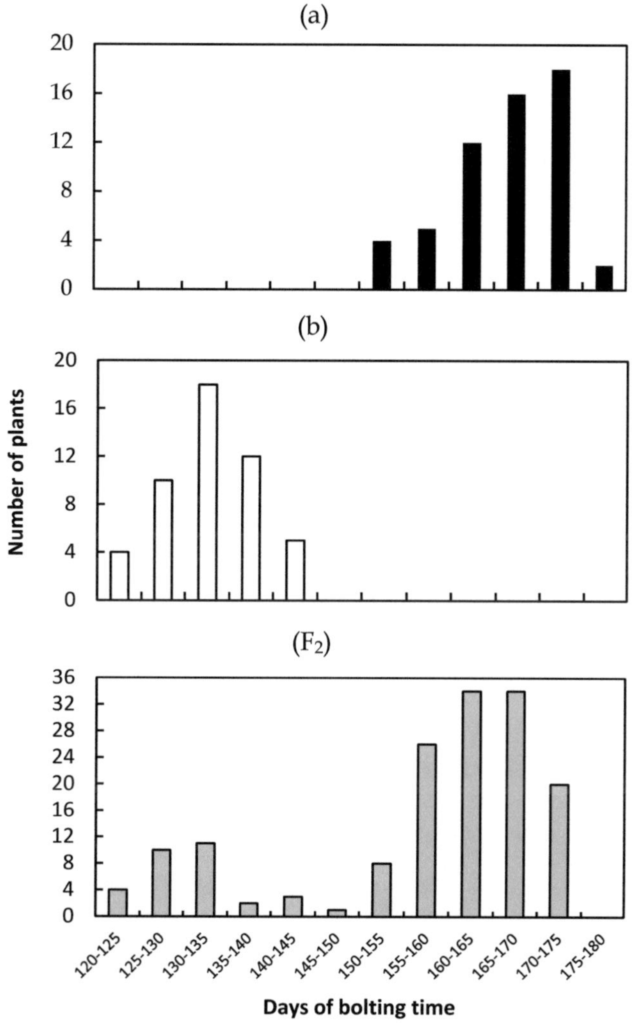

Fig. 1. Distribution patterns of bolting time in F_2 populations derived from crosses between (a) MOS8 (late bolting type) and (b) Guikum (very early bolting type) onion cultivars. These onion cultivars used in this study were inbred lines. The 'days to bolting' time were calculated when 80% of the total population of onion plants had bolted.

Variety	Total	Very early flower bolting	Late flower bolting	Ratio	Test ratio	x^2	P
MOS8	56	0	56			-	
Guikum	48	48	0			-	
MOS8 × Guikum	152	31	121	1:3.9	1 : 3	1.719	0.001

Table 1. Genetics of crossing MOS8 (Late bolting type) with Guikum (Very early bolting type) to identify genes that confer a vernalization response.

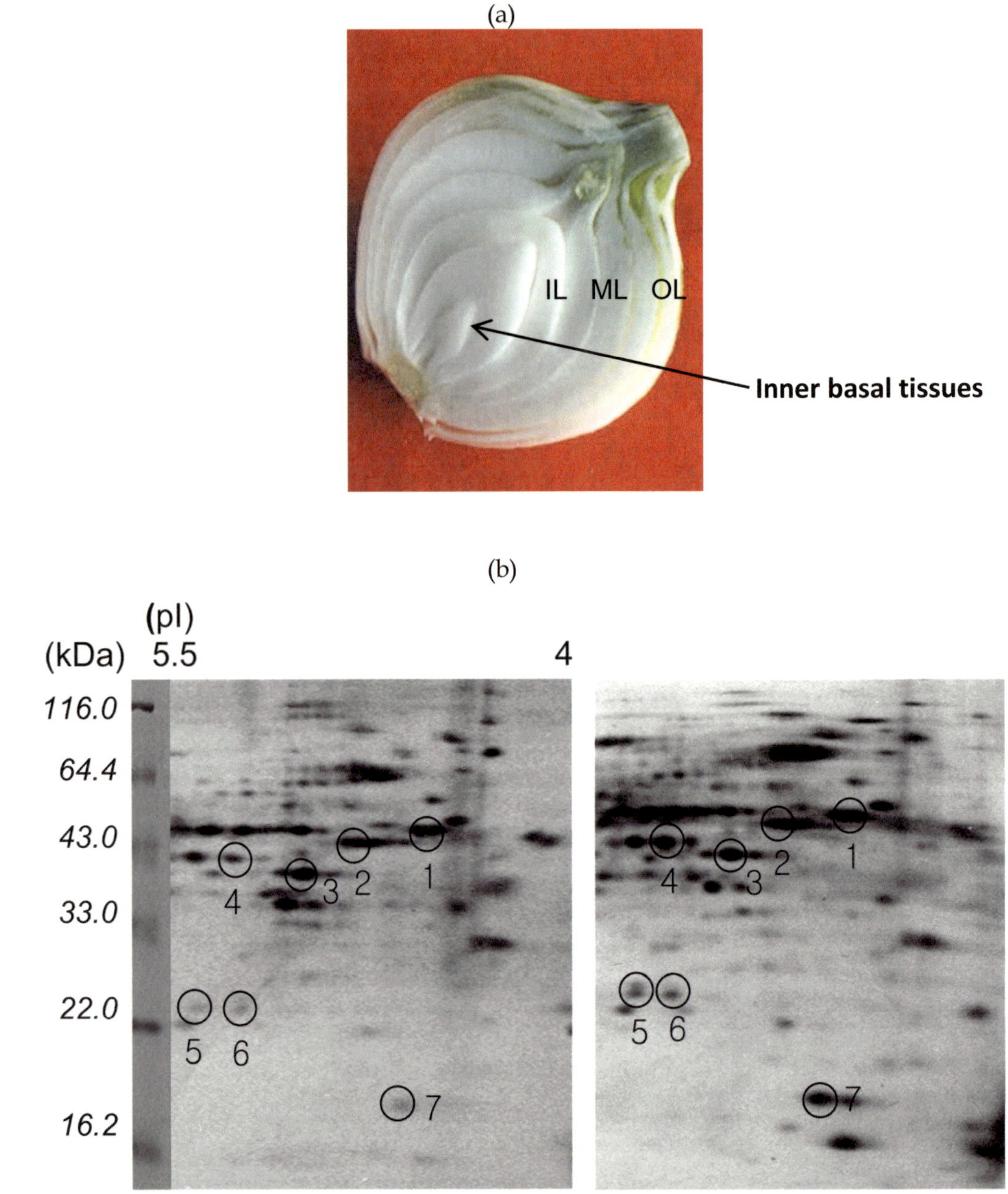

(a) The inner basal tissues of onion bulb used for proteomics analysis. *IL* inner layers, *ML* middle layer, *OL* outer layer.
(b) Protein analysis was performed using medium-range IPG strips with pH range from 4 to 6. The protein spots were identified by protein sequencing and MALDI-TOF MS analysis. Molecular masses (kilodalton) are shown on the left and pI ranges at the top comers of each figure.

Fig. 2. Two-dimensional gel electrophoresis of proteins isolated from onion plants (MOS8 and Guikum).

sufficient amounts from many of these spots for successful protein sequencing. Thus, we chose seven protein spots significantly changed in accordance with the degree of bolting time. The amino acid sequences of the differentially regulated proteins were analyzed by protein sequencing (Table 2). Homology searches were performed using the BLAST search tool. N-terminal sequences were successfully obtained for only one protein (spot 7). The remaining proteins were analyzed by MALDI-TOF MS. Among the other six proteins, three proteins (spots 1, 5 and 6) were not identified, whereas three proteins (spots 2, 3 and 4) were identified as actin, tubulin and keratin.

Spot No.[a]	pI/kDa[b]	Sequences[c]	Homologous protein (%)	Accession No.
1	4.8/46	N-blocked/MS[d]	Not hit	-
2	5.0/43	N-blocked/MS	Actin 1 (96)	P53504
3	5.1/39	N-blocked/MS	Tubulin alpha 2 chain (89)	Q96460
4	5.2/40	N-blocked/MS	Keratin, type II cytoskeletal 1 (90)	P04264
5	5.4/23	N-blocked/MS	Not hit	-
6	5.2/23	N-blocked/MS	Not hit	-
7	4.9/17	N-ARTLQTARRSTGGKAP	Chromodomains of mammalian CHD1 or HP1 proteins (93)	2B2W_D 3FDT_T 1GUW_B 1KNE_P

[a]Spot numbers are shown in Fig. 2.
[b]pI and molecular mass (kDa) are from the gel in Fig. 2.
[c]N-terminal amino acid sequences are determined by Edman degradation.
[d]MALDI-TOF MS.

Table 2. Identification of onion proteins whose abundance varied significantly among onion plants with different bolting time

Interestingly, the amino acid sequence of spot 7 showed significant similarities to several chromodomain regions of mammalian CHD1 or HP1 proteins, though we could not confidently identify an onion protein homologous to this spot in the database because of the short amino acid sequence and poorly characterized onion genome (Fig. 3). The chromodomain appears to be a well conserved motif, because it can be found in wide range of organisms such as protists, plants, amphibians, and mammals (Eissenberg, 2001). Furthermore, proteins with this chromodomain are known as both a positive and negative regulator of gene expression in various developmental processes (Hall and Georgel, 2007). For instance, two tandem chromodomains of CHD1 protein are known to interact with methylated lysines on histones, which include H3K4me, H3K36me and H3K79me, associated with active chromatin, thereby inducing active transcription (Flanagan et al., 2005; Sims et al., 2005). However, the chromodomain of the HP1 protein recognizes and binds to H3K9me for promotion of heterochromatin formation (Jacobs and Khorasanizadeh, 2002; Nielsen et al., 2002). Therefore, chromatin remodeling factors with chromodomains may play an important role in regulating gene expression. Because there is a dramatic change in the chromatin in meristematic regions such as inner basal tissues used in this study.

Onion	ARTLQTARRSTGGKAP_ _ _ _
2B2W_D	ARTXQTARKSTGGKAPRKQY
3FDT_T	ARTKQTARXSTGGKA_ _ _ _ _
1GUW_B	ARTXQTARXSTGGKAPGG
1KNE_P	ARTKQTARXSTGGKAY_ _ _ _
	*** **** *****

Fig. 3. Multiple alignments of amino acid sequences between onion protein spot 7 with other homologous proteins. Identical amino acid residues are denoted by asterisks. 2B2W_D chain D-tandem chromodomains of human CHD1 complexes with histone H3 tail containing trimethyllysine 4, 3FDT_T chain T-crystal structure of the complex of human chromobox homology 5 with H3K9(Me)3 peptide, 1GUW_B chain B-structure of the chromodomain from mouse HP1 beta in complex with the lysine 9-methyl histone H3 N-terminal peptide, 1KNE_P chain P-chromodomain of HP1 complexes with histone H3 tail containing trimethyllysine 9

Consistent with this, lesions in *Arabidopsis PHOTOPERIOD-INDEPENDENT EARLY FLOWERING 1* (*PIE1*), which encodes an ISW1 family ATP-dependent chromatin remodeling protein, result in a large reduction in *FLC* expression, thereby causing the conversion from winter-annual to summer-annual habits (Noh and Amasino 2003). Given that yeast Isw1p, an *Arabidopsis* PIE1 homolog, can bind H3K4me (Santos-Rosa et al. 2003), it might be assumed that PIE1 will bind H3K4me, which is generated by EARLY FLOWERING IN SHORT DAYS (EFS) (He et al. 2004; Kim et al. 2005), and remodel FLC chromatin to allow active transcription. However, an *in silico* search revealed that an onion protein homologous to human CHD1 was not related to the *Arabidopsis* PIE1 gene. This observation raises the possibility that various ATP-dependent chromatin remodeling factors may interact with various methylation states of lysine on H3 to induce transcriptional activation of target genes. Although there is no evidence that this protein spot is relevant to the regulation of bolting time by vernalization, this observation raises the possibility that chromatin remodeling factors may play roles in regulating this process in onion plants.

3.3 *In vitro* HMTase activity assays in onion plants

In order to assess whether histone methylation correlated with bolting time of onion plants, we performed *in vitro* HMTase activity assays using purified protein spots with significant similarities to chromodomains of mammalian CHD1 or HP1 isolated from two onion cultivars (MOS8 and Guikum) with calf thymus histones as substrates (Fig. 4a). Amino acid sequences of the purified spots used in this assay were confirmed (data not shown). The purified protein spots were able to methylate histone proteins in examined onion plants, indicating that the spots are associated with HMTase activity. Furthermore, differences in HMTase activity were observed in onion plants, though equal amounts of calf thymus histones were used in this assay (Fig. 4a, b). However, chromodomains of chromatin remodeling factors like mammalian CHD1 or HP1 generally act as binding modules for methylated lysines on histones. This could be explained by the SET-domain containing histone methyltransferase (Yeates, 2002) being present in extracts from onion cultivars. We cannot exclude the possibility that the purified protein spot is a histone methyltransferase with a chromodomain-like protein SUV39H1 (Brehm et al., 2004; Koonin et al., 1995).

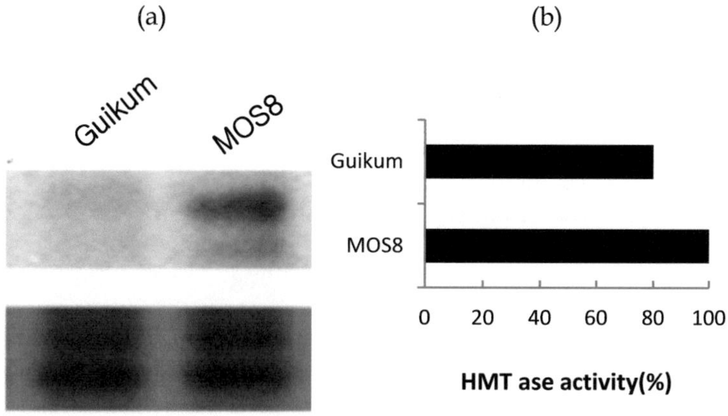

Fig. 4. In vitro HMTase activity in onion plants (MOS8 and Guikum). (a) Fluorography of ^{3}H-methyl calf thymus histone. 200 μg of purified spots containing protein homologous to human CHD1 isolated from two onion cultivars grown for 96 days after transplanting were used in this assay(upper panel). Reaction mixtures were analyzed by 16% SDS-PAGE and autoradiography. Equal amounts of each reaction were confirmed by Coomassie blue stain profiles of calf thymus histones (lower panel). (b) Quantitation of HMTase activity in MOS8 and Guikum

4. Conclusions

Our results suggest that a genetic pathway may be involved in the control of bolting time in onion plants by genetic inheritance, though the regulation of bolting in onion plants may be more complexly governed by several loci. Although it is very difficult to identify confident proteins in onion plants with uncharacterized genome, it appears likely that chromatin remodeling factors involved in histone modification may be conserved in onion plant. Although molecular and genetic analyses of flowering time in *Arabidopsis* have identified several floral promotion and repression pathways, our knowledge of the floral pathways in other economically important crops is limited. Thus, the quantitative trait locus (QTL) mapping and the use of high-throughput experiments such as genomics will provide a better understanding of the regulation of bolting time in onion.

5. Acknowledgment

We thank the Korea Basic Science Institute for their generous technical support of our research.

6. References

Abegg FA (1936) A genetic factor for the annual habit in beets and linkage relationship. J. Agricultural Res 53: 493-511

Ausin I, Alonso-Blanco C, Jarillo JA, Ruiz-Garcia L, Martinez-Zapater JM (2004) Regulation of flowering time by FVE, a retinoblastoma-associated protein. Nat Genet 36: 162-166

Blazquez MA, Ahn JH, Weigel D (2003) A thermosensory pathway controlling flowering time in Arabidopsis thaliana. Nat Genet 33: 168-171

Blazquez MA, Green R, Nilsson O, Sussman MR, Weigel D (1998) Gibberellins promote flowering of arabidopsis by activating the LEAFY promoter. Plant Cell 10: 791-800

Boss PK, Bastow RM, Mylne JS, Dean C (2004) Multiple pathways in the decision to flower: enabling, promoting, and resetting. Plant Cell 16 Suppl: S18-31

Brehm A, Tufteland KR, Aasland R, Becker PB (2004) The many colours of chromodomains. Bioessays 26: 133-140

Clarke JH, Dean C (1994) Mapping FRI, a locus controlling flowering time and vernalization response in Arabidopsis thaliana. Mol Gen Genet 242: 81-89

Eissenberg JC (2001) Molecular biology of the chromo domain: an ancient chromatin module comes of age. Gene 275: 19-29

Eul-Tai Lee, Cheol-Woo Kim, In-Hu Choi1, Young-Seok Jang, Jin-Ki Bang, Sang-Gyeong Bae, Dong-Yun Hyun, Jong-Mo Jung, In-Jong Ha, and Seong-Bae Kim (2009) New Mid-late Maturing F1 Hybrid Onion Cultivar, "Yeongpunghwang". Korean J. Breed. Sci. 41(4) : 587~590

Flanagan JF, Mi LZ, Chruszcz M, Cymborowski M, Clines KL, Kim Y, Minor W, Rastinejad F, Khorasanizadeh S (2005) Double chromodomains cooperate to recognize the methylated histone H3 tail. Nature 438: 1181-1185

Fowler S, Lee K, Onouchi H, Samach A, Richardson K, Morris B, Coupland G, Putterill J (1999) GIGANTEA: a circadian clock-controlled gene that regulates photoperiodic flowering in Arabidopsis and encodes a protein with several possible membrane-spanning domains. EMBO J 18: 4679-4688

Galmarini CR, Goldman IL, Havey MJ (2001) Genetic analyses of correlated solids, flavor, and health-enhancing traits in onion (Allium cepa L.). Mol Genet Genomics 265: 543-551

Gazzani S, Gendall AR, Lister C, Dean C (2003) Analysis of the molecular basis of flowering time variation in Arabidopsis accessions. Plant Physiol 132: 1107-1114

Gendall AR, Levy YY, Wilson A, Dean C (2001) The *VERNALIZATION 2* gene mediates the epigenetic regulation of vernalization in *Arabidopsis*. Cell 107: 525-535

Hall JA, Georgel PT (2007) CHD proteins: a diverse family with strong ties. Biochem Cell Biol 85: 463-476

He Y, Michaels SD, Amasino RM (2003) Regulation of flowering time by histone acetylation in Arabidopsis. Science 302: 1751-1754

He Y, Doyle MR, Amasino RM (2004) PAF1-complex-mediated histone methylation of *FLOWERING LOCUS C* chromatin is required for the vernalization-responsive, winter-annual habit in *Arabidopsis*. Genes Dev 18: 2774-2784

Huang S, Raman AS, Ream JE, Fujiwara H, Cerny RE, Brown SM (1998) Overexpression of 20-oxidase confers a gibberellin-overproduction phenotype in Arabidopsis. Plant Physiol 118: 773-781

Jacobs SA, Khorasanizadeh S (2002) Structure of HP1 chromodomain bound to a lysine 9-methylated histone H3 tail. Science 295: 2080-2083

Jakse J, Telgmann A, Jung C, Khar A, Melgar S, Cheung F, Town CD, Havey MJ (2006) Comparative sequence and genetic analyses of asparagus BACs reveal no microsynteny with onion or rice. Theor Appl Genet 114: 31-39

Kania T, Russenberger D, Peng S, Apel K, Melzer S (1997) FPF1 promotes flowering in Arabidopsis. Plant Cell 9: 1327-1338

Kardailsky I, Shukla VK, Ahn JH, Dagenais N, Christensen SK, Nguyen JT, Chory J, Harrison MJ, Weigel D (1999) Activation tagging of the floral inducer *FT*. Science 286: 1962-1965

Kim HJ, Hyun Y, Park JY, Park MJ, Park MK, Kim MD, Lee MH, Moon J, Lee I, Kim J (2004) A genetic link between cold responses and flowering time through FVE in Arabidopsis thaliana. Nat Genet 36: 167-171

Kim SY, He Y, Jacob Y et al. (2005) Establishment of the vernalization-responsive, winter-annual habit in *Arabidopsis* requires a putative histone H3 methyl transferase. Plant Cell 17: 3301-3310

Kobayashi Y, Kaya H, Goto K, Iwabuchi M, Araki T (1999) A pair of related genes with antagonistic roles in mediating flowering signals. Science 286: 1960-1962

Koonin EV, Zhou S, Lucchesi JC (1995) The chromo superfamily: new members, duplication of the chromo domain and possible role in delivering transcription regulators to chromatin. Nucleic Acids Res 23: 4229-4233

Koornneef M, Alonso-Blanco C, Peeters AJ, Soppe W (1998) Genetic control of flowering time in Arabidopsis. Annu. Rev. Plant Physiol. Plant Mol. Biol. 49: 345-370

Lang A (1986) Hyoscyamus niger. FL: CRC Press.

Lee H, Suh SS, Park E, Cho E, Ahn JH, Kim SG, Lee JS, Kwon YM, Lee I (2000) The AGAMOUS-LIKE 20 MADS domain protein integrates floral inductive pathways in *Arabidopsis*. Genes Dev 14: 2366-2376

Lee I, Aukerman MJ, Gore SL, Lohman KN, Michaels SD, Weaver LM, John MC, Feldmann KA, Amasino RM (1994) Isolation of LUMINIDEPENDENS: a gene involved in the control of flowering time in Arabidopsis. Plant Cell 6: 75-83

Lee JH, Yoo SJ, Park SH, Hwang I, Lee JS, Ahn JH (2007) Role of SVP in the control of flowering time by ambient temperature in Arabidopsis. Genes Dev 21: 397-402

Levy YY, Mesnage S, Mylne JS, Gendall AR, Dean C (2002) Multiple roles of Arabidopsis VRN1 in vernalization and flowering time control. Science 297: 243-246

Lim MH, Kim J, Kim YS, Chung KS, Seo YH, Lee I, Hong CB, Kim HJ, Park CM (2004) A new Arabidopsis gene, FLK, encodes an RNA binding protein with K homology motifs and regulates flowering time via FLOWERING LOCUS C. Plant Cell 16: 731-740

Macknight R, Bancroft I, Page T, Lister C, Schmidt R, Love K, Westphal L, Murphy G, Sherson S, Cobbett C, Dean C (1997) FCA, a gene controlling flowering time in Arabidopsis, encodes a protein containing RNA-binding domains. Cell 89: 737-745

Macknight R, Duroux M, Laurie R, Dijkwel P, Simpson G, Dean C (2002) Functional significance of the alternative transcript processing of the Arabidopsis floral promoter FCA. Plant Cell 14: 877-888

Michaels SD, Amasino RM (1999) *FLOWERING LOCUS C* encodes a novel MADS domain protein that acts as a repressor of flowering. Plant Cell 11: 949-956

Moon J, Suh SS, Lee H, Choi KR, Hong CB, Paek NC, Kim SG, Lee I (2003) The *SOC1* MADS-box gene integrates vernalization and gibberellin signals for flowering in *Arabidopsis*. Plant J 35: 613-623

Nielsen PR, Nietlispach D, Mott HR, Callaghan J, Bannister A, Kouzarides T, Murzin AG, Murzina NV, Laue ED (2002) Structure of the HP1 chromodomain bound to histone H3 methylated at lysine 9. Nature 416: 103-107

Noh YS, Amasino RM (2003) *PIE1*, an ISWI family gene, is required for *FLC* activation and floral repression in *Arabidopsis*. Plant Cell 15: 1671-1682

Park DH, Somers DE, Kim YS, Choy YH, Lim HK, Soh MS, Kim HJ, Kay SA, Nam HG (1999) Control of circadian rhythms and photoperiodic flowering by the Arabidopsis GIGANTEA gene. Science 285: 1579-1582

Putterill J, Laurie R, Macknight R (2004) It's time to flower: the genetic control of flowering time. Bioessays 26: 363-373

Putterill J, Robson F, Lee K, Simon R, Coupland G (1995) The CONSTANS gene of Arabidopsis promotes flowering and encodes a protein showing similarities to zinc finger transcription factors. Cell 80: 847-857

Samach A, Onouchi H, Gold SE, Ditta GS, Schwarz-Sommer Z, Yanofsky MF, Coupland G (2000) Distinct roles of CONSTANS target genes in reproductive development of *Arabidopsis*. Science 288: 1613-1616

Santos-Rosa H, Schneider R, Bernstein BE et al. (2003) Methylation of histone H3 K4 mediates association of the Isw1p ATPase with chromatin. Mol Cell 12: 1325-1332

Schomburg FM, Patton DA, Meinke DW, Amasino RM (2001) FPA, a gene involved in floral induction in Arabidopsis, encodes a protein containing RNA-recognition motifs. Plant Cell 13: 1427-1436

Shindo C, Aranzana MJ, Lister C, Baxter C, Nicholls C, Nordborg M, Dean C (2005) Role of FRIGIDA and FLOWERING LOCUS C in determining variation in flowering time of Arabidopsis. Plant Physiol 138: 1163-1173

Simpson GG (2004) The autonomous pathway: epigenetic and post-transcriptional gene regulation in the control of Arabidopsis flowering time. Curr Opin Plant Biol 7: 570-574

Simpson GG, Dijkwel PP, Quesada V, Henderson I, Dean C (2003) FY is an RNA 3' end-processing factor that interacts with FCA to control the Arabidopsis floral transition. Cell 113: 777-787

Sims RJ, 3rd, Chen CF, Santos-Rosa H, Kouzarides T, Patel SS, Reinberg D (2005) Human but not yeast CHD1 binds directly and selectively to histone H3 methylated at lysine 4 via its tandem chromodomains. J Biol Chem 280: 41789-41792

Sung S, Amasino RM (2004) Vernalization in Arabidopsis thaliana is mediated by the PHD finger protein VIN3. Nature 427: 159-164

Sung S, Amasino RM (2005) REMEMBERING WINTER: Toward a Molecular Understanding of Vernalization. Annu Rev Plant Biol 56: 491-508

Trevaskis B, Bagnall DJ, Ellis MH et al. (2003) MADS box genes control vernalization-induced flowering in cereals. Proc Natl Acad Sci USA 100: 13099-13104

Yang G, Inoue A, Takasaki H, Kaku H, Akao S, Komatsu S (2005) A proteomic approach to analyze auxin- and zinc-responsive protein in rice. J Proteome Res 4: 456-463

Yan L, Loukoianov A, Blechl A et al. (2004) The wheat *VRN2* gene is a flowering repressor down-regulated by vernalization. Science 303: 1640-1644

Yeates TO (2002) Structures of SET domain proteins: protein lysine methyltransferases make their mark. Cell 111: 5-7